I0055208

LEAN REFINING

HOW TO IMPROVE PERFORMANCE IN THE OIL INDUSTRY

Lonnie Wilson

Industrial Press, Inc.

Industrial Press, Inc.
32 Haviland Street, Suite 3
South Norwalk, CT 06854
Phone: 203-956-5593
Toll-Free in USA: 888-528-7852
Email: info@industrialpress.com

Author: Lonnie Wilson
Title: Lean Refining: How To Improve Performance In The Oil Industry, First Edition
Library of Congress Control Number: 2017936848

ISBN (print): 978-0-8311-3612-3
ISBN (ePDF): 978-0-8311-9446-8
ISBN (ePUB): 978-0-8311-9447-5
ISBN (eMobi): 978-0-8311-9448-2

Editorial Director: Judy Bass
Copy Editor: Judy Duguid
Interior Text and Cover Designer: Janet Romano-Murray

books.industrialpress.com
ebooks.industrialpress.com

TABLE OF CONTENTS

ADVANCE PRAISE

"Lonnie Wilson has brought his years of experience and passion in continuous improvement back to his roots in oil refining. This title will galvanize the refinery and continuous process industries to create huge gains in throughput and financial results. With practical and simple language, Lonnie shows how refineries can make huge gains in safety, quality, employee retention, and profits. His guidelines in this work will be followed by the industry's future winners and ignored by those destined to be on the scrapheap of history. It is not just the best book on the subject; it is the only book. It will change the industry forever."

Robert H. Simonis
Founder, KCE Consulting
Director of Rapid Improvement, CEVA Logistics
Lieutenant Colonel (retired), U.S. Army

"Another wonderful resource from Lonnie on implementing lean manufacturing. If you are interested in someone to shine your shoes so you can look pretty at lean manufacturing, keep looking. However, if you are actually looking to Implement lean manufacturing and have a sensei that really knows how to do Lean, that will stand shoulder to shoulder with you and help you actually DO Lean, Lonnie is my recommendation. I wish I could have had the opportunity to have Lonnie teach and mentor me 25 years ago. Thank you, Sensei."

Fred Kaschak
Global Manufacturing Director
The Woodbridge Group

"Here is a book that provides clear direction on how, and why to harness the collective experience and wisdom of the refinery workforce. In complex process plant operations such as oil refineries, raw material is processed with only a few percent profit. The product passes through the plants largely hidden in pipes, tanks and vessels, so the clues to the operation are observed through testing and various sensors. These all produce insight, most of which is difficult for computer analysis, but recognizable by people, especially when an enabling culture is in place. Improved utilization of the people is explained very well by the author in this "how to" for refiners. A "must read" for all refinery employees, these are instructions on how to create a world-class refining team."

RR Kooiman, PE

"Lonnie Wilson merges his extensive knowledge and experience of both lean manufacturing and refining in an easy to grasp way. It reads as a conversation with a friend that is an expert in lean manufacturing and refining and he is on your team to help you solve your most challenging problems. He highlights ideas with specific examples in the refining industry. As a former refinery improvement manager, I can vouch for the values obtained by the implementation of lean in a variety of projects. The next step is for a refinery to completely embrace the lean principles. The refining company that does this will be in a position to set themselves apart from the competition."

N. Brandon Hughes, ChEng & Energy MBA
Optimization & Production Analyst,
YASREF an Aramco JV

"I have found in Lonnie Wilson a rare combination of both passion for, and the ability to clearly communicate, the principles of continuous improvement. And, maybe even more rare, he does so not for his own reputation but to equip and enable others to move out smartly and make a difference!"

Michael J. Wiseman
Lean Six Sigma Black Belt
MJW Lean Consulting LLC

"For 25 years. I've watched Lonnie Wilson unwrap the most convoluted cause and effect relationships within complex systems and identify the opportunity like no other… I really mean like no other. It is simply uncanny how he breaks down complexity to deliver valuable insight and, most importantly, solutions that work. This book presents a snapshot of Lonnie's gift. Enjoy the story telling and reap the benefits of this opportunity to get inside his head!"

Jason Farley
Founder
BridgeGap Consulting

"No matter what your business, the principles identified in this book can positively change your company's operating and financial results—and your management skills! Implementing lean in our company profoundly changed our business, and me personally. Once we perfected these skills, we could take virtually any process and improve it by eliminating waste. This resulted in significant and lasting time and cost savings, while improving output and quality. Lonnie is an early practitioner and expert in lean management. His book offers practical advice on how to successfully implement lean. I highly recommend it."

Eugene J. Voiland
Founding President & CEO (retired)
Aera Energy, LLC

Acknowledgments

Paying Forward and Giving Thanks

Why did I write this book? There are two reasons:

- I wish pay forward with the things I have learned and do justice to the time others have spent in my development.
- I wish to catalyze the use of lean manufacturing, as I firmly believe it will make the world a better place.

As I think about those reading this book, my prayer is that it will be both informative and provocative. I hope to spread new aspects of lean manufacturing that I have learned through my experiences and from my interactions with others. I have been graced to have spent time with some giants in the world of lean manufacturing. I learned directly from Dr. W. Edwards Deming and Dr. Joseph Juran, and was blessed to discuss a variety of topics with them. Two sources of recent knowledge have been Dr. Robert W. "Doc" Hall and Dr. Edward L. Deci. Doc Hall, a very early proponent of lean, is a brilliant thinker and a grossly underappreciated contributor to the lean movement. Dr. Deci is a great social psychologist and one of the thought leaders on the subject of intrinsic motivation. Others I learned a great deal from include my current mentor, Toshi Amino, Dr. David Chambers and Dr. Donald Wheeler, who taught me how to properly apply statistics in industry. Although I only worked with them briefly, both Brian Joiner and Peter Scholtes had a profound effect on me. There have been many others, including several men and women who mentored me in both my career with Chevron and my consulting practice later. In addition, I have learned a great deal from literally hundreds of authors, peers, and clients. To all these people and hundreds more, I owe a huge debt of gratitude, as they taught me endlessly. One way I can pay homage to them and their knowledge is by passing it to you, as best as I can.

As for the information in this book, it is mine, and for you it will be incomplete in many ways. This knowledge I am trying to convey was gained through my experience, and although there is much to be learned by reading, knowledge can never be complete through intellectual pursuits alone. I am hopeful this book will inspire you to action, provoke you to go to the gemba and apply these concepts. Only then, when this knowledge has been tested and refined by the fires of your personal reality, will it become *your knowledge.*

I hope you too will be able to expand this powerful concept of lean manufacturing and pay that new knowledge forward as well.

PAYING FORWARD AS A TEACHER

As a teacher, I make the same agreement with all my students. That is, if you were ever one of my students, you remain a student of mine always, if you so wish. That means you can call me anytime:

- To get advice and ask questions, personal or professional
- Just to discuss a topic
- To teach me something new
- To talk about soccer

The purchase of this book has just made you a student of mine, so you can avail yourself of this option.

Be well and enjoy our book.

AND IN THE TRUE SPIRIT OF TRANSPARENCY

As I write this book and reflect on my experiences in lean and in oil refining, I refer to a lean transformation of an oil refinery as a "target-rich environment." I would like to assist you if you choose to undertake this amazing journey. It is a journey that will lead you and your company to higher levels of profitability, quality, and delivery as well as improved safety and environmental performance, all this as you watch the morale of those in the facility skyrocket. Most people will tell you that "you cannot have your cake and eat it too." That you cannot improve quality without increasing costs, that you cannot improve delivery performance without increasing inventory volumes. Well with a lean manufacturing system "you *can* have your cake and eat it too."

MANY THANKS

This is the third book I have written on lean manufacturing, and it was the most fun to write for several reasons. It is a book about applying lean principles in a petroleum refinery, a topic covered very lightly in the literature, and I enjoy tilling new ground. Second, it caused me to reach back to my early years in manufacturing and recall the instances from over 40 years ago, digging through my files and notes, and that was an extremely enjoyable trip into the past. Third, as I wrote the book, I realized how much of lean manufacturing we had applied, at least the technical aspects, while I was at Chevron. Finally, it was humbling to go back and recount the experiences because I could see more clearly the things I had done very well, but more importantly, the things I would now do much differently.

I don't wish to sound less than humble, but some things come easy to me. I am sure that some things come easy to you as well; we all have our callings. Coaching soccer came easy to me. Leading teams comes easy to me. Teaching comes easy to me. But more importantly, implementing lean manufacturing comes easy to me. Many aspects of lean manufacturing I just do intuitively. Sometimes I even have a hard time explaining why I do them in the way I do, and far more often than not, they work nicely. Long ago I learned that lean manufacturing was a calling of mine, and I was prepared in many ways long before I even started my consultancy.

However, writing is not a calling of mine. I have now written three books, and dozens of my articles have been published. You would think I would be good at writing; not so. Writing does not come easily to me . . . it is effortful. The fact that I can get these documents published is as much a reflection of my editor and the people who have helped me along the way as it is to my authoring skills. In this book, I owe a great deal to the people on my technical review team who read and assisted me with every word. They are Jason Farley, Jose Alejandro Tatay, Randy Kooiman, and Venkatesh Seetharamam, with special thanks to Robert Simonis, Kelly Moore, Michael Wiseman, and Gene Voiland, who in addition to being on the team, each made significant contributions to the content.

Even beyond that, special thanks go to Fred Kaschak, Thom Longcore, Karn Gill, and Brandon Hughes, each of whom, in addition to being on my technical review team, worked directly with me as we transformed their facilities. Although in each case I was their advisor and teacher, each of these four taught me a great deal. My deepest thanks.

This book would not have been a reality without the guidance from my editor, Judy Bass. She not only is supportive but is an inspiration to work with, and I trust we can do more together in the future. "Thank you, Judy, for sticking with me and helping me as you have." And thanks for putting together a solid support team with special thanks to Judy Duguid for editing my words to make a much improved story, and for Janet Romano-Murray for vastly improving the layout and recreating the artwork.

And to my wife, Roxana, who not only is an excellent industrial engineer in her own right, but more importantly, is the love of my life—none of this is possible without your support and patience. I love you.

PREFACE

A MESSAGE TO MY READERS

ABOUT THIS BOOK

Seldom are ideas original. This book is no exception. However, sometimes people are not able to understand how these ideas apply to *their* environment. My hope is that I present these ideas in a manner that will connect with you and your experiences and assist you to a higher level of understanding.

I am sure this book will not answer all your questions; no book can. However, I am equally sure it will answer many of your questions. Additionally, as quickly as you have those answers, they will spawn additional questions you did not have earlier.

Finally, I hope these thoughts will then catalyze your actions so you can put them to use for the betterment of your customers, your employees, your stockholders—and yourself. For as surely as you practice these lean techniques, you, too, will grow and benefit.

ABOUT ME, MY PATH, AND YOUR PATH TO UNDERSTANDING THE LEAN METHODOLOGY

My journey into becoming a consultant specializing in lean was a torturous path with no real plan; rather, it evolved. In the words of Charles E. Lindbloom of Yale University, my path was one governed by "the science of muddling through."

Yet after the fact, although the path was long and torturous, I am thoroughly convinced that this is the best path I could have taken.

No, I did not graduate from college and go to work under the tutelage of Taiichi Ohno. I did not rise through the Toyota system learning as I went, being taught by a master, and then, sometime later, decide I wanted to create a consulting practice to exploit what I had been taught from others. That was not my path.

Far from it.

My path was much longer, much less focused, and, for my money, much better. Much better, because as you will see throughout this book, the path I took is the very path that created the Toyota Production System. It was the path of:

- Planning
- Doing
- Experimenting
- Succeeding and failing; followed by
- Reflection and learning
- Then starting the cycle all over again

Along the way, I had neither formal nor informal mentors of lean manufacturing until the last few years that Toshi Amino has taken me under his wing. However, I would be remiss to say I had no mentors, because I did. Several people took the time from their busy days to spend time with me and nurture my development. The knowledge I got from them—in business, in manufacturing, in management, and in leadership, but mostly in how to be human and relate with people—was simply pure gold. Many people helped me in ways far greater than they could ever appreciate, and my prayer is that this book will give homage to those people.

I have included in all my books "prescriptions" to follow. I do this, as it gives people a definitive starting point and, often, several next steps to guide them. Many people have told me that the steps are the strong points of my writing. Unfortunately, each circumstance is different, and although we may start at a similar point, soon enough the variation inherent in all systems will cause you to diverge from the prescription. You must be prepared for this. And like my learning experience in lean, you too will need to follow the path of "planning, doing, experimenting, succeeding and failing, followed by reflection and learning, and then starting the cycle all over again" as you strive to create a culture of continuous improvement and respect for people. That is what lean manufacturing is all about—that cycle.

That was my path, and it will be yours.

Today, as you embark on this journey into lean, you will have more than I had. First, you will have a sensei who will provide immense guidance, support, and teaching as you proceed. In addition, you have this book, which is focused on refining, and other books to guide you. However, no matter how many books you read, the knowledge you gain from them will be a mere grain of wheat compared with the harvest of knowledge that will come from *doing* lean.

You will need to "plan, do, check, and act" as you embark on this incredible journey. In the plant—dealing directly with the people, directly with the products, and directly with the processes—is where the true knowledge is gained. At the gemba, that is where you will learn. For this there is no substitute. Reading, benchmarking, visiting others, doing outside audits, and using other means of comparison may help your awareness, but awareness alone is not adequate. Awareness that is not followed by experimentation and action is a very hollow form of understanding and seldom benefits anyone.

Initiative is the one litmus test of any transformation. Are you willing to embark on change? Frequently the answer is no, because you may be afraid or have inadequate guidance. But act you must, and frankly if your actions do not have occasional failures, then you are not acting enough. Yes, you must be able to make "smart failures," and you must be able to recover from these failures—both intellectually and emotionally. These failures are not really failures; they are just more knowledge you are acquiring on your way to success. Scientists will tell you they learn 10 times more from their failures than from their successes, so you will need to introduce a new cultural paradigm called "promote fast failures" and then revel in the knowledge gained.

The path to lean is not to copy the path someone else has followed. Although you may get a lot of help from books and others regarding your starting point, soon enough you will be hacking your way through the jungle of problems and issues to address those that will be largely unique to your business, to your facility, and probably to you.

So my advice is to put on your helmet, tighten your chinstrap, and get ready for the experiences of a lifetime—transforming a facility to lean manufacturing. It's worth the effort.

MORE "SLICE-OF-LIFE" EXAMPLES

Finally, if you wish to read more real slice-of-life experiences, go to our website, www.wilsonleanrefining.com, where you can meet Dirk, Jill, Bob, Javier, and many others, whose lives I have touched and whose lives have touched me using this wonderful and powerful methodology known as lean manufacturing.

Lonnie Wilson

May, 2017

Lean Refining

INTRODUCTION

WHAT'S IN THE BOOK

This book is all about lean manufacturing and its application in the business of oil refining. Although we delve into the oil business in general, the primary focus is on refining. And that is due to two very simple observations of mine. First, I know that lean manufacturing will work in the oil business. While I was in the oil refining business (1970–1990), we implemented a great deal of what today is known as lean manufacturing. We achieved huge gains. The gains were not only financial, although those were very impressive. We achieved gains in several other areas including quality, delivery, environmental improvements, safety, reliability, and morale. And we did this while we created greater engagement of the workforce through problem solving. Second, as I have now been a consultant in lean manufacturing for over 27 years, I have employed these same lean concepts in not only refineries but several continuous process industries as well, with great success.

Even though I have looked hard, I have yet to find even one example of any refinery that has transformed to a lean manufacturing system—not one. I suspect that all of them are working at some level on process improvements that they might call continuous improvements, and some are doing some imitations of lean that they call Lean Sigma or Lean Six Sigma. But I do not find even one real facility that has implemented a complete culture-changing continuous improvement process using a fully engaged workforce—and this amazes me. This book will sort through those topics and try to answer:

1. Why is there so little lean manufacturing in the refinery sector?
2. Why should and how can one go about implementing this set of powerful tools in a petroleum refinery?

Just as I am sure that this book will not be the conclusive and definitive treatment of those questions, I am equally sure that it will provide a different and enlightening treatment.

This book is divided into three parts.

Part 1–Lean background, is specific to the topic of lean manufacturing and has little to do with the oil business in general. Three topics are covered, each with its own chapter:

Ch 1. The Story of the Toyota Headrest Cell, the Theta cell

Ch 2. What Is Lean Manufacturing?

Ch 3. Why Is Lean Manufacturing So Misunderstood?

Part 2–Lean and refining, addresses lean manufacturing in the oil business in general and refining in particular. It explains why lean is not the prevalent improvement strategy in refining and why it will not likely appear in one of the major oil companies; rather, it will most likely appear in one of the smaller refinery-focused oil firms. Part 2 also addresses the differences you will see in a refinery application versus the lean stereotype of a tier one auto supplier doing labor-intensive assembly work. The chapters in this part include:

Ch 4. How Much Has Lean Penetrated the Refining Business?

Ch 5. Where Will Lean First Appear in the Refining Industry?

Ch 6. How Can Lean Benefit the Refinery-Focused Businesses?

Ch 7. What Differences Does Petroleum Refining Bring to Lean?

Part 3–Implementing lean in the refinery, addresses the topics you will need to understand to implement the lean transformation in a refinery. It covers in depth the foundational issues, gives you several examples, and explains how to organize and roll out the transformation from the top down. These chapters are:

Ch 8. The House of Lean

Ch 9. Lean Manufacturing Foundational Issue—People

ABOUT THE AUTHOR

I have written this section in two parts. The first section is a summary of my career in refining with Chevron. The second section is "my lean-specific training," and I will cover two time frames. First, I will distill much of the history of my career with Chevron and how this prepared me for a life as a consultant in lean manufacturing, and next I will relate my history after leaving Chevron and how this better prepared me to become a lean consultant in oil refining.

My career in refining

My journey as a lean consultant up to the writing of this book has been simply amazing. It has been filled with lots of reading, studying, and thinking; tons of hard work; and a surprisingly large amount of pure luck. I am fond of a quote from President Lyndon Johnson, who, when he was being chided for most of his success being attributed to pure luck, said in his deep East Texas drawl, "Well, son, it appears that the harder I work, the luckier I seem to get."

I never forgot it and personally find it to be very true.

However, I will give you several examples where I was just plain lucky, no question about it. In fact, there were instances where I did not get what I wanted, and I suspect it worked out far better than if I had gotten what I had originally wished for.

I become a chemical engineer

My path to becoming a lean consultant started in college at Washington State University. I had gotten a scholarship to study either math, a physical science, or engineering. I chose mechanical engineering, as I was told it was the broadest engineering curriculum and led to the most flexible career choices. In my dorm, across the hall from me, a senior, Bill, was in chemical engineering (ChE), and he told me how tough it was. Later I spoke to a few other folks and found out that many students avoided chemical engineering as it was the most challenging. I found this to be intriguing. Always liking a challenge, I wanted to be in the toughest curriculum. I did a little homework, and finally I approached the department chair, Dr. George Austin, and told him I wanted to transfer. At first, he discouraged me. But after another discussion and checking my transcript, he reluctantly agreed. I say reluctantly because he pointed out that there had been 108 students in the ChE 101 freshman introductory class, but that number had shrunk to 62 in the second semester in ChE 102. And of those 62, only 33 were promoted to ChE 201. His point was not only that it would be tough, but that there might be resentment because, though some of those who started had quit of their own accord, a rather large number had been asked to not return.

Dr. Austin told me that the department had more students than places for them in the junior- and senior-level classes (there was only room for 18), and hence there was a very strict grading policy. An F meant you were dismissed from the school of ChE; the department gave no Ds, and no classes could be retaken. So if you could not get a C or better, you could not remain in the curriculum.* If I remember correctly, there were 16 in our graduating class.

It turns out that not only the education I received, but my interactions with Dr. Austin and the rigors and competition in the ChE department had set me up well for industry. Being a chemical engineer armed me with a powerful set of tools that helped me greatly when I went to work. And they help me greatly in my current role as a consultant to this day. I want to thank Bill, Dr. Austin, and others for this random series of events that led me to become a chemical engineer.

*You need to recognize that at this time, 1965–1969, the Vietnam War was in full escalation and student deferments were available to only those students making progress toward a degree in 4 years. In addition, unlike today, colleges were overcrowded and expanding like crazy to educate the "baby boomers." In my freshman year, Washington State University had 8,600 full-time students; when I graduated in 1969, there were 14,800 full-time students. Colleges, with demand exceeding supply, could comfortably afford to be selective at that time.

I go to work and make a dumb decision

After four years at Washington State University with Dr. Austin, I accepted a very prestigious job offer from DuPont. I was among a very small group of engineers that had been hand-selected from a nationwide pool. We were chosen to work in the Engineering Service Department. However, in an incredibly immature move (I was 22 at the time), less than a year later, I quit DuPont. Possibly the dumbest career move I have ever made, but I guess that is what you do when you are 22.

I go from hero to zero . . . but still land at Chevron on the rebound

I was not too worried, as I had numerous job offers when I graduated. And I decided I would go to work for Chevron (at that time it was Standard Oil of California). I had been a summer intern for the company and had done well, and Chevron courted me strongly. In fact, I had nine different job offers from the company, including one in the Engineering Department, one in Chevron Research Corporation, three in different refinery locations, and four in different locations in exploration and production. I was confident I was a commodity in high demand. Dumb me!

After I left DuPont, I called Chevron and told the people there that I was on the market. The recruiter said the company had no positions but she would check it out. I was crushed; I had gone from hero to zero in less than a year. However, a couple of days later I got a call, and the recruiter said that based on my experience, there was a "fit" for me at the company's El Segundo Refinery. Although I wanted to go to the refinery in Richmond, the recruiter made it very clear that El Segundo was the only option. That experience sobered me considerably. I had gone from an engineer in very high demand to a mere commodity that might "fit in" because of my prior experience with Chevron. I later learned that the only reason I was given a job offer was that the El Segundo Refinery had not met its hiring goal for new engineers. Now I knew for certain I was just a number; and in all honesty, I was lucky El Segundo was not very good at hiring young engineers. Things worked out very well for me there.

Good things happen to me at El Segundo

While at the El Segundo Refinery, at least four major serendipitous things happened. First, I was assigned to a designs engineering group that supported the Isomax Operating Division. The Isomax complex was new and still under construction, which was well behind schedule, so the complex was also in start-up mode. The eyes of the refinery management and the corporation were on this complex and how it was doing. Hence only the best engineers were assigned there, and consequently my boss was one of the better engineering supervisors. The urgency of the work, the quality of my supervisor, and my competitive nature made for an exceptional learning environment. In addition, as a chemical engineer, I got deeply involved in not only the processing operations but the plant and equipment limits as well. Some of us became bottleneck management experts. I learned at an amazing pace and soon was more knowledgeable than engineers with twice my experience. Since the pressure was on, we could not fail to perform. Many engineers came to our group but were rotated out if they failed to produce. I did not recognize it until much later, but I was clearly a high-producing engineer. Many managers noticed that, and so they were willing to overlook my flaws (I was rough around the edges, to put it politely). Soon I was promoted to supervisor. So fate had put me in a group that not only learned quickly but had great visibility and had to work under a great deal of pressure, all of which accelerated my growth.

Second, I met my mentor there. He was the superintendent of the Operating Division and chose me to be his student. He did not advertise it as such, but his actions were clear. He worked hard toward my development and spent a great deal of one-on-one time with me—coaching, guiding, and challenging me, as well as seriously chewing me out on occasion. He taught me the Supervisory Development Tools and made sure I used them. He was intently focused on developing people and inculcated that value in me during my first supervisory assignment. He was gruff and seemingly harsh, but at the end of the day, he not only preached but lived what he said. He told me frequently, "Train and develop your people, and they will take care of you." Or another of his favorites was, "Keep your eye on the work, but take care of your people, and they will take care of everything else." At Chevron we had all the "tools and programs" for developing people; unfortunately, not everyone used them. I was lucky: I had a mentor, and he not only used these tools and programs; he lived them as well.

Third, while in El Segundo, I had to go through strike training. Every two years the refinery labor contract was up for renewal, and three times while I was at El Segundo, there was a walkout. The refinery prepared for the potential of a walkout by training all the supervisory and staff personnel to run all the plants and equipment. I did this three times. In so doing, we went through all the operator training including studying the documentation, walking all the lines, studying the control schemes, taking the tests, and performing all the potential situations, "the situationals," including the emergency procedures. We did everything in the training manuals—everything. We had to spend time on the day, afternoon, and graveyard shifts. This training was invaluable. But even better training occurred when the workforce walked out. We ran the plants. I actually ran the plants. This was amazing training. It was no longer observing and critiquing; it was actually making the products, on time and on spec. Twice I ran plants in the Isomax Division, and later I was an operator in the Sulfuric Acid Alkylation Plant. The shortest walkout was about 10 days, and the longest lasted over a month. This was the ultimate "go to the gemba" experience. I viewed refinery work significantly differently after each of these three strike opportunities. In addition, I was also a better engineer, a better supervisor, and a better manager after each of these three experiences.

Fourth, after only three years, I was promoted directly to a supervisory position. At this time in El Segundo, few people got promoted from engineer with less than five years' experience. At the time of my promotion, my time in service raised some eyebrows, as did my new job.

I get promoted and learn about cultural change

I was promoted to the job of the head draftsman for the refinery. For 60 years in this refinery, no one had become the head draftsman without first being a draftsman and then a lead draftsman; I had been neither. There were two very interesting aspects of this promotion. First, my mentor had been promoted to the refinery chief engineer, in charge of the Engineering and the Maintenance Divisions. He orchestrated my promotion with the Refinery Personnel Committee. Second, it was his belief that these draftsmen had far more potential to help the refinery than in the way they were currently being utilized, that is, almost exclusively as draftsmen and not much else. After all, that is how it had been for the last 60 years. If you were to start as a draftsman, and we had 24, you would retire a draftsman, with the slight possibility that you could become a lead draftsman; or if really, really lucky, you could become the head draftsman. This attitude was cultural in nature. Most of the refinery management thought of draftsmen as only draftsmen—nothing more. Plus, only a few of the draftsmen sought to be more. They were quite content.

That was soon to change.

My mentor told me their attitude was as a result of two things. First, all the previous head draftsmen were stuck with these paradigms and passed them to their people. They could not shake them even when challenged. Second, draftsmen were a part of the refinery technicians' pool (a pay grade, not a job) along with engineering technicians and equipment inspectors. The total pool of technicians was about 60, and the draftsmen were force-ranked predominantly in the bottom third. My mentor believed they were more talented than that. He was in a distinct minority with that opinion; neither his subordinates, peers, nor his superior could see the draftsmen as anything other than draftsmen. It was a rigid refinery paradigm my mentor wished to change. My job was to find out what kind of potential we had among the draftsmen, and if I determined they were in fact better than advertised, my job was to prove it.

I went about the task of finding that out. The first thing I did was to ally myself with my two direct reports, the two lead draftsmen. That was easy. Second, we started running the drafting room like a business with a business plan and a meaningful schedule of work. We expected the draftsmen who put together the schedule to meet it; it was a matter of simply asking them to be responsible. Third, we created a much stronger approach to employee growth and development. To the existing appraisal program, we added:

- A five-year growth and staffing plan to be used by the lead draftsmen and myself so we could plan for hiring and replacement of the men and women we expected to promote, even though, at this time, no one wanted our draftsmen
- Individual personal development plans
- Five-year growth and development plans for each individual

Next, we changed the surroundings. The drafting room was noisy, dreary, and poorly lit. The floor was two shades of gray tile, the walls were drab green, and the ceiling was a 12-foot elevated ceiling with recessed lighting that I was sure Thomas Edison himself had built at Menlo Park. The luminosity at the drafting table level was only 30% of standard, and each draftsman had at least one auxiliary lamp that he had purchased himself. Part of the problem with noisiness was that all the refinery drawings were kept in the same room with the draftsmen. This meant that besides the draftsmen rummaging through the files to find drawings, the engineers also came to search the files, and the combination of the tile floors and people's leather-heeled shoes only made it noisier. Finally, the most demeaning thing of all, for all 24 draftsmen and a varying number of contractors—there was one phone, only one stinking phone.

We changed all that.

We change the facilities . . . and prepare for progress

We replaced the floor with light, bright tile and installed carpet runners to cut down the noise. We installed a phone at each drafting table, moved all the drawings to the basement, and hired a clerk to search for drawings. We installed a new dropped ceiling with lighting that exceeded current standards. And we did about 50 other upgrades.

The story of the drafting room revamp has an amazing story line:

- We showed these draftsmen that we cared and that they mattered. They responded, and the improvement in morale was immediately measurable.
- We used the most imaginative and creative ways to get around all the bureaucracy. It was no exaggeration to say that it was easier to start up the fluid catalytic cracking unit than it was to change the office manual (yeah, we had an office manual, and it was 8½ inches thick). Through some hard work coupled with creative accounting, some cajoling, and some fracturing of—but never breaking any—refinery rules, we finished in record time. We were not so interested in revising all the paperwork and procedures as in making this a much better work environment. We violated no laws and hurt no one, but we did cut a few bureaucratic corners—quite a few. It was both fun and rewarding.

Time to spread the word and fight for rankings

I visited two managers: the head of Contract Construction for the Refinery Major Projects and the maintenance manager. I told both we had numerous draftsmen who could fill roles in their organizations and asked them to keep an eye out because some of these draftsmen were very talented. They listened.

Soon it was time for me to participate in the semiannual forced ranking of all technicians. It was tough. My group was largely in the bottom third, and I felt this was a massive error. Not one supervisor had more than 4 technicians, whereas I had to represent 24. With each supervisor fighting for his own people, it was no surprise I had few allies. My job was to advocate for placements as I saw them. I did my homework; I had my facts regarding the performance of my people. During the forced-ranking session, I was fighting for my people. I was convinced many were placed too low in the ranking. During the session I fought a number of battles, and armed with relevant data, I won more than my fair share. At times I was mad, I was frustrated, and I was a little brusque—generally, as I said earlier, I am a little rough around the edges; I am terrible at playing the political games and don't like them at all. But in the end, we made great progress.

I was totally spent after the ranking session, and I had ruffled a few feathers. But some draftsmen in the middle third had risen into the top third, and many in the bottom third rose to the middle. We had made significant progress in getting these draftsmen viewed in a more objective light, and we had some fine raises to give out as well. More importantly, the group of engineering supervisors who participated in the forced-ranking sessions started seeing these draftsmen in a different light.

Soon enough the draftsmen started seeing themselves in a new light as well. We were making progress. I was elated, and when my mentor checked in with me after the first ranking session, he said he had seen the ranking and was pleased to see his bias confirmed. I recall his words, "I just knew these men were better, and you can tell if you just listen to them." Soon enough I was approached by the head of Contract Construction; he wanted two construction reps. We already had started discussing the five-year growth and development plans with all the draftsmen (we were the only group that did this formally), so we already knew who was interested. We approached the two top-performing draftsmen, whom we knew were interested, and had them apply for the job. Soon enough we had moved two draftsmen. They both got promotions with significant raises along with their new jobs. I had been in my job for less than a year, and we had, although I did not know it at the time, significantly changed the culture.

I often reflect on this experience and recall a number of culture-changing activities that we performed without truly realizing the depth and breadth of our actions at that time. In the ensuing six months we hired several new draftsmen, among them the first women draftsmen in the refinery (they preferred that term over draftswomen), and we trained them. Another six people were promoted, including one of the lead draftsmen.

I get another promotion . . . and learn even more

After two years in the job, I was moved to a supervisory position in engineering, and four significant things happened. First, my stock was at an all-time high; I was in demand as a supervisor, and lots of people knew that. Second, without fully understanding it as such, we had made a huge cultural change in the drafting department. Third, I had learned a million lessons in leadership and management—especially how to handle people. In addition to the growth experience described above, I was sent to a series of training sessions including the refinery's first school for first-line supervisors, classes on motivation and discipline, and a weeklong seminar by Bill Oncken on managing management time (later made into a book; I highly recommend it to all) and several other topics. But fourth and most importantly, we took a group of employees that were not thought of very highly and not only changed the way they were perceived by others but changed their actual performance as well. We changed how they thought about themselves. In so doing we created development and growth opportunities these individuals never had before and provided the refinery with a new talent pool. I was pleased, my boss was pleased, my mentor was pleased, and the refinery in general was pleased.

It was at that very time I learned that:

- The vast majority of skills needed to vault your enterprise into becoming a lean enterprise are already inherent within your organization.
- The key to success is largely an issue of unleashing these skills.

- The major and very critical tool needed to unleash these skills, and hence to become a lean enterprise, is to learn and employ the skills of lean leadership at all levels of the organization.

These are 3 of the 15 key beliefs that guide quality consultants in all we do.

These beliefs have been reinforced numerous times over my 47-year career and have guided me as a manager early in my career and later as a consultant. They are the very foundation on which my consulting practice is built.

All of this came about because some random recruiter had sent me to El Segundo in spite of my desire to have a job at the Richmond refinery. Also, to some degree this happened because I had first chosen to work at DuPont, and in an immature move, I had quit. Even if I had known what was available, I never would have made the choices that were made for me; some random, some not so random.

My stock is high. I go to Perth Amboy

From here, it did not take too much time for me to be transferred to Chevron's refinery in Perth Amboy, New Jersey. I started out in the Engineering Group, where I supervised two groups of designs engineers, but my primary task was to design and implement an engineering hiring and training program. Since I had done similar work in El Segundo, I accomplished that quickly. We hired our first batch of engineers, and they were incredibly successful, outstripping all expectations. Previously the Perth Amboy refinery had not been able to attract any engineers except those from local schools. Now we were getting engineers from some of the finest schools, including Stanford, Northwestern, and Princeton, to name a few. No one within Chevron believed we could attract such strong talent, but we did. Corporate execs came to talk to me to see how we were so successful.

I become a section head

Quickly I was moved to Operations as a section head, a nice promotion but more so a clear sign I was on my way up. I was in charge of the crude units and some lesser plants, with about $400 million worth of investment and an operating expense budget of over $100 million a year. I had 5 shift supervisors, 6 head operators, and 36 operators in my section. Although I was very familiar with the processes, I was not familiar with our equipment here and the nuances of our section. I knew I had a great deal to learn. Luckily, at Perth Amboy, we had the finest operator hiring and training program I have ever seen, at Chevron or any other refinery. I set up a training plan for myself, and for the next four months I came in two hours early to read, study, and take the tests and situationals that all operators must take. I passed them all and felt far more armed as a section head. Operations was a totally consuming job. It seemed there was a crisis every day, with both processing and people challenges all over the place. It was hard to plan any day; it seemed the plant ran us, rather than the opposite. But it was "action city;" we were never bored, and we learned at an exponential rate.

As I had done in other jobs, I approached this a little differently than most new section heads. First, I spent time doing the training; second, I decided to attack all the chaos we were living through each day. I noticed two aspects of operation that I found disturbing, which appeared normal in all parts of the refinery and had been normal at El Segundo as well. I was disturbed that we would have numerous live alarms flashing on the alarm panel and take little or no action in response. Also many automatic control loops were on manual. Some had been on manual since start-up several years earlier. I quickly decided that we needed to change that. However, within Chevron and at Perth Amboy in particular, having alarms "in" and controllers on manual was an accepted practice—it was a way of life. Intuitively I knew it was wrong, but I also knew I had an upward battle. I enlisted the help of the shift supervisors and the instrument/electrician supervisor, and we began work to ensure that the alarms were responded to properly and the control loops were set on automatic. As soon as we got going, the gains were noticeable. We made less off-test product, were able to approach the specification limits on products, reduced product give-away, and increased yields; and the plants generally ran much better.

I get credibility

Despite these gains, the effort lost momentum. I was struggling to get the momentum back when, just by luck, something happened that would shake things up. I went to the control room and saw that the large atmospheric column to the No. 2 Crude Unit was in distress. After only a little questioning, I learned that the overhead pressure controller was on manual. All the side cuts were in trouble. The heavy straight-run draw had been reduced to avoid sending high end-point product to the Rheniformer, and the No. 2 oil draw had been reduced to avoid end-point problems as well. The plant was struggling, and so were the operators.

I asked why the controller was on manual. The shift supervisor said that a maintenance person "had looked at it," which was not very helpful. We asked the mechanic, and he confirmed that he had looked at it and "could see nothing wrong."

I was sure that the investigation had not been adequate, so I asked the shift supervisor and the maintenance supervisor to get dressed; we were going to check it out. It was the middle of January, the temperature was around zero, and no one in his right mind wanted to go visit the pressure transmitter, which was at the top of the atmospheric column. I strongly suspected the mechanic had not gone up there to investigate, although he said he had.

We climbed the column, the three of us. It was bitter cold, and the wind was blowing to make it even worse. After a bit, we got to the top of the tower, and I could immediately see the problem. The insulation on the steam tracing had failed, and the differential pressure (dP) cell leads were exposed. The leads to the dP cell had obviously plugged. The maintenance supervisor used some insulation tape to get the steam tracer in place and cover the leads. He then said, "Let's get the hell out of here," and my reply was, "As soon as we get the instrument on automatic."

We dropped down to a lower platform, out of the wind, and waited. After for what seemed an eternity, but was probably about five minutes, the leads to the dP cell cleared, and the head operator radioed us that he had the pressure control on automatic and it was responding. We descended the column and thawed out. But that activity was the talk of the control room for the next two weeks. People were convinced I was crazy as a loon. No manager in a suit and tie ever went to the top of the atmos column, ever—much less in the freezing weather of January. But they also understood what I was willing to do to get the work done, and furthermore they now knew I would not listen to a bunch of lies.

From that day forward, I was viewed differently by everyone in that control room, especially the operators; they knew they had an ally in their day-to-day battle to run the plants. That experience alone created so much credibility for me with the operators. I doubt they were willing to jump on a hand grenade to save my life, but they clearly appreciated the work I was willing to do to give them the tools they needed to do their job properly. They trusted me. I was flattered. At Chevron and at Perth Amboy in particular, trust was a commodity in scarce supply.

I did not appreciate it at the time, but when I read about the Toyota concept of "going to the gemba," this is the classic example I think of. It shows the manifest power of going to the gemba to grasp an understanding of the real situation.

We set crude run records

With all this work on alarms and control loops, in addition to some other operational moves, in a few months we were setting new records for crude rates. For nine consecutive months we exceeded the historical record for each of those months. We set new records on both crude units and ran more crude than the refinery had ever run before. At the same time we also set records for the lowest "slop creation," and off-test product was significantly reduced. With better controls, we ran longer between decokings, and overall crude rates increased even further. We began to get real cocky and comfortable and pushed the units even further—right up to the point where we screwed up, and I mean *really* screwed up.

We push too hard, all hell breaks loose . . . and luckily it ends well

As we pushed the units harder, our limit became vacuum bottoms spec, which was a minimum viscosity, as asphalt was our primary product. To make the viscosity, we reduced the wash oil in the bottom portion. We pushed so hard, and even though we had all kinds of process experts helping us, we ran out of liquid in one downcomer, coking it up completely, and we ended up shutting down the whole crude unit. It was a screw-up of galactic proportions and cost the refinery millions. It was not much of a feather in my cap. However, as it turned out, there were so many experts watching us that it was hard to pinpoint a guilty party. Then, after an expensive shutdown, we were up and running.

Since I was in charge of this disaster, I felt compelled to solve the root cause of the problem. And with a lot of hard work and a little bit of luck, I was able to fully understand precisely what had happened. And by precisely I mean *precisely*. By carefully reviewing the design and scrupulously reviewing the data prior to shutdown, we were able to model almost perfectly the failure mechanism. Hence, prior to start-up we installed what was an elegant solution to our problem and could further push the crude unit and do so safely. It involved better instrumentation on the lower part of the column and was incorporated into all new designs. Surprisingly, in the fullness of time we became known as the place that came up with the design that allowed all Chevron vacuum units to increase capacity and reduce bottoms make. The financial gains from this, Chevronwide, were huge.

The process had come full circle. We had shut down 70% of the refinery's capacity and for some time were pure "zeros." However, after we were up and running and others could implement this elegant solution, with the huge resultant gains, once again we were "heroes"—and all was forgiven.

I go to El Paso

The dust had hardly cleared on our start-up when I was promoted to chief process engineer in El Paso. And voilà, we were off to Texas. This was a real feather in my cap. I was now certain that I was one of the anointed ones, as it was almost unheard of for anyone to get to a Group 1 (a pay classification) in less than 10 years, and I had done just that. When I arrived, the refinery manager immediately advised me that he was very disappointed in the refinery's ability to hire new designs and process engineers and that we needed several. There were several support functions he was concerned about. In particular, we were under some scrutiny by the local and state environmental enforcement agencies, but worse, our Corporate Environmental Group felt we were not in full compliance with the corporate environmental policy. He asked me to do a full study of the process engineering support functions and recommend any needed changes. In addition, he asked me to take over responsibility of the refinery laboratory, a large group of union-represented people who had previously reported directly to him.

It took me three months to get acquainted with the refinery personnel and perform the evaluations requested. I found that the day-to-day process engineering support to the refinery was simply first rate. I also found that we were short on IT support, that we were way behind in environmental compliance, and that the refinery had excessive give-away in both vapor pressure and octane. Almost immediately we took two actions. First, we implemented an engineer hiring and training program, as we had implemented in Perth Amboy; and second, we allocated engineering support to the environmental compliance issues. Shortly thereafter we reorganized the process support function and created two lead process engineering positions. Later we expanded the IT group, began to provide blending technical support, and started an Environment Group with its own lead environmental engineer. As our hiring program took effect, we hired eight new, very talented engineers, and in less than a year the group structure had changed dramatically with four supervisors reporting directly to me—the two process lead engineers, the newly appointed lead environmental engineer, and the laboratory supervisor. My group had expanded to over 50 people.

I discuss my future

At about this time, I had a fateful discussion with my boss, the refinery manager, Jim. Jim had been relegated to El Paso some 15 years earlier when he embarrassed an obnoxious vice president and did so in public in front of the man's peers. Those there said Jim was right, but he had used the wrong venue to express his concerns. Anyway, it turned out that Jim liked El Paso, and although he did not have the job challenges he wanted, he was comfortable in El Paso. A real plus for Jim was that he was viewed as a strong teacher, and many up-and-coming managers sought to work in his organization. Having a tour of duty with him was an obvious plus on your résumé. In addition, his recommendations bore some real weight, and his peers and superiors sought his advice on personnel matters. Although pigeonholed prematurely in his career, he held tremendous respect among his peers and superiors alike. Not surprisingly, he liked this type of recognition, and it took the edge off the fact that his career had been severely and prematurely sidetracked because of one obdurate yet powerful man.

I too was happy in El Paso. I liked the refinery, I liked the town, and my whole family liked El Paso as well. In particular, the refinery provided me with a tremendous learning opportunity. El Paso was small enough that you could see the operations of the entire refinery from crude receipts to shipped product and everything in between; you could wrap your mind around the whole of a petroleum refinery. In the larger refineries, to get this exposure might take a whole career, and even then you would likely have sizable knowledge gaps.

Since my family and I were happy in El Paso, and I had gotten deeply involved in a variety of community activities, especially my kids' sports activities, I was in no hurry to move. And life at Chevron was good. Chevron was growing. A new coker project and general refinery upgrade had been approved for El Paso. I knew growth in El Paso was a very likely possibility, so I had a very, very sensitive discussion with my boss. Normally you would never have this discussion with your boss at Chevron—not at that time, unless you were willing to be pigeonholed for your career. However, Jim and I had become friends, and I thought I could be honest with him. I told him I appreciated all that Chevron had done for me, and though I did not want to truncate my career, I would like to stay in El Paso for as long as possible, for both personal and professional reasons. He understood fully and told me he would orchestrate my career as such, with one exception—he would let me know if any amazing opportunity should present itself somewhere else. He kept his word, and six months later I was promoted to refinery chief engineer in charge of engineering, maintenance, and reliability. Three years later we reorganized the refinery, and I was promoted to technical manager in charge of all technical functions; this was basically my old job of chief process engineer along with the engineering, equipment inspection, and reliability functions. Jim had kept his word.

Things change, with me and at El Paso

I had now been in El Paso for over four years; I came in with a promotion and had gotten two more promotions, and my stock was very high. However, about this time I started feeling a little bit uncomfortable about work in general. I could not put my finger on it, and I was not sure what it was, but at that time, I had just reread a book by Dr. W. Edwards Deming, *Out of the Crisis*. It went a long way toward opening my eyes to many of the weak practices we had at Chevron, which, at Chevron, were considered perfectly normal. To name a few:

- We had no mission, no vision, and no values statements.
- Waste was considered normal, with blend tests, reblend retests, and re-reblend re-retests on finished products, and huge volumes of off tests were routinely produced.

- Year after year, mistakes were repeated with no real effort to solve the root cause of the problems.
- Quality was something managed by someone else; our job was to make volumes.
- Waivers on products were obtained simply because the products were difficult to make to spec, especially aviation and jet fuels.
- There was a top-down style of leadership with little real employee engagement.
- Local leadership was weak, with no refinery-specific plans other than to meet the demands requested by marketing.
- This was a multibillion-dollar business with no profit and loss statement. As all other Chevron refineries, we were only a cost center.

Based on the way I had been treated, I thought Chevron was blind to many business practices that could and should be dramatically improved. I read increasingly and got new and different ideas. I took the teachings of Dr. Deming to heart, and we started using statistical process control (SPC) in the entire division. Wow! That was a real eye-opener. As a young and up-and-coming manager, I was sent to some very special training sessions that allowed me to discuss many of these topics with a more-open, less-biased group of employees. These training sessions, along with a much-accelerated reading program, discussions with others, and a significant amount of both prayer and reflection, caused my awareness to skyrocket.

Although Chevron had treated me quite well, I was now sure it had some major performance areas to address.

In addition, to the surprise of many, including me, the coker project and the refinery upgrade were canceled. I recall commenting to Jim that the cancellation of the coker project sealed the fate of the El Paso refinery. With the lost upgrade capacity afforded by a coker, the refinery could not financially compete in our distributive marketing area. The folks at the home office were nonplussed by the whole event;

I was mad as hell. I now knew the refinery was not a strategic cog to Chevron. In addition, at Jim's request, I recapped the capital budgets for the refinery for the last five years and noted that 92% of all money had been spent on refinery replacement and environmental projects; virtually nothing had been spent on the ability to upgrade product, reduce operating expenses, or improve the financial position in any manner. This was yet another eye-opening experience.

These realizations got me thinking. I was becoming aware that Chevron may not be the nirvana I once thought it was, El Paso certainly was not a key location, and maybe refining was not even a key activity within Chevron. For the first time, I seriously began to realize that my future may not be with Chevron.

Things change even more at Chevron

Then the merger of all mergers, at least to me, occurred. Chevron purchased Gulf Oil. I was deeply involved in a great deal of postmerger activity. The El Paso refinery perked along as usual, but something significant happened. Jim, the refinery manager, was selected to go to one of the old Gulf refineries and "get it back on track," as he related. It was a great opportunity for Jim, but I had lost a friend and mentor, and with him went the substance of the discussion we had several years earlier regarding my career wishes. I did not realize the impact of this, because it was some time until Jim's replacement arrived. By then much had changed.

The Gulf managers needed to be assimilated into the Chevron system, and Chevron was now awash in mid-level managers like me. Chevron tried to handle this with two early retirement programs. Meanwhile a whole series of activities complicated matters, as Chevron tried to spin off some facilities while retaining the core businesses it wished to retain. All these assimilations and spin-offs created a chaotic environment, and the net result was a significant reduction in management opportunities and movement.

And then it happened. We got a new refinery manager. And unfortunately for me, he did not like me at all. We started out with a toxic relationship, and it never improved. It was surprising to me, as we had worked together earlier. Although every section of my division was performing better than those in virtually any other refinery, he was hypercritical of the most minor issues. We were hiring, training, and not only filling our own promotions but sending many people for promotional opportunities to other locations. We met or exceeded every aspect of capital budget management. Refinery reliability was up, and our blending support group was generating millions of dollars by reducing product giveaways. He failed to recognize our large contributions but would laser in on seemingly minor issues. For the first time, I was faced with a supervisor who did nothing to help me in any way. I finally concluded that he simply did not like me, and I accepted that.

About two years later, I was asked to transfer to one of the old Gulf refineries. It was not a promotion, I would be going to a refinery that I knew Chevron intended to sell, and it was in a location in which I did not want to live. I was told I had no option and that was where my next check would be. I reminded my manager of my earlier discussions with Jim, of which he was previously aware, and he simply dismissed them as no longer being relevant. This action caused me to take some vacation time and discuss the situation with the family.

Before the week was out, I handed in my resignation and retired after 20 very interesting and fun-filled instructive and developmental years.

My "lean-specific" preparation to become a lean consultant

How my work at Chevron had prepared me to be a consultant in lean manufacturing

At the time of my retirement, I did not have a plan of any sort. I did not have a client, nor did I have a business plan. However, I did have a certain level of confidence born of my history, and I knew I was willing to work hard to make things work. So with a little trepidation and a lot of enthusiasm, I embarked on my next career; I was going to be a consultant.

With Chevron I had spent significant stints in three major refineries. In addition, I was dispatched to several other refineries to consult and solve challenging problems including operational, maintenance, and environmental issues. In my 20 years I held positions in designs engineering, process engineering, IT, facilities planning, oils planning, operations and operations management, maintenance management, and environmental management. In addition, on many occasions I was sent as a corporate troubleshooter to many locations for one-week to three-month assignments. You see, problem solving was one of the blessings bestowed upon me.

I have always read books, maybe 10 or so per year from the time I graduated from college. As you will recall, I was made aware of the writings of Dr. W. Edwards Deming and reread *Out of the Crisis*. It had a profound and enlightening effect on me and accelerated a change in my life. Soon I was reading books by Walter Shewhart, Joseph Juran, Kaoru Ishakawa, Taiichi Ohno, Shigeo Shingo, Masaaki Imai, Richard Schonberger, and Robert "Doc" Hall, David Chambers, Peter Drucker, Donald Wheeler, Sir Ronald Fisher, Abraham Maslow, Douglas McGregor, John Sanford (the Episcopalian priest and Jungian analyst, not the mystery writer), Thomas Merton, and many others. They opened new thoughts to me in not only management but problem solving, statistics, motivation theory, and just being human. These readings also quickened my interest in learning more about working in business and being a manager. My readings expanded; soon I was reading 40 to 50 books a year, which I continue to do.

Sometime later, I was introduced to Peter Scholtes, much by accident. I happened to be visiting another refinery when the refinery manager, a personal friend of mine who knew of my interest in Deming and his 14 Obligations of Management, approached me and told me to cancel my schedule and attend this class the refinery was sponsoring. It was entitled "Team Problem Solving." It was incredible. I took home a three-ring binder of the materials from the class.

This refinery had hired Joiner and Associates to teach the class. Joiner and Associates was a firm that followed the teachings of Dr. Deming. Brian Joiner and Peter Scholtes were the instructors in this class on solving problems using a team methodology. I cannot tell you how much this training by Peter Scholtes and the materials in the three-ring binder profoundly changed my life. I could fill another book with just that story, but I can tell you it was the first real teaching I had received that allowed me to facilitate the engagement of a broad range of employees—real employee engagement. It was the true and deep engagement that is taught by Toyota. Chevron had one or another "employee involvement program" from time to time, but all were filled with bureaucracy replete with complicated forms and lots of management reviews and approvals; they operated at a snail's pace, and all eventually died of their own weight. However, using the team-based problem-solving techniques taught by Peter and found in the raft of information in his three-ring binder, I was able to experience, firsthand, a simple working methodology to get true employee engagement. Peter played a huge role in my development as a lean consultant, and to him I am forever thankful. Later, the three-ring binder was published as a book that is now in its third edition and its zillionth printing, named *The Team Handbook;* it's a classic. Get it, read it, and apply it; you will not be disappointed. (In fact, do a book search and get the first edition. It is Peter's original work and superior, in my opinion, to later editions.

Well, the rest is history as they say. Soon I was teaching statistical process control and group problem solving, using a team approach, to my entire division. We were solving tough problems and using SPC to make a ton of money for the refinery. And soon enough we expanded to using cross-functional teams including people from other divisions. The rewards were amazing. It was this success, more than anything else, that fueled my interest and guided me in starting my consulting practice in 1990. It did not dawn on me at this time, but the key motivating factor for me to start my consulting practice was that for the first time, and by using these techniques, we were getting *true employee engagement* and that engagement transcended refinery divisions. We were working as a team to solve refinery problems. For a company that has not only silos, but thick-walled, steel-reinforced silos, working together in this fashion was truly paradigm busting at its finest. For those working on these teams, the excitement was obvious, and the high level of energy was noticed by all. In so doing we made the refinery a much better place to work, and the morale of these groups was noticed by everyone.

At that time, there was very little interest in SPC, group problem solving, or the teachings of Dr. Deming at Chevron's refineries. In addition, after a little nosing around, I could tell that outside of Chevron, there was probably even less interest. In our refinery here in El Paso, there was not a lack of interest; there was disinterest. The refinery manager, my boss, told me with distinct disdain that "this statistics garbage is just looking at history. Can't learn much from that.

Just by luck, I was empowered to embark upon and continue these efforts because the VP of manufacturing, had a growing interest in the work of Dr. Deming. Not so much interest that he would recommend any changes at his level, but he was intellectually interested. He would come to the refinery three times a year, and on each visit, he would want to see what we had been doing "with statistical methods and Dr. Deming's work." This became a large part of his visit as we would go over our accomplishments while we discussed the operating expense or the capital budget plan. His simple question empowered us, and since we were uncovering great opportunities, it was hard to stifle our efforts. From basic SPC we expanded into measurement system analysis and designs of experiments, and we used a number of statistical techniques along with group problem solving to resolve a number of refinery issues.

How working as an independent consultant further prepared me to be a lean consultant in oil refining

With this as a backdrop, I retired from Chevron and started my consulting practice so I could spread the word of Dr. Deming. However, his 14 Obligations of Management were never packaged in any way that attracted much interest. I tried a couple of approaches but then settled in with my two staples: statistics and statistical problem solving using small groups. My business flourished. I was drowning in work, and my name had gotten around. In the 27 years of my consulting practice, even my first few years, I never had to advertise; and I was able not only to keep afloat but to prosper nicely.

Along the way, I was asked to work on other projects that brought me further notice. I started teaching the elements of the Toyota Production System following the teachings of Taiichi Ohno. I was lucky to land large consulting jobs with two Toyota firms and became exposed not only to the actual activities in Toyota plants, but interestingly enough, to the subject of value stream mapping long before *Learning to See* was published.

In addition, Toyota published *The Toyota Way, 2001*, and I found that many of my beliefs of the Toyota Production System, which frequently clashed with the beliefs of my consultant friends, were confirmed. It was as much about respect for people as it was about the technical stuff, embodied in the "continuous improvement concept."

This was both eye-opening and heartwarming as I realized the importance of many of the teachings from my mentors regarding leadership, management, and just proper handling of people. I recall my mentor saying, "As a manager-leader, you need leverage. Take care of the people, and they will take care of everything else. That is your leverage." This same philosophy was right there in one of the few documents that was written about Toyota by Toyota, not what was written by outsiders. Right there in *The Toyota Way, 2001* was the proper balance to explain the importance of the human element in lean manufacturing.

My skills with SPC allowed me to develop the "three-part inventory system" and back it up with statistical logic (see *How to Implement Lean Manufacturing,* 2nd ed.). In my early years as an engineer with Chevron, I got very skilled at bottleneck analysis and management. Also, all Chevron engineers were quite accomplished at working with process flows, process designs, process management, process diagnosis, and process mapping. We also spent long hours working on shutdown planning and time management, on the design and management of large and small capital projects, and on critical path scheduling for both the projects and large maintenance activities. In short I was comfortable with many of the technical aspects of the Toyota Production System early in my career.

However, it was always a bit of a struggle to explain what I did. My title was "Statistical Consultant." Problem solving has always been my forte, and my clients would say, "We call Wilson when we have a really big issue." So I guess I was "the statistical consultant that solved really big problems." Unfortunately, that was not much to put on a business card. Although I never struggled for work, I struggled for a title. It was hard for me to explain to most others what I did. Then Jim Krafcik saved the day for me. When Womack, Jones, and Roos's book *The Machine That Changed the World* was published and I came upon the term "lean manufacturing," coined by Krafcik, voilà, I had a title! I was now a lean manufacturing consultant.

I was traveling more than ever, and work was incredibly interesting. Then it happened. I was diagnosed with cancer. My life changed a great deal. For about a year I did almost no work as I took on this new challenge. But as life will give you challenges, sometimes there are other opportunities buried inside those challenges.

Earlier I had been approached to write several books. First, one on SPC, as my clients thought I presented it in such a usable way. Next I was asked to write a book on measurement system analysis (MSA) for a large international automobile tier one company. Later I was asked to write a book on designs of experiments (DOE) and teach the class at UTEP's School of Mechanical and Industrial Engineering. Well, previously when I had been asked, I was always too busy and so begged off each time. Now, with this cancer gnawing at me and with my work life on hold, I had no excuse to not write a book. Unfortunately, my interest in SPC, MSA, and DOE had changed. Although those were still topics of great passion to me, there was already good material on the market.

I had already written several large documents on lean manufacturing and decided to write a book on this topic . . . mostly because I now had the time to write (that should give you pause as you read any author and ask how is it he or she has the time to write!). Anyway, I wrote my first book, and it sold like hotcakes, which means McGraw-Hill made money, but it gave me what I was looking for—credibility. The book created an interest in my work, and I became very busy.

Meanwhile, several things had happened. I had defined myself as a consultant in lean manufacturing and furthermore had decided that these lean techniques were exactly the same techniques we had used in my refining days. I became anxious to utilize these techniques in my experiences in refining. I consciously sought out clients in the refining and continuous process industries and in the next 10 years had six major clients from the continuous process industries. Two clients were petroleum refineries implementing some version of what they called Lean Sigma . . . more on that later. I could vividly recall from my earlier days with Chevron how easily we had made progress—and money—using this lean philosophy, and I wanted to make my mark in the refining business. I was convinced the opportunity was huge.

But in the refining sector, no one was buying; interest was practically zero. I tried to cultivate interest. I became involved in numerous forums in the oil and refinery sectors. I spoke at three conferences focused on the oil and refining business, and I was convinced that if we could only get a foothold in a refinery setting, we could show the power of lean and the world would be ours for the taking. All my efforts at blogging and commenting at forums, as well as speaking at conferences and conventions, did not seem to be going anywhere. I was not sure what to do next to get lean into the refining sector.

I get my opportunity

And then it happened. Not too long ago, a young engineer looking for a mentor for his company's Lean Sigma effort in a refinery, contacted me. I was ready, and we jointly took on a 20-month journey that was, by any measure, incredibly successful. We were completing projects and making some real money. There were several both large and small projects. And there were projects to improve planning, improve loss prevention, and improve environmental and safety performance as well as projects to improve reliability. We had the full complement of refinery tasks being executed using the refinery's Lean Sigma methodology. Of course, the projects with the most exposure were those with large financial gains. One project had just booked over $1.5 million in one-time savings. Another blending project was saving over $1.2 million per month, and yet another was transferring $900,000 a month to the bottom line. There was a second blending project whose potential was over $5 million per year and a third worth $1.3 million per month for the six winter months. Not only these but other projects were being completed, belts were being trained, and refinery engagement was very high. By any objective measure the project was a success.

The messages I learned from this engagement were clear, For one thing, my work before I left Chevron and my work with both refineries and other continuous processing plants after I left Chevron confirm beyond any doubt that the principles that make lean work in the assembly industries and in the auto industry are the same principles that will make a lean transformation in a refinery work. For another thing, with a deep background in both lean manufacturing and refinery management, I felt very prepared to transform a refinery from its conventional means of operation to one using a lean manufacturing philosophy.

But the most startling finding had to do with the progress made during the intervening years since I had retired over 25 years earlier. I had expected to see many positive cultural changes in refining operations. I did not. What I did see was that:

- The overall culture in a refinery had changed very little. Overall direction, including accomplishing a mission and attempting to attain a vision, was virtually nonexistent. The refinery was rife with silos, and each group was operating with its own agenda. Individual goals would normally trump group goals, and group goals would trump refinery goals. Teamwork was neither promoted nor rewarded and hence virtually nonexistent. The focus was not so much on achieving excellence as on making sure people made no major mistakes. Creativity, imagination, and an entrepreneurial spirit were not encouraged. On the rare occasions that someone would try to improve things, almost always that person felt criticized by the effort, and these change agents were then not viewed as "team players." "Getting along" was much more important than "getting good"; and this was often reflected in the individual performance appraisals of employees. Finally, I found a lot of little fiefdoms with highly protective owners—so protective that individual subordinate development was not occurring. And in the end, many people were not being prepared for promotion or transfer; doing this was viewed as weakening the little fiefdoms.

- All the plant process information was now electronically available in real time, which was not the case 25 years ago. This should have been a huge asset; however, in many, many instances, it worked to the detriment of the refinery. With this information readily available at your desktop, the need to go to the plant, to go to the gemba, seemingly is reduced. But more often than not, failing to go to the gemba is an enormous mistake. In lean we go to the gemba to grasp the real situation, which means gathering information beyond just the pressures and the temperatures. Of course, the "real situation" includes the process information, but it also includes the situational information occurring at that time. It provides product information and ancillary information that may not be transmitted. However, most importantly the "real situation" includes information directly from those involved, and that can only be obtained at the gemba. So now engineers, planners, and operations managers, to name a few, could access the process information from the quietude of their offices—and they did. In some cases, this helped with the efficiency of the problem solvers, but my overriding observation was that many of these people stayed in their offices not because they should but because they now *could,* and the problem solving suffered. You will learn that "going to the gemba" is one of the Five Guiding Principles in *The Toyota Way, 2001,* the cultural guidance booklet for the entire Toyota Corporation.

- The bureaucracy had increased dramatically. There were more procedures and more forms with more needed signatures, and worst of all there were a lot more lousy meetings that were almost universally poorly managed and sporadically attended. In addition, there were tons more staff personnel with coordinators all over the place. There were maintenance coordinators within the Maintenance Division and maintenance coordinators within the Operations Division. There were operations assistants to coordinate the work of the engineers as well as coordinators for the environmental groups. There were people on top of people and people to keep track of the people who were on top of the people. The bureaucracy was both stifling and demoralizing. It was disturbing.

What has all of this taught me about lean in an oil refinery?

First, it will work nicely, possibly better than in the tier one auto suppliers you read so much about. Second, the gains will be large. There will be financial, quality, and delivery gains, as well as improvements in environmental performance, safety, and morale. More than that, you can create an environment that feeds all the intrinsic motivators, and your plant will be a place that people want to work in. Not only does business performance improve, but relationships improve and business-created stress is reduced.

Most people will tell you that you *cannot* improve quality without increasing costs and that you *cannot* improve delivery performance without increasing inventory volumes, especially in an oil refinery. What I have learned is that with a lean manufacturing system, you *can*. Lean has the key qualities that allow you to unlock the potential in your business—and in your people—so you get a whole lot more and for a whole lot less, and you *can* have your cake and eat it too.

Finally, lean manufacturing is all about the people as well as the process. This fact has been missed by most of the companies implementing lean, and I have made that point clear in many places. In addition, this book has several case studies and examples that show how being part of a lean transformation can have a very positive effect on a more personal basis.

Enjoy the journey.

Part 1 Lean Background

CHAPTER 1

THE STORY OF THE TOYOTA HEADREST CELL, THE THETA CELL

Aim of this chapter—*The story of the Theta cell tells in graphic detail what the concept of kaizen is—in practice. It is no understatement that many of my Japanese management friends, and sensei, refer to lean as the "kaizen mind-set." "Lean" is a term coined in North America; and many auto company and Japanese firms are hesitant to use the term "Toyota Production System" (TPS) for a variety of good and bad reasons. Yet whether you call it lean or the TPS, all agree that the basis is a kaizen mindset. Here is the story of how to create a kaizen mindset. In it, you will see an accurate implementation of the famous TPS.*

BACKGROUND

The Theta cell was designed to produce a headrest for Toyota. This cell was one of five value streams in this plant. Some value streams were tier one and others were tier two suppliers to the auto industry. This cell was U-shaped in design. It was a tier two value stream and was currently staffed with 16 workers. The design of the cell was such that at modest defect rates, the profitability was very labor sensitive. Target profitability required the productivity be 100 headrests/person/shift. Following PPAP and a brief start-up period, the cell leveled out at 65. (PPAP is an abbreviation for the production parts approval process, which is a standard in the automotive industry. The PPAP is a major part of the process and product validation, prior to the initial shipment of products.)

After two years and numerous efforts to improve, including visits from the home office support staff, the workers still struggled to reach a productivity level of 80; the cell was losing money. This is where the story of the Theta cell begins.

This plant, located in Biloxi, Mississippi, with its five value streams was part of a much larger international corporation that had decided to implement lean manufacturing across the corporation. The executives selected a steering committee, primarily from the C-suite and the next level of reports, the Business Unit VPs. In addition, they chose a lean support team (LST) that reported to the chief operating officer (the COO, not the CEO). To support this team, they created two positions, managers of process excellence (MPEs), and filled them with two mid- to high-level managers. One was most recently a division manager over six plants, and the other was formerly a plant manager of a recently closed plant. Both had training in previous programs of process excellence, both were very knowledgeable of the production process, and both had proved to be excellent managers and excellent problem solvers. The ex–plant manager had some experience with lean manufacturing and the Toyota Production System, but both were far from being experts in the field.

To augment the LST the executives hired an outside consultant; that was me. Although I disliked the term, they called me their "sensei." My task was to fill the expertise gaps and to provide guidance and balance to the lean transformation effort, including the steering committee and senior management. In addition, as they had a strong, rigid, and in many ways highly dysfunctional culture, I also, as always, had the task of being the "outside observer," as I was not "enculturated." With this as a background and with very little (like practically zero) training of the management team, very little preparation, and a weak plan—and contrary to my advice—the steering committee nonetheless decided to dive into action. The first major effort was to launch two lean tools, 5S and kaizen, across the entire corporation. Shortly thereafter, the two MPEs were shuttling across the world to the 40 plant locations teaching 5S and kaizen.

At the recommendation of the sensei (me), the steering committee decided to do some pilot testing to gather the information needed to plan a full corporatewide transformation. Six sites were chosen as alpha sites; this plant was one of them. The alpha site work consisted of using the LST to implement the transformation model at these six facilities and evaluate the strengths and weaknesses of the transformation model. The plan for the alpha sites was to visit each site once per month for the next six months; this was done. Recommendations were then made to the steering committee, and the transformation was started. Although a great deal was learned from the alpha sites, only minor changes were made to the previous plan. Hence the plan for the corporatewide rollout was still very weak, with only general goals and no measurable milestones. However, one very positive outcome was that two beta sites were selected to test new techniques and further refine the lean strategies, tactics, skills, and standards. The new strategies, tactics, skills, and standards would then be rolled out to the rest of the corporation. The steering committee chose the Biloxi site as a beta site based on our recommendation. We were very impressed by the plant manager, a "salty, seasoned" man who had shown a remarkably keen interest in lean and proved to be a voracious learner during the alpha site work.

THE PATH TO LEAN

A six-step process whose purpose is to create flow and drive major sources of waste and loss from your value stream.

As the first step, before going to the plant, we wanted to do all we could to "onboard" the members of the senior leadership. Consequently, we discussed our rollout plan with them. That included the plant manager, the director of U.S. operations who had management responsibilities for three plants, and the VP of the Worldwide Business Unit. The plant manager, Dirk, was more than enthusiastic; he could not wait to get started. The director was highly interested and wanted to get going; he said he would be at any meetings or trainings we needed; just give him a day or two notice since he worked at the home office 400 miles away. The VP, on the other hand, was somewhere between indifferent and opposed to the entire effort. The truth is, as it was eventually revealed, he only supported it because the COO was pushing this transformation. This VP, who literally knew nothing about lean or lean concepts, nor did he even attempt to make a superficial effort at learning about lean, was politically astute enough to appear to support the lean transformation. He learned the jargon and attended the steering committee meetings, and that was about all. This was more than disconcerting. Nonetheless we proceeded.

Our first order of business was to advise the three managers of the results of the alpha site work. We shared the strong points and the weak points we had cataloged as part of the alpha site work. The managers, like those at the other alpha sites, showed an almost total lack of industrial engineering skills, and we recommended they hire two industrial engineers. The VP balked totally, as he said this would add substantially to the plant's overhead costs. After a good bit of discussion, pointing out the progress the plant had made with the LST, and with some help from both the director and the plant manager, the VP reluctantly agreed to add one engineer. An industrial engineer from a nearby college was hired and arrived four months later following his graduation.

> **THE MEANS TO LEAN**
> To problem solve your way to the ideal state through the total elimination of waste using a fully engaged work force.

At the Biloxi plant, we continued the pattern of meeting monthly for one week at a time. One member of the LST was assigned to be the mentor-teacher for the plant. The role of the LST was to provide the guidance and the technical support needed so that the facility could make a true lean transformation, including the cultural change that is so vital to sustain the efforts. The primary task of the LST was to train the plant manager, as he was the key change agent. Later you will learn that this method of teaching—training the plant manager and having him train his subordinates, etc.—is absolutely critical and is the key to proper deployment as well as both sustaining the gains achieved and continuing to improve. At the plant, a lean guidance team (LGT) consisting of Dirk and his eight direct reports was formed. This group would meet each time the mentor arrived. The week would start with a reflection on recent efforts followed by creating an agenda for the week's activities. Each week's training and activities were led by Dirk, with the mentor providing support and guidance as needed. Frequently the mentor would hold specific skills training sessions for those implementing kaizens and other change activities; that is, JIT training. In addition, during each visit the mentor would spend time with several key plant members as well as make independent observations. Each day would end with a wrap-up, and the subsequent day would start with a reflection on the prior day's activities. Following this pattern, a great deal of work was accomplished, but for the sake of brevity, we will focus on one value stream, the Toyota headrest cell, the Theta cell.

THE KAIZEN EVENT

Our plan for the Theta cell was to go to the gemba and observe. First, we provided some very focused training on observing and performing time studies. After this we introduced the *path to lean*:

1. Synchronize supply to the customer (external synchronization).
2. Synchronize production (internal synchronization).
3. Create flow.
4. Establish pull-demand systems.
5. Standardize and sustain.
6. Improve and drive to the ideal state.

(For a detailed discussion with several examples of the path to lean, refer to How to Implement Lean Manufacturing, *2nd ed.)*

At this time we also introduced the concept that the *means to lean* is to problem-solve to the ideal state through the total elimination of waste using a fully engaged workforce. Then we got organized and assigned responsibilities and proceeded to the floor. The LGT was to do spaghetti diagrams and studies on each of the 16 operators, plus make general observations. In addition, each of us was to write down as many possible kaizen opportunities as we could find. We spent the entire afternoon on the floor observing and gathering data.

The information we gathered was revealing. The spaghetti diagrams looked just like the plant, a total mess of disorganized activities. The utility operators (there were three) would move into areas that were falling behind and "help them." Operators would frequently leave their workstations to chase both parts and tools. The time studies were particularly revealing. Cycle times were all over the map, with many workstations showing +200% variation from the mean.

Also, we found that the work content at each workstation might change with the workers. Some workers liked to do certain activities, and as they moved around the cell, frequently they would bring the work with them. The work at each station, although not defined by lean standards, was well enough defined that this was obvious. In addition, WIP inventories would build up at a workstation; a utility operator would respond to assist the worker and, once done, would then move to the next accumulation. It was total chaos—a target-rich environment for lean activities.

The Theta cell has a lead hand, a nonsupervisory worker with only a few defined duties. One duty was to review rejected headrests and rework them or scrap them as necessary. In addition, the lead hand was the material stocker for the cell, but if he or she were busy, someone else would go to the warehouse and acquire the parts, and guess what? Yes, you got it! Subassemblies would then accumulate in front of that vacated workstation, and one of the utility workers would fill in the gap, that is, if a worker was available. It was all I could do to wait until we got together to start straightening out this "target-rich lean opportunity," but wait I did! Amid all this chaos, there was some really good news: this was a motivated group! The members worked hard and really wanted to do better; they just did not yet know how to do it. Early in the evening we adjourned to the conference room to understand our data and observation. We combined our thoughts, plotted our measurements, and reviewed our notes and spaghetti diagrams until very late in the evening, at which time, tired but optimistic, we adjourned for the evening.

THE INITIATIVE MANTRA

- Start where you are
- Use what you have
- Do what you can

The following morning, while the cell was operating in its "normal mode," we made a quick visit to the floor and then met to reflect on all we had seen. After a review of the data and combining the list of improvement activities, we embarked on a design review. We had over 90 improvement ideas, and we knew we could not do them all.

I reminded them of the Initiative Mantra, and we decided to follow the implementation outline in *How to Implement Lean Manufacturing* (2nd ed.); this was a key part of the rollout plan.

Consequently, we began by implementing the "path to lean" for the Theta cell:

1. We first calculated takt. We then adjusted takt for what we believed was the OEE and calculated the cycle time at each workstation (synchronized externally).

2. With a great deal of help from key members of the LGT, we synchronized internally. First, we analyzed our time studies for each workstation to obtain the total work, as it was now defined. Then, on paper, we reallocated the work until each person had about the same amount of work and it was about 80% of takt time. While doing this, we were able to combine work elements, and consistent with our takt time calculations, we could eliminate two workstations. This allowed us to proceed with two less staff and tighten up the cell. We wrote some new job instructions detailing the changes in work assignments. Since we had not changed the work, only reallocated it, this was mostly a cut-and-paste activity.

3. Next, to facilitate flow and create pull, we introduced one more rule, and we worked at this for the next 18 months. That was, "Do not build ahead." Each workstation had a "finished work" location, a kanban square, and when it was full, the worker needed to stop producing subassemblies. The workers had a serious problem with this. Standing still was not part of their culture; they clearly were workers. Over the years, their management, like so many other managements, had spent so much time making their people efficient, they forgot to make them effective. Working hard and staying busy was part of their culture, and they did not adjust well to a directive that told them that "when your kanban square is full, stop working."

In addition, we:

1. Eliminated the need for the utility operators and assigned them to another cell. This was a key point, as the utility workers effectively "hid" the problems.

2. Redefined the work of the lead hand to first make sure all stations were stocked with both parts and tools. Second, he or she would look for problems, especially materials accumulations, and begin to problem-solve the issues. Finally, the lead hand would do rework on a fill-in basis.

3. Told all workers to stay at their station, and if they needed something to ask the lead hand.

4. Explained the need for each worker to do the work at his or her assigned station and no other work.

Looking at the spaghetti diagrams, we could see that we needed to relocate some equipment. That night we moved and reanchored the equipment. We put down some temporary lines with tape to start to make the cell "visual." This included floor placement for containers and pallets plus the finished work kanban squares mentioned earlier. We completed that work by midnight, and we were ready for the day shift.

The next morning, when the operators arrived, the production manager trained them along with their supervisor, and we started producing with the new design. By the end of the shift, we were producing at a rate well over 100 headrests/person/shift. The plant LGT met and did an informal on-site reflection, and we made some minor changes prior to the second shift's arrival.

Upon arrival, the supervisor from the first shift, along with the production manager, met with the Theta cell team for the second shift. He told the team of the redesign and summarized the changes including the new work rules, and the second shift started up without incident. The lead hand, Derek, kept all stations stocked, and work was flowing smoothly.

A few hours into the shift, I was talking to Derek, who was a very competent guy but equally grouchy. I asked him, "How do you like this new layout?" His reply I cannot print; simply put, he did not like it at all. He felt the layout had slowed production such that he would have to be working this Saturday and he had planned to go deer hunting instead. I dug into what he meant, and finally he said, "We are just going too slowly." I told him we were producing at least 50% more than this cell had ever produced. Again, I cannot print his reply, but in effect he told me I was crazier than a loon. Calmly I challenged him to count pallets as we produced them and calculate the rate of production. We did this together, and when he saw the cell was producing 90% more than it

THE SIX QUESTIONS OF CONTINUOUS IMPROVEMENT

A six-step process of structured problem solving and decision making to guide thinking so that both PDCA and hypothesis testing are incorporated into the thought process.

had ever produced, he was flabbergasted. His reply was, "But nobody is working too hard." He simply did not believe the production count and said he was going to recheck his numbers.

Derek did not believe we were producing more because he had inappropriately equated motion with work and effort with productivity. I returned at the end of the shift, and Derek and his team had produced at a rate of 158 headrests/person/shift, roughly double their average. Derek and I discussed it, and he was able to give me numerous examples of waste we had designed out of the system. He was now a believer . . . a bit skeptical still . . . but we had definitely gotten his attention. At that time, I told Derek to oil up his musket, get his gear ready, and warn the deer, because he would be hunting by Saturday. In the next three days we made some minor modifications to the flow, but by the end of the week we were producing at 188 headrests/person/shift. I left for El Paso feeling really good about what the team had accomplished, and we were just getting started. And Derek went hunting.

KAIZENS GALORE

The next week I got a call from Dirk. He was concerned, as the operators were now overproducing so much they could get by working only six hours per shift, and he did not know what to do. We gave him some instructions and following that, Dirk did some rough time studies, rebalanced the workstations and was able to reassign one person to another cell. Soon enough, after the opearators had implemented more kaizens and people got used to the new work rules and flow, once again the cell began to overproduce. Again Dirk called, and after doing some calculations and considering some options, he decided to shut down production at the seventh hour and assign kaizen activities to the team for the final hour of each shift. The operators implemented many kaizens, including upgrading their 5S standard and building a production-by-hour (PBH) board. At this time, we gave the assignment to the lead hand to initiate problem solving when production differed by +/– 15% from the target on the PBH board, but if production was in that range, to just keep producing—and smile. We effectively introduced the concept of "normal" versus "changed." Production smoothed even more, and the overall rate increased. Each time that we reduced the staffing in the cell, we rebalanced the work and retrained the workers. Later we introduced TWI training and instituted standard work for each workstation and used standard work combination tables to assist in problem solving. (TWI training is a methodology developed as part of the Training Within Industry service.)

When I arrived the next month, the operators consistently produced more than 225 headrests/person/shift, and the productivity was rising. They would make improvements, rebalance the work, and reassign the released worker to another cell. At that visit we reviewed the activities at the Theta cell but really focused on general plant improvement efforts that could be applied plantwide.

Before my next visit, I reminded Dirk of the training he had received during the alpha site phase, specifically the Six Questions of Continuous Improvement:

1. What is the present condition?
2. What is the desired condition?
3. What is preventing us from reaching the desired condition?
4. What is something we can do, right now, to get closer to the desired condition?
5. When we do this, what should we expect?
    a. What will happen?
    b. How much of it will happen?
    c. When will it happen?
6. What have we learned?

I asked Dirk if he felt comfortable training his staff members on this technique and if he felt they could train their subordinates. He enthusiastically said yes, and when I arrived for my next visit, people were deeply into structured problem solving. I went to the floor and asked the lead hand about his problem-solving activities. He proudly showed me a "whiteboard" he and his crew had made to do the problem solving. They had continued to improve and once again had about 45 minutes at the end of each shift to do group problem solving.

I continued to visit the plant, and the process continued to improve. After the second visit we really had no more "kaizen events" where the engineers and supervisors locked themselves up in a room and redesigned the flow. However, each month the number of improvements (kaizens) increased. Two of the kaizens were relocation of pallets for ease of pickup and relocation of workstations to reduce the cell size; we moved workstations closer to one another, to reduce transportation and WIP. Other kaizens included changes in work procedures, redesign of fixturing for ease of use, and reduction of batch sizes of incoming parts. The kaizens were too numerous to mention all of them here, and most were never formally documented. But they were all discussed within the cell, and before any changes were made, they were put up on the whiteboard to solicit the input from all three shifts and then executed.

It was a system that had a bias for action. The plant did not have a bias for documentation, nor did it have a bias for the ritual of bureaucratic form-filling and higher-level review and approval. The supervisors had an opportunity to review the whiteboard and get in their input, if need be. The vast majority of the kaizens were implemented via the Nike principle, Just Do It. Furthermore, if the kaizen required some engineering assistance, the workers would contact the engineer directly, get his input, and then proceed. Everyone was focused and aligned on continuous improvement at this cell, and it was being done through problem solving, elimination of waste, and a fully engaged workforce. The line people were engaged, the support staff was engaged, and the supervisors and managers were engaged.

My favorite two kaizens were both at the final inspection and pack-out workstations. The first kaizen involved the relocation of an entire workstation, and the other kaizen was the relocation of the pack out containers at the work station. The first kaizen occurred at the end of the line, where there were two workstations for inspection and pack-out. The parts at these stations approached the operators differently as they advanced on the conveyor. For one work station, the parts were presented to the operator "right hand first," while the parts for the other operator were presented "left hand first." An operator noticed that for the "right hand first" acquisition, this operator then transferred the part to his left hand so he could use his right hand to "guide his eyes," which was part of the inspection standard. This added almost 2 seconds to the cycle time, which was around 20 seconds. To capture this opportunity to reduce the cycle time, the workstation needed to be moved to the other side of the conveyor, necessitating a complete redesign of the workstation. The team did it during a routine kaizen. Up to this point that type of redesign was left to the engineers or supervisors. But this showed we had now begun a new phase of work, and this was clearly a cultural change for this plant, which, by the way, had a strong and formerly very antagonistic union presence.

The second kaizen was much subtler but even more impactful. At these same workstations, the pack-out containers were directly behind the operators necessitating a full 180 degree turn of the body, to insert the parts in the containers. One operator did an ad hoc trial of his own. He placed the containers at his side and had the lead hand do a time study. They found they could save over 3 seconds/part because the operator could now pack parts without moving his feet to turn around. He could reach the containers by just rotating his trunk a little. It presented a problem, however, as it now made it harder for the forklift to access the parts. During one of their 1-hour kaizen periods, they addressed this issue, and not only did they redesign this workstation, saving around 14,400 seconds/day (4 hours of work), but this became the new standard for the plant and was implemented at over 24 workstations. These improvements, operator initiated and operator implemented, are a true sign of a kaizen mindset on the floor and a true sign of genuine employee engagement.

In the ensuing months, literally hundreds of kaizens were done, and nearly all were done solely by the line workers. Eleven months after the initial kaizen event, this headrest production was discontinued when the new model was introduced. In its last month of production, the cell productivity was 304 headrests/person/shift.

IS KAIZEN AN EVENT, A PROCESS OR . . . ?

The story of the Theta cell is an example of the difference between doing kaizens and doing kaizen events. Kaizen events, kaizen bursts, kaizen blitzes, kaizen projects, kaikaku, jishuken, or workshop blitzes, whatever you wish to call them, are management-led and engineering-intensive improvements. Kaizen events are distinguished from kaizens, which are normally smaller projects done directly by the line worker. To create a kaizen mindset, neither the kaizen nor the kaizen event is sufficient; yet both are necessary. In the Theta cell we redesigned the cell and made a huge initial improvement from 80 headrests/person/shift to 188; this was a kaizen project. It was a large and necessary improvement, and had we turned it over to the line workers to execute it likely would have failed. They did not yet have the skills to design and execute a change of this magnitude and complexity. However, over the next 11 months, the line workers and their supervisors executed over 100 kaizens, and the productivity rose to 304 headrests/person/shift. By the same token, the management and engineers often did not have the intimate process knowledge to guide those smaller kaizens. It takes the ability to do kaizen events, as well as kaizen, to make the improvements necessary to create a kaizen mindset, which then will allow you to achieve world-class performance.

IF WE DO KAIZEN EVENTS AND GET OUR FLOOR TEAMMATES TO IMPLEMENT KAIZENS, IS THAT ENOUGH?

If you looked on the floor, that is what you would have seen: one major event done in one week and then a whole bunch of smaller kaizens done over the next 11 months. That alone would have impressed you, and yet those two activities cannot stand alone. And that is what many companies try to do. They initiate kaizen events using management and staff subject-matter experts to make process improvements—and they predictably will make progress. Then they roll out the concept of kaizens to make process improvements. What predictably happens is while the managers are busy breaking their arms patting each other on the back, the plant's "improvements" slowly deteriorate, and the workers, who have been disempowered for the last 30 years, are now predictably confused. You see, they don't really know how to use their newfound "empowerment."

As surely as this approach will be effective initially, these gains will wither and die in the fullness of time. The problem is that the third part of the "kaizen mindset of management" is missing. Not only must the members of the management team be preaching kaizen, but they must be teaching kaizen; and furthermore they must be practicing kaizen themselves. They must become lean leaders to complete the kaizen mindset for the facility.

So what did Dirk and his team do to create this third leg of a kaizen mindset? It is simply said but far less often done, and that is they got fully engaged as a management team and provided what we now call "lean leadership." Let me introduce you to the model Dirk and his team used on each step, large or small, as they navigated through the murky waters of changing the culture in the plant. He and his team transformed this cell from one that lost money for over two years to one that became a huge moneymaker for the plant. They did this while they improved productivity, quality, safety, and on-time delivery and made a huge improvement in morale as well.

Dirk started in each case with the Six Questions of Continuous Improvement. This meant that at each step, he had a clear vision of his objective. Next he went to the floor to get real-time data, and then he compared the current state with the desired state. In the old days, we called this a "gap analysis." Next, with the appropriate members of his team, he asked, "What is preventing us from reaching the objective, the desired state?" Following this, the team came up with "countermeasures" (these are the so-called tools of lean), and the team members were asked, "When we do this, what should we expect?" regarding their predictions. Finally, the process ended with a reflection based on the question "What have we learned?"

And to show their management commitment, Dirk and his team did this with their management problems as well, those not on the floor but in the office. They attacked such issues as:

- How do we deal with high employee attrition?
- How can we implement the TWI training model?
- How do we improve morale?
- How do we teach the six-question format?

And in so doing, they modeled the correct behavior by using the six-question format. They did this *before* they expected their subordinates to use the six questions and hence became role models as well as teachers of this process. In addition, they studied the Six Skills of Lean Leadership (see below); then they practiced those skills assisting and critiquing each other as they jointly learned as a team. As a unified team they actively changed their style of leadership to the style defined so well in *How to Implement Lean Manufacturing* (2nd ed.), as explained on the following page:

What Model of Leadership Should We Change To?

First, leadership is not some manager sitting in his office presuming that he knows it all, and "all of it" can be expressed in a spreadsheet that he can receive on his PC and then, from the comfort and solitude of his office analyze this information and make a calculated decision that will drive his business to prosperity. This model must be discarded. It simply does not work and it will never work. I call the new model of leadership—well, it is not really new; the Japanese have been using it for 70 years that I know of. It is called lean leadership. It has six basic qualities (the Six Skills of Lean Leadership), which are as follows:

1. Leaders as superior observers: They go to the action—they call it the Gemba—to observe not only the machines and the products but also to spend significant time with the employees. They strive to be aware of not only the products and the processes but more importantly the people. They also are in contact with their customers. A much overlooked leadership skill they have in abundance is the ability to be an empathetic listener.

2. Leaders as learners: They do not assume they know it all. Rather, they go to the floor to learn. They are in "lifelong" learning mode. They are masters of the scientific method. They learn by observing and doing, but most importantly they are superior at asking questions, they learn by questioning.

3. Leaders as change agents: They plan, they articulate, and sell their plans, and they act on their plans. They are not risk averse, yet they are not cavalier. They do not like to, but are not afraid to make mistakes.

4. Leaders as teachers: They are "lifelong" teachers. When something goes wrong, their first thought is not "Who fouled up?" but "Why did it fail?" and "How can I use this as a teaching opportunity?" They teach through the use of questioning rather than just instructing.

5. Leaders as role models: They walk the talk. They are lean competent. They know what to do and they know how to do all of

Lean that is job specific to their current function in the organization. There is no substitute for this. NONE.

6. Leaders as supporters: They recognize they mainly get work done through others, so they have mastered the skills of "servant leadership."

Taken from How to Implement Lean Manufacturing, *Second Edition.*

Next, as Dirk and his team "problem-solved their way to the ideal state," or simply put, implemented the kaizen mindset, they found dozens of obstacles along the way. These obstacles were largely the answer to Continuous Improvement question 3, "What is preventing us from reaching the desired condition?" As they found these obstacles, they pulled in the necessary countermeasures so they might make progress. They first learned the techniques from their sensei, then they used and practiced these techniques, and finally they taught them to their subordinates. Often these countermeasures are referred to as the "tools of lean." Call them what you like; they are a means to an end, and that is "better, faster, and cheaper." A partial list of the countermeasures Dirk and his team implemented included:

1. Cellular manufacturing, hybrid cellular-flow line manufacturing
2. Producing to takt
3. Line balancing, time studies
4. Stable flow at takt
5. Standard work
6. The seven wastes, TWO DIME
7. The path to lean
8. Lean leadership
9. Leader standard work
10. Leading from the front
11. Cultural basics

12. A culture of continuous improvement
13. A culture of respect for people
14. The Five Cultural Change Leading Indicators
15. The Five Tests of Management Commitment
16. Best methods
17. OEE with quality losses, cycle time losses, and availability losses
18. EPEI, a planning algorithm
19. Reactive versus preventive versus predictive maintenance
20. The means to lean
21. The Cultural Mantra
22. Learn-do-reflect
23. The Continuous Improvement Mantra
24. The Initiative Mantra
25. Problem solving to reach the ideal state
26. Value stream mapping
27. Value-add versus non-value-add versus non-value-add under the current conditions
28. Target conditions
29. PDCA
30. The Six Questions of Continuous Improvement
31. Sustaining the gains, the Five-Step Model
32. How to standardize
33. Kepner-Tregoe problem solving
34. Toyota practical problem solving

35. The Five Whys
36. Statistical process control and statistical problem solving, and we created one Black Belt for the facility
37. Hoshin kanri (HK) planning using catchball with consistent hourly, daily, weekly, monthly, annual, and five-year goals
38. The visual factory reflective of HK plans
39. TWI training with a TWI support structure (Training Within Industry services)
40. Job cross-training
41. Individual job training
42. Individual career plans
43. Calculating lead time, lead-time reductions
44. Three-part inventory system
45. MTI, MTO, and FTO systems (make to inventory, make to order, and finish to order)
46. Job succession planning
47. Five-year individual growth planning
48. Five-year facility staffing plans
49. Group dynamics
50. Meeting management skills
51. How to brainstorm
52. Team leader facilitation skills
53. Layered audits
54. Kamishibai audit system

This is a rather awesome list of countermeasures (the lean tools), to say the least. And Dirk and his team were by no means thoroughly skilled in all these tools. For some tools, such as TWI training, OEE, standard work, and the Six Questions of Continuous Improvement, plus line balancing and time studies, they not only were skilled, but were capable of teaching others. For other tools, such as the three-part inventory system, SPC, leader standard work, and hoshin kanri planning using catchball, they were applying and rapidly learning. For other skills such as Kepner-Tregoe problem solving, statistical problem solving, job succession planning, and the five-year facility staffing plans, they were just getting started. Still, Dirk and his team were well on the way to learning both how to apply these tools and how to integrate them into a cohesive system that mimics the Toyota Production System.

In a nutshell, with the help of their sensei, they were practicing their way to competency while making huge progress in plant performance.

To put the plant performance in perspective, in this, the second year of the transition:

- The plant had its best safety year ever.
- Morale was visibly improved and confirmed in the Annual Engagement Survey. On a visit from the CEO, after talking with a variety of line workers, he was visibly impressed, and he said, "Now we finally have a place we can bring our customers to and they can see for themselves our huge improvement and our lean progress. Talking to the people you can just feel it in what they say . . . and how they say it."
- On-time delivery was practically 100% with a 30% reduction of inventory.
- All quality metrics, both internal measures and external measures, had improved.
- Most telling of all is that the gross profit of the plant had risen from 5% at the beginning of the year to 14% at year's end.

All this while the *demand had decreased.*

Now you can see the plant has the third element of a kaizen mindset, an engaged management team with the skill set of the lean leader and the plant performance to prove it. Once again, we reaffirmed our beliefs that:

- The vast majority of skills needed to vault your enterprise into becoming a lean enterprise are already inherent within your organization.
- The key to success is largely an issue of unleashing these skills.
- The major and very critical tool needed to unleash these skills, and hence to become a lean enterprise, is to learn and employ the skills of lean leadership at all levels of the organization.

As you read this book, we will frequently refer to this example, so dog-ear the page or put in a bookmark; we'll come back here often.

CHAPTER SUMMARY

The Theta cell is an example of "lean done right." There are four very strong lessons to learn from this example. First, even though this is an example of just one value stream, it is also the model to follow in implementing lean in an entire facility. Second, it explains in detail the concept of a kaizen mindset, especially the need to do large projects, the kaizen events, as well as small, operator-driven kaizens. Third, it points out the need for ongoing support and skill training for all the people involved and for the method of that support, which is to train the people on the management team and have them train their subordinates. You did not see even one example of classroom training or "death by PowerPoint." Fourth, it is a very representative example of the "means to lean," which is to:

- Problem solve your way to the ideal state
- Through the total elimination of waste
- Using a fully engaged workforce

CHAPTER 2

WHAT IS LEAN MANUFACTURING?

***Aim of this chapter**—It is interesting that the very definition of lean manufacturing is still a huge discussion point, even among the lean professionals. This I believe to be a rather large problem for those wishing to embark on the transformation to a lean manufacturing system; people should know what they are to accomplish before they embark on a task. As Yogi Berra once said, "If you don't know where you are going, you might end up somewhere else." So let's start out with the term "lean" and then get into just what lean manufacturing is; and then we will end with a useful, working definition.*

LEAN SOUNDS SIMPLE . . . DANGEROUSLY SO. HENCE PROCEED WITH HUMILITY

Interestingly, the fact that lean sounds so simple is a large problem for the lean professional who is teaching the subject of lean manufacturing. Since it sounds simple, much like being a parent sounds simple, many people underestimate the effort it takes to become proficient in lean. Well, it is a problem at first. Slowly but surely, as a teacher of this fine technique, I find that my students start to "get it" as they practice it—and not before. I cannot be any clearer on this topic: to understand the lean philosophy and the lean methods, you absolutely need to practice them. There is no substitute . . . none. And in so doing, you take the journey described by Justice Holmes when he said, "I don't give a fig for the simplicity on this side of complexity, but I would die for the simplicity on the other side."

You see, in learning about, and understanding *to the point of being wise*, almost any subject of substance, we all take the same multistep, *thought-action-reflection, thought-action-reflection* repetitive path as outlined as follows:

1. At first, when we know very little, it is normally very simple to understand, or so we think.

2. So we try to generalize. We find that those generalizations, while initially seem very clear in our mind, are often not quite so clear when we teach them or even discuss them with a colleague. All too often we will find that those seemingly well-articulated thoughts and ideas can be quickly defused by a simple question or two from someone with more knowledge, or sometimes by someone with significantly less knowledge.

3. And the final test of our ideas is, Will they face the refining fires of reality? Will they work when tested in the field? What happens when we really act on these ideas?

4. All too often, in part or in total, they fail this "reality check." We then conclude that it is not so simple as we once thought. And if we persist in action, we learn more, and soon it gets more complicated; as we practice and execute, learning even more about the subject matter, it often gets very complicated.

5. Through careful and introspective reflection, we find that our earlier ideas and generalizations that we once thought were pretty simple are now "not so simple;"rather, they are pretty complicated.

6. We now have many questions we had not even thought of on day one.

7. And then if we test our concepts, practice more, persist, and pay close attention, we will find that we will both succeed and fail.

8. And if we reflect upon and learn from both the successes and the failures, we will again gain more wisdom, and we might begin to see some "common threads of thought."

9. Then, in a cold, hard, dispassionate analysis, if we are humble, introspective, and thankful for the successes, as well as being

reflective, persistent, and challenged by the failures, at this time we are in danger of entering the rarefied air of becoming wise about our subject matter.

10. When we repeatedly go through this entire process of thought, action, and reflection and put in this deep physical, intellectual, and emotional effort, very likely we will come to understand the basic underlying principles of our subject matter . . . and voilà, once again it becomes simple.

11. Maybe, just maybe, we have reached "the simplicity on the far side of complexity."

12. We will then be able to simply explain these complex concepts such that they have meaning to the neophyte and the expert as well.

13. But this never happens without a long, thoughtful, action-based, reflective, and persistent journey filled with both successes and failures.

And your journey to become proficient in lean manufacturing will take this path—that is, if you "think-act-reflect" and do it repeatedly and do it with depth. All such learning efforts are "ongoing experiments," as almost no one can chart a precise and successful path very far in the future. In addition, unless you are divine, you can expect to make some mistakes along the way. However, if you rigorously follow the thought-action-reflection process manifest in the Six Questions of Continuous Improvement and learn from those successes and mistakes alike, they too will guide you to an improved direction and an improved destination.

WHAT IS THE ORIGIN OF THE TERM "LEAN MANUFACTURING"?

In their book *The Machine That Changed the World,* authors James Womack, Daniel Jones, and Daniel Roos used the term "lean manufacturing" when they were referring specifically to the Toyota Production System (TPS). First they defined "lean production":

After World War II, Eiji Toyoda and Taiichi Ohno at the Toyota Motor Company in Japan pioneered the concept of lean production. The rise of Japan to its current economic pre-eminence quickly followed, as other Japanese companies and industries copied this remarkable system.

and then they credited John Krafcik with coining the term:

Lean production (a term coined by IMVP researcher John Krafcik) is "lean" because it uses less of everything compared with mass production—half the human effort in the factory, half the manufacturing space, half the investment in tools, half the engineering hours to develop a new product in half the time. Also, it requires keeping far less than half the needed inventory on site, results in many fewer defects, and produces a greater and ever growing variety of products.

Well, from there it is no stretch to change the word "production" to "manufacturing," and we have created a whole new concept that, if used properly, will change the world; that concept is *lean manufacturing*.

A SIMPLE, USABLE DEFINITION BUT WITH QUESTIONS UNANSWERED

A simple and fairly accurate, although not very detailed, definition of lean manufacturing is "Problem solving your way to the ideal state through the total elimination of waste." This definition raises the questions:

- What is the problem solving that needs to be done?
- How do you go about this problem solving?
- What is the ideal state?
- What is waste?

- Just what is the uniqueness of this manufacturing system?

And not surprisingly, once we delve into the answers to these questions, dozens more are raised. And the key question that remains in probing this definition is, "Just what is the uniqueness of this manufacturing system?"

Almost all businesses have some type of ongoing problem-solving effort in place. Hence, with a clear conscience and with conviction, they likely will say, "We are doing that already." They will readily agree that neither do they focus on quantifying the seven wastes, nor do they speak about the "ideal state;" but they are working toward their annual improvement goals and trying to reach the year-end objectives. So regardless of the jargon used, they will claim that their behaviors are the same as this definition of lean manufacturing. And that is a decent argument when applied to this commonly used definition of lean manufacturing.

However, the problem is that this response, upon scrutiny, is misleading if not outright wrong. First, what they call problem solving is more appropriately called firefighting. Second, they are not striving for the long-term goal we call the ideal state. Third, while they may be removing waste, there is no concept of "total removal" with process improvement; rather, it is "Let's remove enough to make the bottom line look better," and the total focus is on results. Lean manufacturing goes much deeper than what most people envision it to be.

Finally, there is one major, very major aspect completely missing. Just *who* is doing the problem solving? More, much more, on that later.

WHAT DID OHNO SAY?

Taiichi Ohno was the architect of the Toyota Production System. Perhaps we should listen to what he said. Although he never gave a specific definition of the TPS, in his writings he was clear.

His most notable writing on the subject is his book *Toyota Production System: Beyond Large-Scale Production*. In it, Ohno makes three key statements, which when taken together define his TPS:

- "The basis of the Toyota Production System is the absolute elimination of waste."

- "Cost reduction is the goal."
- "After World War II, our main concern was how to produce high-quality goods. After 1955, however, the question became how to make the exact quantity needed."

Taken together, we could then write a definition such as "The TPS is a production system which is a *quantity control system,* based on a foundation of *quality,* whose goal is cost reductions, and the means to reduce cost is the absolute elimination of waste."

A PARADIGM SHIFT OF LEAN: TO REMOVE, NOT TO ADD

The overall approach to lean is a revolutionary concept to most Western thinkers. When we work to improve a system, almost automatically we think we need to add things. We think about adding controls or adding robots to make things better. Almost reflexively we also complicate things and add extra layers of both review and approvals. The TPS is just the opposite. It is the removal of wastes. It is largely a process to simplify and shrink things—something we Westerners do not normally relate to. All too often we feel we need to add some stuff, so we can show someone and, with that approval-seeking look, ask: "See what I did to make this better?" On the other hand being amazed by the absence of things, no matter how productive that may be, is just not very exciting to many people.

The Seven Wastes

It is more than interesting to note Ohno's focus on wastes, which was pure genius. It created a focus and alignment that permeated the entire Toyota Corporation. He categorized the wastes as:

1. *T*ransportation (of products and intermediates)
2. *W*aiting (of people)
3. *O*verproduction (including not just making more than you need, but making it way too early as well)
4. *D*efects (and rework)

5. *I*nventory (anything above what you need to protect sales)
6. *M*ovement (meaning movement of people)
7. *E*xcess processing (doing more than necessary to meet the customer's needs)

We teach these categories using the mnemonic TWO DIME, making it easier to recall.

This definition, derived from Ohno's teachings, is a clarification of the previous definition. Although most organizations who focused on process and product improvements tacitly attacked these wastes, none had created a systematic, focused, management-driven approach to waste reduction. Most firms only attacked the large wastes biting them in the butt at that moment. This is a characteristic of Japanese thinking I will clarify later and is a significant shortcoming in most Western thinkers: that is, we Westerners all too often only go after the obvious and the big wastes and do so using short-term thinking.

Look anywhere you would like and you can see:

- Thousands of dollars circling the drains
- Products and projects not meeting lead times
- Quality deteriorating

That is because the seven wastes are more than hyper-resident; they are institutionalized in your plant operations and your operations support systems. It is a problem of epidemic proportions. As a bare minimum, there are huge opportunities to be captured by the transformation to a lean manufacturing system . . . for those willing to undertake that effort.

Its focus is quantity control

In Mr. Ohno's definition, there is the seemingly cryptic topic that the TPS is not a *quality control* system; he hardly mentions it at all. On the surface this makes no sense since Toyota made its inroad into the automobile market by leveraging quality. Ohno is not saying that quality is not important; he is just saying that the company had *already achieved* high levels of quality, and it was no longer a manufacturing constraint.

By 1978, Toyota's high levels of quality had become an extremely strong point for both marketing and sales. To put this in perspective, let me give you an excerpt from my earlier book, *How to Implement Lean Manufacturing* (2nd ed.):

The TPS is not a complete manufacturing system. In fact, it is only a part of a manufacturing system. To better understand what part of a manufacturing system it is, or rather what it is not, we need to return to Ohno's book for a moment. While discussing flow as the basic condition, he writes:

> "After World War II, our main concern was how to produce high quality goods and we helped the cooperating firms in this area. After 1955, however, the question became how to make the exact quantity needed." (T. Ohno)

These two sentences are so simple that their significance is missed by almost everyone. However, the implication to these two sentences, especially to those wishing to undertake a TPS initiative, must be thoroughly understood. For example, let me paraphrase it a bit: "From the end of WWII until 1955, we had focused our attention on improving the quality of our goods. By 1955, we thoroughly knew how to provide quality to our customers. We could discuss the key quality concerns with them, we could determine how to supply quality and we could provide at a very high level. We used a long list of tools to achieve these communications skills, but the most important two were the simple customer quality questionnaire—which we statistically analyzed, of course—and we also became very proficient at Quality Function Deployment (QFD). Long ago, we ceased using inspection, especially visual inspections by humans, as a means of achieving quality. Instead, we moved to process control as a means to make the process more robust. To do this, we first became proficient in data gathering and analysis using such techniques as those that Ishikawa outlines in his writings. Now the vast majority of our data is used for process improvement rather than product evaluation. In addition, we became very proficient in a wide range of statistical techniques so we could analyze and make better decisions with our data.

The four fundamental statistical techniques of Measurement System Analysis (MSA), Statistical Process Control (SPC), Designs of Experiments (DOE), plus Correlation and Regression are widely understood at even the supervisory level in our plants. We also made all levels of personnel responsible for root cause problem resolution, which means we trained them in various levels of problem solving—the "5 Whys" being the cornerstone technique. Another significant effort was the transition in our quality, process, and product data. Initially, the majority of our data was attribute data on the product. We moved from a high percentage of attribute quality characteristics of the product to variables characteristics of the process. We recognized early on that high levels of quality could not be achieved if we used attribute data, so this meant we needed to correlate the attribute defects to process parameters, and so we became skilled at this very early in our quality efforts. We had been very committed to providing quality products and had been working very hard on quality. With the help of Deming and a unified effort pushed by JUSE, we could supply excellent quality and our costs and losses associated with quality were very low. We had what most Westerners would call a very mature manufacturing system that could consistently produce high-quality goods and deliver them on time for a reasonable price. Quality was no longer a production problem. Now we needed to look at the losses caused by producing the wrong quantities—especially the wrong quantities produced and delivered to the wrong places at the wrong times."

If I were Ohno, that is what I would have written, because that is where they were as a manufacturing company. So what Ohno built, the TPS, had a foundation of quality, but his TPS is not a quality system. Yes, it had jidoka, but we will learn that jidoka is there to support JIT. In addition, and as an example, Ohno makes nearly no mention of Cp and Cpk which are the two accepted measures of process capability or "process goodness." Nearly every book you read on manufacturing and process quality reduces the concept of quality to Cp and Cpk for all measurement data, yet Ohno hardly mentions it. One has to ask why? Well, it is because of what he said—they simply could supply high-level quality, and quality improvements were not what they needed to focus on to reach a higher level of manufacturing excellence. Their focus, as he says, was on quantity. The implication of this information, to some company that wishes to embark on the journey into Lean, is usually quite sobering. Ohno says they just spent seven years—seven very focused years—learning how to deliver quality, after which they embarked on the journey of quantity control. I can say with assurance that most companies who entertain the option of mimicking the TPS do not have the sound foundation that Ohno describes in his writings. After all, that is what made them Toyota and the TPS only took them farther along. Most think they can bypass this step and are always disappointed to find that there are no shortcuts. If you want the benefits of the TPS, the foundational issues must be addressed. It does not mean you should not embark on the journey. The foundational issues must be addressed, or your effort will be in vain. However, it is possible, with good guidance, to attack the foundational issues as well as the quantity control issues, simultaneously.

However, my criticism of this definition, by Ohno, is serious. Although what he uses for "his" definition is not inaccurate, it is, however, incomplete. It fails to emphasize a key point of the TPS, maybe *the key point of all key points,* the human element—the respect-for-people pillar made so obvious in *The Toyota Way, 2001,* as you will see later in this chapter.

Ohno did not miss this point. In *Toyota Production System: Beyond Large-Scale Production,* he made the strong statement that "there is no magical method. Rather, a total management system is needed that develops the human ability to its fullest capacity to best enhance creativity and fruitfulness, to utilize facilities and machines well, and to eliminate waste." Unfortunately, it is buried in a section entitled "Cost Reduction Is the Goal," and he gives it an "oh, by the way" treatment. It is anything but an "oh, by the way" aspect; it is at the heart of the TPS and hence at the heart of lean manufacturing. For reasons only known to him, he did not emphasize these points. However, in this and other writings by him, he frequently makes reference to "challenging workers," "well-trained workers," "the power of the individual," and "the power of teams." So people and how they worked, how they were supported, and how they were respected was clearly a part of his thinking, even if he did not use it in his definition.

LET'S SEE WHAT TOYOTA, THE CORPORATION, HAS TO SAY

Numerous books have been written about Japanese manufacturing methods in general and the TPS in particular. However, to the best of my knowledge, the only book that was written about Toyota and *by Toyota* is *The Toyota Way, 2001*. It is a 13-page booklet affectionately referred to as "The Green Book." Toyota wrote it when it made its goal to become the largest auto company in the world and thought it best to write down the company's values and beliefs.

In the introduction of *The Green Book,* Fujio Cho, then the CEO of Toyota, said: "Since Toyota's founding we have adhered to the core principle of contributing to society through the practice of manufacturing high-quality products and services. Our business practices and activities based on this core principle created values, beliefs and business methods that over the years have become a source of competitive advantage. These are the managerial values and business methods that are known collectively as the Toyota Way."

What the company has compiled in *The Toyota Way, 2001* not only provides guidance for the Toyota Production System. Much more broadly, it is Toyota's effort to describe its culture. I define culture as the "the combined actions, thoughts, values, beliefs, artifacts, and language of any group of people."

If you read Fujio Cho's quote, you will see that he is articulating the culture, summarizing it as their:

"business practices and activities based on this core principle created values, beliefs and business methods that over the years have become a source of competitive advantage. These are the managerial values and business methods that are known collectively as the Toyota Way."

The Toyota Way, 2001 is summarized on the Toyota website as follows:

The Toyota Way 2001

Toyota Motor Corporation, April 2001

Illustration from Toyota Motor Corporation Environmental & Social Report 2004, page 75

Interview with Mr. Fujio Cho, President, Toyota Motor Corporation July 2003

Question: What is the relationship between the Toyota Way and Toyota's management?
Answer: The Toyota Way, which has been passed down since the Company's founding, is a unique set of values and manufacturing ideals.

Clearly, our operations are going to become more and more globalized. With this in mind, we compiled a booklet, The Toyota Way 2001, in order to transcend the diverse languages and cultures of our employees and to communicate our philosophy to them.

Toyota Motor Corporation Annual Report, 2003, page 19

Since Toyota's founding we have adhered to the core principle of contributing to society through the practice of manufacturing high-quality products and services. Our business practices and activities based on this core principle created values, beliefs and business methods that over the years have become a source of competitive advantage. These are the managerial values and business methods that are known collectively as the Toyota Way.

Mr. Fujio Cho, President, Toyota Motor Corporation
From the Toyota Way 2001 document

Figure 2.1 **The Toyota Way, 2001** *seen in Figures 2.1, 2.2, and 2.3.*

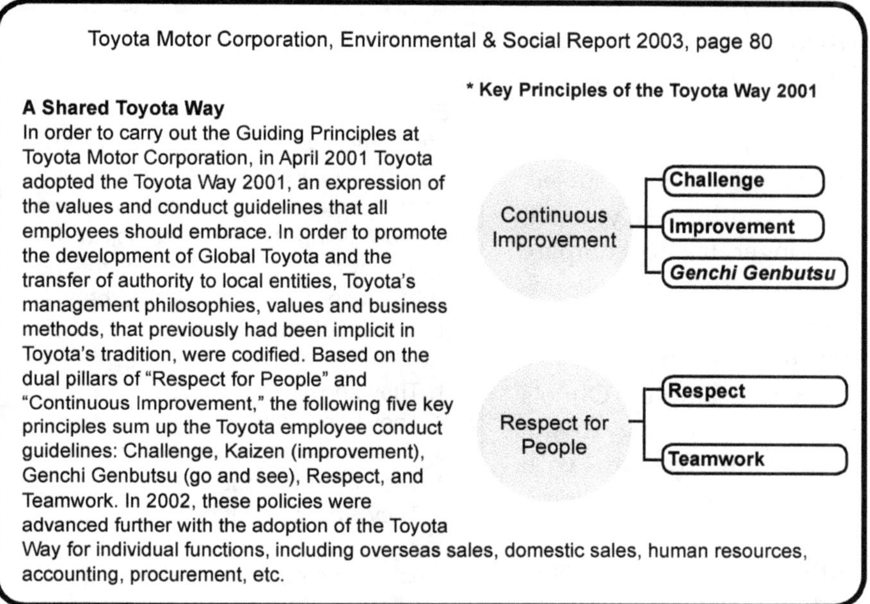

Toyota Motor Corporation, Environmental & Social Report 2003, page 80
A Shared Toyota Way
In order to carry out the Guiding Principles at Toyota Motor Corporation, in April 2001 Toyota adopted the Toyota Way 2001, an expression of the values and conduct guidelines that all employees should embrace. In order to promote the development of Global Toyota and the transfer of authority to local entities, Toyota's management philosophies, values and business methods, that previously had been implicit in Toyota's tradition, were codified. Based on the dual pillars of "Respect for People" and "Continuous Improvement," the following five key principles sum up the Toyota employee conduct guidelines: Challenge, Kaizen (improvement), Genchi Genbutsu (go and see), Respect, and Teamwork. In 2002, these policies were advanced further with the adoption of the Toyota Way for individual functions, including overseas sales, domestic sales, human resources, accounting, procurement, etc.
* Key Principles of the Toyota Way 2001
Continuous Improvement
Challenge
Improvement
Genchi Genbutsu
Respect for People
Respect
Teamwork

Figure 2.2 The Toyota Way, 2001—*Key principles.*

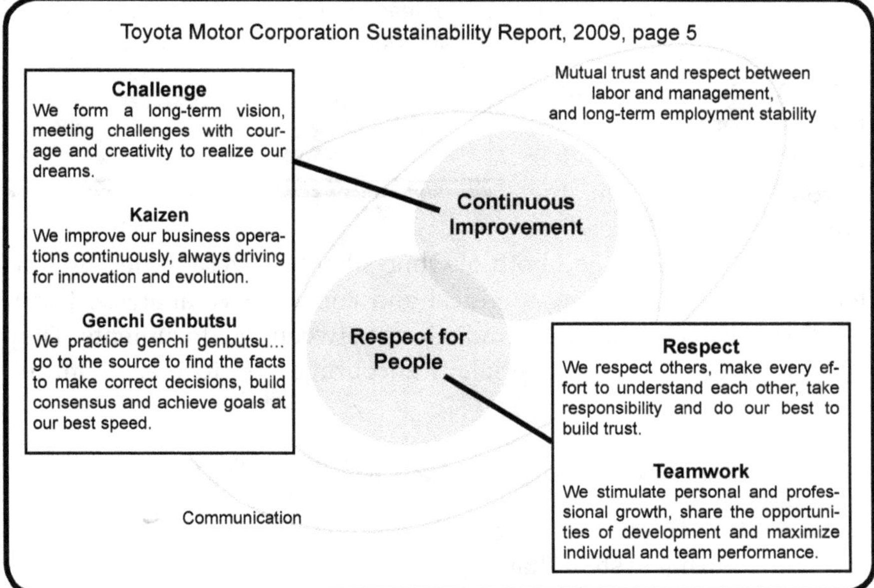

Toyota Motor Corporation Sustainability Report, 2009, page 5
Challenge
We form a long-term vision, meeting challenges with courage and creativity to realize our dreams.
Kaizen
We improve our business operations continuously, always driving for innovation and evolution.
Genchi Genbutsu
We practice genchi genbutsu... go to the source to find the facts to make correct decisions, build consensus and achieve goals at our best speed.
Mutual trust and respect between labor and management, and long-term employment stability
Continuous Improvement
Respect for People
Respect
We respect others, make every effort to understand each other, take responsibility and do our best to build trust.
Teamwork
We stimulate personal and professional growth, share the opportunities of development and maximize individual and team performance.
Communication

Figure 2.3 The Toyota Way, 2001—*Sustaining the key principles.*

Many get confused when they try to compare *The Toyota Way, 2001* with the writings of Ohno and others. However, once you realize the different subject matters addressed, you will no longer try to compare them as if they are both addressing the same subject; they are not.

For example, Ohno's writings cover the Toyota Production System with a intensive engineering view, and he is very light both on the human side and on the culture. *The Toyota Way, 2001*, however, is a document describing the Toyota cultural ideal. It does so with no particular emphasis on production, no more than on sales or marketing or any other Toyota function. On the other hand, after reading *The Toyota Way, 2001* and both of Ohno's books several times, and after outlining and making a critical and comparative analysis, I cannot find any substantive contradictions between the documents. They simply cover different materials; I encourage you to read them all.

ENGAGEMENT

Engagement is a simple concept, yet often misunderstood by many. Engaged workers are not just hurrying around at a feverish pace. They are actively doing the right things, in the right ways, at the right times, for the right reasons. There are five behavioral traits of engagement:

1. They know what to do.
2. They know how to do it.
3. They have there sources to do it.
4. They want to do it.
5. They want to do it better.

I seldom find people who don't want to do a good job, nor do I find people who naturally do not want to improve. So as you see, if people are not engaged, it is for reasons 1 through 3. And these are all obligations of management.

(There is a great deal more on engagement, including several examples at our website, www.wilsonleanrefining.)

MY DEFINITION OF LEAN MANUFACTURING

My definition of lean manufacturing is now:
A manufacturing system that:

1. Has respect for all its stakeholders, including customers, employees, stockholders, and suppliers. It is the duty of lean

management to engage all and create an environment where all will flourish.

2. Is built on a strong foundation of process and product quality.
3. Has a focus on quantity control to reduce cost by eliminating waste.
4. Is fully integrated.
5. Is continually evolving.
6. Is perpetuated by a strong, flexible, and appropriate culture that is managed consciously, continuously, and consistently.

I refer to this as the "sixfold definition" of the TPS—and of lean manufacturing by extension. It is a much better characterization of lean manufacturing. But unfortunately to those who are practicing lean manufacturing, this is old news, while to those on the outside, this definition may not be very clear at all.

As recently as four years ago, my definition of lean contained the final five points in the list, and it was only recently that I added the sixth point, item 1. It occurred to me that not including this was a blind spot of mine, a realization that came about when a client and I were working on job succession planning (one of five personnel plans I normally help develop). We were developing a cross-training program for line workers.

Five Supervisory Tools

1. *Individual performance plan.* This plan addresses the growth and developmental needs in the current job. If the subordinate would have this discussion with his or her boss, the performance plan would be the documented answer. The discussion would start, "Boss, in about a year you will evaluate my performance, and by that time I want to be the very best employee you can imagine. To get that evaluation from you, what must I do in the next year?"

2. *Five-year business growth and staffing plan.* This is a tool that lets us look at the facility's strategic plan over the next five years and ask ourselves, "Just how will we staff that in the

future? What will our structure be, and what kind of skills will we need?"

3. *Job succession planning*. In addition, we look at job-specific attrition that may be caused by retirements, promotions, and even those who might quit. We ask, "If Margarita left her job today, who is a capable replacement?" For each job, we create a list of potential candidates with the categories (1) Ready now, (2) Ready in one year or less with more job-specific skill development, and (3) Ready with more job-specific skill development; greater than one year of development is required.
4. *Individual five-year growth and development plan*. This is just what the title implies. And in this challenge the supervisors address how they can assist each subordinate to improve, oftentimes integrated with job succession planning (what can or should the supervisor do?). In addition, we challenge the individuals so they grow at the maximum rate through self-development (what can or should the individual do?).
5. *Leader standard work and daily feedback systems.*

(There is a great deal more information on these plans, including numerous forms, at our website, www.wilsonleanrefining.com.)

The plant manager, Dirk, and I had spoken about all five personnel plans, but we started with job succession planning. As we were working on this, he remarked, "I don't see the personnel development so strongly presented in your book as you and I are now working on it."

At first, I immediately and almost reflexively went into denial and explained to him the value of people and searched in my book for references to the development of people and good human relations in general. There are plenty. But then he said, "You have referenced several times the strength of the 'respect-for-people' concept in *The Toyota Way,* but I cannot find it in your definition of lean. Why is that?"

Holy enlightenment, Batman! He was right, and I needed to change! I had given it short shrift and needed to change my definition because I was saying one thing, yet practicing another. I had missed this in my definition of lean; I had not missed the actual work of people development and the full utilization of the individual in actual practice. In every lean engagement, we would teach, train, develop, and improve on all five supervisory tools; however, for some reason I never included it in my definition. That is somewhere between ironic and embarrassing in that I have been very critical of Ohno for not discussing this more openly, and yet I was doing the very same thing. I guess we all have blind spots even after lots of practice. A final comment: I have reflected on why I could miss such an important and basic point, because even though I had been doing it since my early days with Chevron, over 40 years ago, I failed to write about it. It just shows that often those things you are very good at, you simply take for granted, much as Ohno took a lot of things for granted in his writings. It shows the need to listen, question, and reflect on all you do.

WHEN THE STUDENT BECOMES THE TEACHER

Instances like my exchange with Dirk, where the student becomes the teacher, should be an "avalanche of diamonds" to a wise teacher. What it means is that your students are starting to think on their own and they can challenge your knowledge base. This should be great news to you, but unfortunately, most teachers react like I did, reflexively, and go into denial. Often, they have enough technical muscle to dance around the issue and save some face. On the other hand, the best teachers listen, take a deep breath, reflect on what has been said, and then deal with the reality of the situation. A teacher who can do that will continue to receive these avalanches of diamonds. It is the ultimate compliment to you as a teacher.

I have spent a great deal of time developing, modifying and refining this definition of lean manufacturing. It is my belief that this definition best embodies the concept of lean manufacturing as practiced by the Toyota Motor Corporation. Hence, it is my hope that as you read this book, this definition will grow clearer.

But be warned, and be very careful, as it will not have real meaning until you have practiced lean manufacturing for some time. And then if, after practicing it for some time, you feel it needs modification, call and let me know how to improve it. Keep in mind that you are still a student of mine, and as such, I am always highly complimented when the "student becomes the teacher."

> **THE IDEAL STATE**
>
> The ideal state is a process that will produce:
>
> - Zero defects
> - 100% on-time delivery
> - One-piece flow
> - No wastes
> - A safe, secure work environment
>
> The ideal state is more of a concept than a present-day reality; nonetheless it will guide all our problem solving.

We have discussed in depth many definitions of lean manufacturing, and believe me, there are more. Unfortunately, most of the time the words really do not properly carry forth the context of what lean manufacturing actually is as practiced by Toyota. My hope is that as you embark on your lean journey, you will keep in mind the sixfold definition of lean and refer to it always. Repeated reviews of it will cause you to reflect and recenter yourself, and maybe, just maybe, you can find a way to explain it even better in the future.

Unfortunately, a true and deep understanding of the sixfold definition, like all lean concepts, can only be mastered through experience. Hence it is very likely this definition will not resonate with a critical mass at the beginning of a lean transformation. Although in time, it will provide you with excellent guidance.

SO HOW DO WE TEACH IT SO OTHERS WILL UNDERSTAND FROM DAY ONE?

We will introduce an interim definition that will resonate with almost anyone, and this will be very descriptive early in your transformation and for quite some time after.

- Lean is the creation of a culture of continuous improvement and respect for people.
- The "means to lean" is to problem solve your way to the ideal state through the total elimination of waste using a fully engaged workforce.

This short definition above is a good starting point. It is useful, easily remembered, enlightening, and helpful to further your understanding of this powerful concept. And, as you mature in your lean knowledge, I expect you to revisit the sixfold definition of lean frequently, and it will then have both a new and deeper meaning. At the right time, well into your journey, you will know when it is right to adopt the deeper sixfold definition.

PRACTICALLY, HOW DOES THE "MEANS TO LEAN" REALLY SERVE YOUR BUSINESS NEEDS?

Let's further examine bullet point two, the "means to lean," that we just read in the interim definition of lean.

Problem solving. Upon discussing this with the typical manager, he or she might say, "We already do that." And the person usually is partially correct. Upon scrutiny of the problem-solving tactics being used, I normally cannot find a good problem-solving methodology that is taught to all and used on a routine basis by worker, engineer, and manager alike; the typical plant paradigm is that structured problem solving is a supervisory and engineering tool. More often than not, I find a business that is very adept at what I call "firefighting," which is a euphemism for "temporary symptom removal without real root cause problem resolution."

The concept of the "ideal state." This basically means we will strive for perfection, and even more importantly, we will always think and problem solve for the long term as well as for the short term. Since perfection is not normally achievable today, we will set interim goals and interim operating positions, which we define as "target conditions." In addition, there is very likely waste in the product as well as waste in the process; both are addressed. There is no end to the level of problem solving or improvement we wish to achieve. And even if we achieve our goal today, there is still more to do. Quite frankly, there is no longer a discussion of "What is the goal?" The answer to that question is defined by the ideal state. The remaining questions include "How much of it we

LEAN IS THE SEARCH FOR CAPACITY

". . . and the creative use of that capacity." This is a grossly underappreciated aspect of lean, but it goes to the very heart of what makes lean so powerful. Increased capacity is as close as you can get to a cure-all for all that ails a manufacturing facility.

will achieve?" and "When will we achieve it?" And then there is the basic planning question, "Who will do what and by when?"

The total elimination of waste. All waste, not just the big problems but all waste, is a target. The elimination of waste is a very strategic approach; it has a "capacity-creating potential." Another way of stating this is that lean is "the search for capacity and the creative use of that capacity." Every business can operate with greater efficiency and greater flexibility *with more capacity.*

In addition, unused capacity is a huge weapon in the acquisition of future business. With spare capacity, relative to future business, you can bid lower or have higher margins, reduce the cost of capital, and get to launch more quickly. It is a win-win-win feature.

And by eliminating waste, whether it is scrap production you reduce, equipment capacity you free up through greater availability, or labor productivity gains achieved, these all represent increased capacity to produce.

A fully engaged workforce. Our interim definition specifically does not say, "through the engagement of engineers and managers." It says "fully engaged," not just physically engaged, and it uses the term "workforce." This encompasses the line as well as all support functions—including all managers. It includes everyone from the board of directors down to the hands-on-tools workers and everyone in between. It includes the C-suite, senior management, middle management, and the supervisors. It includes the salaried and the hourly and the union-represented workers. It is everyone; there are no exceptions.

Taken together, as you use your totally engaged workforce to create capacity, which is then creatively utilized, not only will you improve profitability, safety, product quality, and product delivery, but you will also satisfy your customers, your employees, and your stockholders alike. Concurrently, you will create a culture of continuous improvement and respect for people that will allow you to not only survive today but also prosper tomorrow.

LEAN IS NOT A SPECTATOR SPORT

In a lean transformation, all are involved, all are committed, and all are engaged. It is "All hands on deck" and "No spectators allowed." All need to come prepared to engage fully with their mind and their body. No one is on the sidelines.

OK, SO JUST WHAT IS THE UNIQUENESS OF LEAN AS A MANUFACTURING SYSTEM?

Well, there are four "conceptual uniquenesses."

First, and most importantly, is the "respect-for-people" concept. And the biggest element of that is the *engagement of all employees* in the continuous improvement effort. Everyone becomes a problem solver, and everyone is not only a body to do some physical work, but a mind to be utilized for the intimate process knowledge it contains. It is the mandate of lean management to create an environment where all will flourish. You will notice that all six qualities of lean leadership have a keen focus on the second pillar of the lean culture: respect for people.

Second is the *continuous improvement effort,* and it has six distinct differences:

1. In a typical refinery, projects are assigned to managers, who then assign them to engineers and supervisors for completion. It is normally a top-down, one-way, autocratic strategy. In a lean facility, hoshin kanri planning is employed using the process of "catchball" (more on that later), where managers discuss with their subordinates *what* needs to be done, while the subordinate imagines *how* it could be done. It is a meshing of the results, the "whats," with the means, the "hows." This is done throughout the entire structure, getting input from all and achieving *facility-wide alignment as well as levels of both focus and commitment* never seen in a top-down, one-way, autocratic mode.

2. The approach to improvements in a typical refinery is a series of projects done almost exclusively by engineers and other staff. These projects have a beginning and an end. Lean projects have a beginning, but the end is not dictated by some management bias; *rather than treating it as an event, the personnel are continuously trying to reach the ideal state.* Consequently, mid-project if they see a course that will yield 50% greater benefits, they are not only encouraged but required to change course.

3. A refinery project frequently has a postproject review that focuses on the results only. Likewise, a kaizen project has a reflective postmortem. However, *both the results and the process of the kaizen are reflectively reviewed*; this is called hansei. (The U.S. Army has a very similar process called AAR, After Action Reports.)

4. In addition to evaluating the results and the process of the project or kaizen, *each project is a designed learning experience.* The people are handpicked, and so not only does the kaizen improve the process, but through learning in a supported, mentored environment, the people grow as well.

5. All kaizens, large or small, are *done at the lowest possible level.*

6. Likewise, kaizens *have the support of a sensei* who will provide help and guidance as needed.

Third, the typical improvement effort is focused solely on the results; in fact it is often called management by results. There is nothing wrong with focusing on results. The problem arises when the focus is *only* on the results. That ignores a reality of manufacturing: to get good results, you must have good processes. So the third difference in a lean manufacturing system is that *all improvement efforts focus on the process improvements as well as the results improvements.*

Fourth, the culture of Toyota, built into the TPS, makes perfect use of the intrinsic motivators. The famous Dr. Edward L. Deci of Rochester University, who has done extensive research on motivation and especially intrinsic motivators, and coauthor Richard Ryan describe in

TERRY'S DELI

I would frequently eat lunch at Terry's Deli, where they had the best Italian sub I ever ate. One day Terry asked me, "What do you do?" I told him, and he asked me the fateful question, "Sounds interesting, but just how could that apply to my business?" I asked Terry to gather a little data for me and told him I'd see him in a week. We redesigned his "sandwich manufacturing business," and Terry's revenue increased $700 a week. I never had to pay for another sub at Terry's. For the complete story, read about Terry's Deli in the Appendix—Chaper 2.

an article, "The 'What' and 'Why' of Goal Pursuits: Human Needs and the Self-Determination of Behavior," the intrinsic motivators as "innate psychological nutrients that are essential for ongoing psychological growth, integrity and well being."

There are five motivators, and they include a sense of:

1. Control
2. Meaningfulness
3. Accomplishment
4. Community
5. Growth

As you embark on a lean transformation and get the engagement of the workforce, some wonderful things happen. As you utilize the concept of delegation, which allows decisions to be made at lower levels, you find an increased *sense of accomplishment* as workers see that they are meeting the company goals of quality and production. As the workers are completing kaizens and improve their workplace, they get a significant *sense of control* over their work environment. As they work together to solve problems, they feed not only their *sense of control* and *meaningfulness,* but also their *sense of community*.

TRANSPARENCY

Transparency is a concept in which any given situation is obvious and easily diagnosed by the worker, supervisor, and manager. The metaphor is the "scoreboard effect." That is, in a sporting event, you can enter the stadium, look at the score-board, and in two seconds understand if your team is winning or not. The question is, can you look at a process or a workstation and in two seconds evaluate if things are normal or abnormal? If so, you likely have good transparency.

In a separate article, "Self-Determination Theory and the Facilitation of Intrinsic Motivation, Social Development, and Well-Being," Dr. Deci notes: "Yet, despite the fact that humans are liberally endowed with intrinsic motivational tendencies, the evidence is now clear that the maintenance and enhancement of this inherent propensity requires supportive conditions, as it can be readily disrupted by various nonsupportive conditions."

As you go deeper into the transformation, you will find that the very nature of lean manufacturing is geared to incessantly feed and support these intrinsic motivators—or using Dr. Deci's words, to provide the "maintenance and enhancement of this inherent propensity."

Most manufacturing systems look to control and frequently overcontrol the workers' activities and work methods. Lean asks the workers to participate in this activity. Most manufacturing systems bring in external problem solvers to work on problems after the fact, often arriving at inappropriate solutions that are always late. Lean asks the workers to solve the problems in real time. Most manufacturing systems share production data as if it were a secret and most often only when the production falls short and the message comes as negative feedback. Lean strives to have transparency on the floor and keep the operators "in the know." This is a unique aspect of lean manufacturing and possibly one of its strongest aspects, as it vigorously feeds the intrinsic motivators.

THE PALETERIA

While working in the yard, our new neighbor, Willard, stopped by to thank my family and me for helping his daughter find her lost dog. Willard, too, asked me, "What is it you do?" Later he asked, "OK, how can that help me?" Willard owned a paleteria, a business that sells paletas, the Mexican version of popsicles. I spent time with Willard and taught him how to apply lean principles to his paleta manufacturing, which was basically a mixed-model production line. He was able to significantly reduce labor, raw materials, and inventory. Willard, like Terry, paid me in food; this time it was paletas.

AND FINALLY, WHAT ABOUT APPLYING LEAN PRINCIPLES TO OTHER ASPECTS OF THE OIL BUSINESS? AND TO BUSINESSES THAT ARE NOT MANUFACTURING BASED?

Much of the documentation on lean as a concept, including books and Internet materials, focuses on lean applications in manufacturing. And the vast majority of substantial data is highly focused on the assembly operations, particularly supporting the auto industry. Hence, when people ask, "How will this work in my world?"they often have difficulty imagining how the lean principles might apply.

My focus in this book is clearly on manufacturing and even more narrowly on oil refining. However, within an oil refinery, and within other businesses as well, we have applied lean principles very broadly. For example, in oil refining we have used lean principles to leverage the business in the manufacturing sector as well as in accounting, finance, human resources, oils and facilities planning, designs and process engineering, safety and emergency response, environmental design and compliance, and even the administrative services. The principles that drive lean apply very broadly in a manufacturing environment—in *any* environment. Likewise, we have employed these lean principles in nonmanufacturing businesses including doctors' offices, hospitals, schools and universities, lawyers' offices, employment services, banks, one sandwich deli, and one paleta (Mexican popsicles) manufacturer.

Should you wonder if lean will apply to your situation, ask yourself the following set of questions:

1. Does your business use processes to produce a product (or service)?
2. Is the success or quality of your product dependent upon the success of your processes?
3. Do your processes have weaknesses or contain wasteful activities?
4. Are these weaknesses or wastes a problem to the business?
5. Are the people who are using these processes able to reduce these wastes via problem solving?
6. If these wastes were reduced, would yours be a "better, faster, cheaper" business?

If you can answer yes to these six questions, then yours is a target-rich environment for lean applications. Lean will help you become a better moneymaking machine, create a more secure work environment for all, and be the supplier of choice to your customers.

CHAPTER SUMMARY

It seems odd that after nearly 30 years we have not agreed upon a definition for lean manufacturing. Maybe we just have not worked with it long enough to truly understand the depth and breadth of is potential power.

We found the origin of the term and reviewed what Taiichi Ohno, its creator, said. More importantly we explored what he did and put it in perspective. This provided a good treatment of the continuous improvement pillar of lean. Next, we reviewed what Toyota actually said about its culture, and this gave us a much better perspective on the second half of the equation for lean, the respect-for-people pillar. I doubt the definition is finished, but it is far more mature than it was five years ago. Consequently, I have given you a simplified although very usable definition: Lean is the creation of a culture of continuous improvement and respect for people. And the means to lean is to problem-solve your way to the ideal state through the total elimination of waste using a fully engaged workforce.

Later, as you "learn by doing," you can use and teach the full six-part definition.

The lean manufacturing model is unique in that there is an inordinate emphasis on the human element creating engagement of all; the planning model creates an alignment and commitment of this engaged workforce; and as people work to continuously improve the business, they focus on the process as well as the results. And maybe most importantly, lean techniques supply natural feedback to the intrinsic motivators.

And by the way, lean is a very broad manufacturing system and has applications far beyond the auto assembly business, far beyond classical manufacturing itself . . .even at businesses like Terry's Deli.

Chapter 3

Why Is Lean Manufacturing So Misunderstood?

***Aim of this chapter**—There is a growing belief that lean transformations are failing and that lean itself is flawed as a manufacturing system. In this chapter, we will explore the most basic reasons why this belief has been perpetuated. And we will explore why any transformation must go beyond the "tools of lean" into the role of lean management and finally the cultural change that is required to sustain the gains of the lean transformation. However, for this perception that the lean concept may be failing, the key question remains, "Is this a failing of the concepts of lean, or are the lean concepts sound, and it is simply a failure to execute the transformation properly?"*

WHAT'S IN THE BLOGS AND ON THE WEB?

There seems to be a lot being written about the shortcomings of the lean transformation. There are a number of discussions on the blogs and forums citing problems with lean and many stories of "gains not sustained." Mostly it just seems like the "air is leaking from the lean balloon," and the optimism and energy surrounding lean is flagging. I have spoken to many managers and lean geeks alike, and many tell me I am overly concerned. I am not so sure.

I have yet to see anything that represents a significant study; rather, there are some anecdotal studies, mostly self-reported data, done in a purely nonscientific fashion. In addition, there are several articles that represent the opinions of many of the lean community. I, for one, agree that many of these lean transformations are falling far short of initial expectations. In fact, our firm, Quality Consultants, specializes in working with companies that are struggling with lean or have tried lean and failed.

The anecdotal studies

I know of two studies, both citing self-reported data, that were handled in a rather nonscientific fashion. Both were written by companies that are associated with the lean community.

The first is the AlixPartner's "Senior Executives Survey: Manufacturing-Improvement Programs: Effective?" published in October 2011. The strength of this evaluation is that the authors used some actual data and interviewed some C-suite personnel. Their conclusion was that "nearly 70% of manufacturing executives reported their manufacturing-improvement efforts led to a reduction in costs of 4% or less."

The weakness in the study is that lean programs were intermixed with Six Sigma and other improvement models, so the findings do not specifically point to lean. In addition, very little follow-up was done to dive deeper into the explanations for these shortfalls.

The second study was reported in the 2007 Industry Week/MPI Census of Manufacturers. This study noted that "nearly 70% of all plants in the US are employing Lean Manufacturing as an improvement methodology." It went on to say that "only 2% of companies who responded to the survey have fully achieved their objectives and less than a quarter of all companies (24%) reported significant results."

Neither study tried to determine if the failings of these improvement efforts were due to a fundamental mismatch of their improvement models with the business needs or if it was simply a matter of poor execution.

However, following the publication of these surveys, the Industry Week/MPI Census of Manufacturers generated a lot of action on the web. In 2008, Rick Pay, a management consultant, wrote an insightful article for *Industry Week* entitled, "Everyone's Jumping on the Lean Bandwagon, but Many Are Being Taken for a Ride." In his article, he cites four major reasons companies fail to achieve benefits:

1. Senior management is not committed to and/or does not understand the real impact of lean.

2. Senior management is unwilling to accept that cultural change is often required for lean to be a success.

3. The company lacks the right people in the right positions.

4. The company has chosen lean as their process improvement methodology when a different process improvement program—or none at all—would have been a better choice.

Simply put, he finds that three of the four reasons (items 1, 2, and 3) are related to the execution, and none point out any fundamental weakness with the lean model.

Jeffrey Liker and Mike Rother dove into the fray with an article published at the Lean Enterprise Institute. In it they say:

> We both have concluded from our different journeys and experiences with companies that people have a fundamental misunderstanding of what the Toyota Production System is in practice. We mistook lean solutions for the process that leads to what we see in a Toyota plant. We need to look more deeply at the human thinking and the processes that underlie specific practices we observe.

They are simply saying that the lean tools (lean solutions) were mistaken for the improvement process.

Liker and Rother reference the "human thinking and processes that underlie the practices," which when dissected is equivalent to "just how they do things in their plants." And "Just how we do things around here" is the operational definition of the culture within these plants.

The authors summarize by saying:

> When we look at lean this way (that is, as a culture-changing mechanism) it is not a set of techniques for eliminating waste, but a process by which managers as leaders develop people so that desired results can be achieved, again and again.

In summary they are also saying that it is not so much the "tools" such as problem solving, heijunka, or kanban; rather, lean is more of how we go about doing these things, and it is a matter of culture change; it is a matter of how it is executed. And I concur fully with that concept.

MANY STILL THINK OF LEAN AS AN APPLICATION OF THE "TOOLS OF LEAN"

When *The Machine That Changed the World* was published in 1990 and the Lean Enterprise Institutes (LEI) were founded in 1997, a global awareness was accelerated that had a profound impact on the awareness of the manufacturing world. That was the good news—really good news.

Largely through the efforts of these institutions, the concept of lean manufacturing was brought into the mainstream in a manner not accomplished earlier. And with this awareness, the belief was that if you implemented the "tools of lean" and used the countermeasures of takt, standard work, value stream mapping, heijunka, kanban, poka yoke, 5S, and others, you could get the gains achieved by the vaunted Toyota Production System (TPS). That was the bad news.

The TPS was that machine that had changed the world, and now many of its "secrets" had been made public. Furthermore, the LEI began an aggressive campaign to provide the resources that it believed would further the expansion of the TPS, including training classes and books galore. Most of what it published focused on the tools of lean. Consequently, over the next decade, it became known that to "become lean" was to implement the tools of lean.

Unfortunately, this was wrong, terribly wrong.

Later, this error became obvious to the leadership at LEI, and to his credit, LEI founder James Womack wrote one of his many e-letters, entitled "From Lean Tools to Lean Management." He said, "I think of the period from the early 1990s up to the present as the Tool Age of the lean movement." He continued:

> The attraction of tools is that they can be employed at many points within an organization, often by staff improvement teams or external consultants. Even better, they can be applied in isolation without tackling the difficult task of changing the organization and the fundamental approach to management. I often say that managers will try anything easy that doesn't work before they will try anything hard that does, and this may be a fair summary of what happened in the Tool Age. . .

> Fortunately, I now see signs that the lean movement is finally tackling the fundamental issues of lean management. I've recently had a number of conversations . . . with senior managers who realize that they need to think more about lean management before thinking further about lean tools.

Just what is a tools approach?

When a facility decides to convert to a lean manufacturing production system, it often puts together a rollout plan. Those who choose a tools approach say something like, "For this year we will teach all facilities three topics: lean basics, 5S, and kaizen. Next year we will teach standard work, kanban, and cellular manufacturing. In year three we will teach TWI-type training, three-part inventory controls, and leader standard work." Rolling out the "tools" in this fashion, then, becomes the implementation plan. It is really very simple.

So what's wrong with a tools approach?

This is a common question and highlights a large misunderstanding of the purpose and power of the tools of lean. First, this is a good way to shortcut real problem solving, by which progress is really made. And in some cases, by learning some of the tools, you will be able to make some short-term progress, but continued progress and sustaining of the gains is doomed to failure.

The lean answer

In our definition of "lean" from Chapter 1, the "means to lean" is "to problem solve your way to the ideal state." Think of the ideal state as perfection: no waste, one-piece flow, with 100% on-time delivery of a competitively priced product with 100% quality yield, produced in a safe environment where morale is sky high and job security is virtually guaranteed; that is our ultimate goal. Since we are not there yet, we must problem-solve until we are. And for most businesses, that means we need to problem-solve forever; after all, we are developing a culture of continuous improvement. Keep in mind that a problem is some deviation from the desired state, often called a gap, that is worthy of our attention. So we problem solve to remove or reduce the gap.

However, in the world of lean we seldom have solutions since "solutions" imply some permanent fix. More often than not, we have countermeasures. The word "countermeasure" is chosen over the term "solution" since countermeasure implies that more can be done—which is always the case. So in "lean-speak" we "implement countermeasures" to improve processes and products; and all the tools of lean are countermeasures to problems. For example, kanban is a countermeasure to inventory and delivery problems; takt is a countermeasure to overproduction; heijunka is a countermeasure to excess inventory and delivery issues; SMED is a countermeasure to lost time in changeovers causing excess inventory and lost capacity; 5S is a countermeasure to waiting for tools, parts, and finished goods and walking around looking for them; and so on. Each of the tools of lean mitigates some type of a problem that is an impediment as you attempt to reach the ideal state.

LEAN... OUR LEARNING MANTRA

Lean skills are learned by doing... and in no other way

When the approach to your lean transformation is to teach all the "solutions" up front, you literally have the cart before the horse. Recall that the Six Questions of Continuous Improvement (enumerated in Chapter 1 and detailed in Chapter 9) start with an assessment of the current condition ("What is the present condition?"). The second question ("What is the desired condition?") asks to seek an understanding of the ideal condition (or target condition) and to see if any gaps exist. If a gap exists, then you can ask question 3, "What is preventing us from reaching the desired condition?" Not until these three questions have been asked and answered is there an understanding of what the possible countermeasures might be. And again, these countermeasures are the tools of lean. So you see, there is no point in having solutions (countermeasures) unless they match a problem. This seems so obvious, and yet it is completely missed by those embarking on a tools implementation approach to a lean transformation.

A practical metaphor

Imagine you are going to start an automobile repair shop. For the sake of argument, let's say you hire some bright but inexperienced people to be your mechanics. You now need to train them. With a garage full of cars to work on, would you train them on voltmeters and then say, "Here is a voltmeter. Now go find an electrical problem to work on? Or, "Here is a tire iron and a jack. Now go find a car with a flat tire and fix it?"

This is exactly what a tools approach forces the transformation to do; that is, "We have taught you the countermeasures. Now go find a problem to fix." As ludicrous as that sounds, that is exactly the impact of using the tools approach.

So just what should happen? How do people learn the tools?

The "roll-out-the-tools" approach, as described above, is the classic example of a "push system" whereby tools are pushed out to the implementation team and the members of the team are expected to utilize them. The opposite of this is a "pull system." In a pull system, problems are first determined and prioritized, and then when improvement opportunities are discussed, the problem solver goes through the Six Questions of Continuous Improvement. When the problem solver gets to question 4, "What is something we can do, right now, to get closer to the desired condition?" he or she immediately begins to think of the possible countermeasures to pull in to solve the problem.

This creates an issue in that there are literally hundreds of countermeasures and very likely there are a myriad of countermeasures that this problem solver has not yet mastered. Consequently, this person asks, "Just what countermeasures are available to me?"

The short answer is, "That is exactly why you have an experienced sensei helping to lead this effort." Initially and maybe even finally, no one has all the answers, but the knowledge of the lean tools is only learned in one way—and that is by using them to solve problems. And that is why our lean mantra is "Lean skills are learned by doing . . . and in no other way." In his book *Attaining Manufacturing Excellence,* Robert "Doc" Hall said, "The best understanding comes from doing, because manufacturing excellence is really a *philosophy of doing*. Just reading and talking about it is 'Mickey Mouse'."

What about learning the tools in the classroom?

And to those of you who wish to train the lean skills through rigorous classroom training, forget it; it won't work.

Classroom training is adequate to introduce a topic and even to gain some intellectual understanding; but to learn it, you must do it . Think about it for just one second; exactly how many world-class golfers learned their trade by going through a motivational speech, some informative lectures, a couple of PowerPoint presentations, and some classroom simulations?

So how do our implementation team members learn the tools? *Simple, they must use them on the floor in real-time problem solving—and with the guidance and support of an expert.*

SO WHY DID IT TAKE 15 YEARS FOR THE LEAN COMMUNITY TO NOTICE THIS—AND OTHERS STILL HAVE NOT?

Kaizen events work

First, a staff or external consultant-driven program that is performing kaizen events will almost surely get you early and often large results. In Chapter 1 you read the story where the sensei and some managers and engineers performed a "kaizen event" at the Theta cell. The gains were enormous. That surely gets management attention. Make no mistake—this approach will get you short-term gains. However, as you read on in this example, you will find the gains were not only sustained but improved upon by using a fully engaged workforce that executed literally hundreds of "kaizens," creating a continuously improving process. However, to execute and sustain those gains, not only do you need a fully engaged workforce, but you also need a great deal of additional skills, as well as the full engagement of the management team. Note the large number of lean skills taught over this one-year period. Equally important was the way they were taught. First the sensei, using real data, taught the management team. Once the members of the management team demonstrated competence, they then taught these same skills to their direct reports and so on, right down to the line worker. All were taught using real-time problem solving on real-time problems. Classroom training was hardly used at all; it was JIT training and JIT problem solving, done at the gemba. It was *lean done right,* and the gains were humongous.

So by using external consultants, or staff experts who already know the lean tools, short-term gains can easily be achieved. Every company has a lot of low-hanging fruit. A set of trained eyes with some willing workers can easily harvest it; there is no secret to that. This type of kaizen event done in isolation is no different from a Six Sigma program in that it is done by a select group of individuals working in isolation on a specific problem. Unfortunately, when they leave, the knowledge and experience go with them. Yet those who do nothing but kaizen events often claim they are "doing lean" or they are implementing the Toyota Production System when but a cursory review will tell you that:

- It is not a continuous improvement.
- It is not problem solving to the ideal state.
- It is not done through the use of a fully engaged workforce.

Often for these firms that employ kaizen events alone, the only resemblance to the Toyota Production System is that the engineers and management team know the jargon of lean. Unfortunately, at the behavioral level they are still doing things as they have for the last 30 years; it is business as usual with some Japanese terms thrown around as window dressing.

The allure of the tools approach is just too appealing

Once a company has been successful with a kaizen event or two, interest soars. The people at the company are convinced that the first dynamic is to begin using the tools of lean. Soon a variety of managers become familiar with some of the tools and begin to use the jargon—not the tools, just the jargon, especially the Japanese terms—in their daily discussions. The whole facility takes on an "air of competence" in lean manufacturing. The people in management as well as the entire team really begin to believe that they can add a kanban here, install a pull production system there, and finish it up with the three-part inventory system, and they have the whole issue of implementing a lean manufacturing system under their belt. After all, a kanban will assure JIT delivery to the production line, a pull production system will avoid overproduction, and the three-part inventory system will compensate for systemwide variation while it assures 100% customer service. It all seems so easy and clear.

At this point, buoyed by these early and short-term successes, the members of the management team believe a second dynamic. That is, they can simply "bolt on" these tools to their existing production system; that is all they need. They fail to see:

- What support is needed to teach the tools, such as training and maintenance
- What problems are created by using the tools, such as how do they interact with your ERP (Enterprise Resource Planning) system
- What effect these tools have on other aspects of the entire system, such as inventory management, inventory controls, and delivery procedures
- What interactions, both positive and negative, occur as a direct result of implementing these tools

But worst of all, using this "bolt-on" theory, they fail to address the aspects of their management and production systems that are currently inadequate. In addition, they fail to address how they will handle the issues that arise by using these lean tools. Recall that the first aspect of all lean tools is that they create "transparency" and cause problems to surface. Even if you just bolt these tools onto your system, they will cause other problems to surface. That is one of their strengths.

For some reason, these managers assume they can just add these new tools and nothing else will need to change.

That concept is simple enough; that's probably why it is so appealing. Unfortunately for those who believe this bolt-on theory, it is incorrect. Not only is it incorrect; it is a belief that, will completely destroy your ability to effect a lean transformation. It is another litmus test for failure.

Sustaining the gains takes a much deeper approach

The problem with the kaizen event is just that—it is an event. Lean, on the other hand, is a continuous improvement dynamic. It is continuous improvement because new ideas come about all the time, and some things deteriorate. Left alone, any system, even one of a world-class performer, not only will deteriorate but will also fail to keep pace and eventually will no longer be world class. To create a culture of continuous improvement is not just a worthy task; it is a difficult one as well. It must be led by the management team using a *teaching-learning-doing model*, the model of lean leadership.

As you read about the story of the Theta cell, you saw a learning dynamic executed by the management team. First, the people on the management team learned from the sensei, and then they, in turn, taught all their subordinates. You will learn that these are two key characteristics of lean leadership: learning and teaching.

BEWARE OF "THIS SIDE OF COMPLEXITY"

This process of achieving a few short-term successes and believing "you have it all mastered" is what I call being on "this side of complexity" (see Chapter 2). Frequently people on the "this side of complexity" are quite confident of their knowledge base even though they may be wrong, very wrong. That is why the section is entitled "Lean Sounds Simple . . . Dangerously So. Hence, Proceed with Humility." Lean requires that you "learn-do-reflect" and you may cycle through that process several times before you have truly captured the intimate process knowledge it takes to make the system perform to a lean standard.

Often this is a long journey and yet another reason you need a sensei, who can guide you away from being overly impressed with short term gains, and provide the balance that will lead you to the "far side of complexity" and its inherent wisdom.

And the managers must change . . . first

This requires that most managers must dramatically change many things they now do. And this fact oftentimes is enough to squelch any thoughts of a lean transformation. All too often I find that the managers are not in active learning mode, nor do they perceive themselves as teachers; they would rather delegate those skills. Seventy years of the TPS has taught us that we should not delegate these skills.

So what does the management team normally do? Either delegate the skills or just not implement the transformation.

Return to the quote from James Womack when he says:

"The attraction of tools is that they can be employed at many points within an organization, often by staff improvement teams or external consultants. Even better, they can be applied in isolation without *tack-*

ling the difficult task of changing the organization and the fundamental approach to management. I often say that managers will try anything easy that doesn't work before they will try anything hard that does, and this may be a fair summary of what happened in the Tool Age." Or paraphrasing, "Management tries the easy stuff and avoids the difficult tasks," especially the difficult part which includes the concept that "they, the management must change." It is much easier for management to dictate that everyone else must change.

So in a nutshell, if kaizen events are all the leaders do, they will have some early gains and some rather large gains, but those gains will not be sustained over time—and the company will not have a lean enterprise.

Meanwhile . . . what is getting published?

Fortunately, many of these stories of the early successes get published in the e-zines and blogs by both the companies and the consultants who assisted them. And unfortunately, the long-term issues are never published, as the companies do not like to admit the failures, and the consultants do not want to be associated with these "failures in the fullness of time." The net effect is lots of "happy talk" about the short-term gains . . . and the long-term issues show up dramatically differently.

Just how do these long-term "failings" manifest themselves?

They show up as articles and surveys which profess that—"this lean thing really does not work"—and lots of negativity in the blogs, forums, and conventions. And that is pure poppycock. I have studied many implementations, and without exception, where lean has failed to make a facility or a value stream a better moneymaking machine and a more secure workplace, as well as the supplier of choice for its customers, it has been a matter of poor execution rather than a fundamental weakness in the concepts of lean. The 2007 Industry Week/MPI Census of Manufacturers mentions at least three major aspects of poor execution and yet is silent on "weaknesses" inherent in lean manufacturing as a concept.

I have categorized these "weaknesses in execution" in my earlier book, *How to Implement Lean Manufacturing* (2nd ed.). In Chapters 2 and 3, I catalog the "Six Roll Out Errors" and the "Ten Lean Killers." If you can successfully execute the countermeasures to these 16 issues, you can have a successful lean transformation.

Let's reflect on all this and the story of the Theta cell

If it has been a few days since you read Chapter 1, I suggest you go back and reread that chapter; the message is crucial to your understanding of lean manufacturing.

Let's make a high-level review of the cell's progress using the key metric of productivity:

- Initially the cell productivity was 80 headrests/person/shift (h/p/s)
- A management-led kaizen event was executed, and in one week the cell productivity increased to 188 h/p/s
- Over 100 employee-led kaizens were implemented, and over the next 8 months, the cell productivity improved even further to 304 h/p/s

Just how is the story of the Theta cell instructive as a means to understand what lean manufacturing truly is?

The first phase, a large kaizen event

First, the management-led, engineering-intensive kaizen event was implemented. This kaizen event made use of a large number of lean tools. Takt calculations, line balancing, standard work, kanban and pull systems, flow implementation, product leveling, OEE, 5S, transparency, and statistical problem solving were all used. And a kaizen event, using this number of lean tools coupled with a productivity improvement of 235% in just one week, would easily qualify as an excellent example of lean manufacturing—to most people.

This is exactly what you found in the articles you read in the magazines, the posts you read in the blogs, and the presentations you heard at the conferences. And quite frankly, as far as the plant was concerned, we might have stopped there because we had already exceeded, by over 88%, the level of productivity needed to be profitable (100 h/p/s). A great improvement for a few days' work, done in less than one week—that sounds pretty impressive, and it is. That is why articles and blog posts like these get written, published, and read. And that is the good news.

The bad news about this "kaizen event" is threefold. First, this is only a start, and there is much more to be improved. Second, these gains are not sustainable. This process will need maintenance, and also entropy will take its toll soon enough and the deterioration of these gains is a foregone reality. Third, this is not what lean manufacturing is all about. Do not be confused. If you stop at this point, you will be very disappointed, as much of the surveys and blog discussions have shown.

The second phase, lots of little kaizens plus some other "stuff"

The second phase, employee-led kaizens, is equally instructive. First, the productivity gains from 188 to 304 h/p/s were achieved by the people on the floor, those workers actually executing the process. The nature of these kaizens was such that they drew on the intimate process knowledge of the worker, rather than the intimate process knowledge of the engineer and manager. Furthermore, the kaizens were both taught and supported by the management team and the workers had available the technical support needed to succeed. Finally, these workers were not empowered, but their "disempowerment" had been removed and they were allowed to become engaged—once again.

Table 3.1 shows a side-by-side evaluation of these two approaches to lean manufacturing: the "kaizen event" only and the lean approach using both the "kaizen event and ongoing kaizens," comparing each with the definition of lean manufacturing.

What's the bottom-line impact of these two types of kaizens?

The kaizen event alone

What is the bottom-line impact of this strategy in the Theta cell? Recall that the workers needed to get to 100 headrests/person/shift to meet the target profitability. Since the labor was over 30% of the product cost, any impacts on labor productivity would be felt directly on the bottom line. At a productivity of 100, they could achieve 9% gross profit (GP). With the improvements from the kaizen event alone, to 188, GP would increase to 23%, a hearty and healthy increase. Now that's the good news.

Table 3.1 Comparison of the Two Approaches to Lean Manufacturing

Lean Concept	Management-Led Kaizen Event Only	Management-led Kaizen Event, plus Employee-Led, Smaller Kaizens
Better, faster, cheaper	All factors improved productivity by 235%	All factors further improved productivity by 380% in total
Culture of continuous improvement	Improvement stopped when the kaizen event ended; it was not continuous	The employees continuously improved the process
Culture of respect for people	Few were involved	All employees—floor workers, managers, and staff participated
Problem solve to the ideal state	We stopped problem solving when we were successful	We continued to improve all aspects of operations. No end was in sight to the improvements
Through the total elimination of waste	These activities focus on removing wastes	These activities focus on removing wastes
Using a fully engaged workforce	Here we only used supervisors and engineers; there was only partial engagement of the workforce	Not only were all the people and workers involved in the improvement activities, but they were also actively and aggressively taught a number of skills; their needs were readily addressed

The bad news is that often people see these huge gains after the kaizen event, and they stop there. Then they move these process-improving resources to another task, so the continuous improvement aspect is never achieved. There is even worse news, and that is, with no process maintenance and no sustaining mechanisms in place, these gains would soon shrink. It is not unusual to see a 50% loss in gains in the subsequent three months and a total regression to the original state, or worse, in less than one year. What happens is unfortunate, logical, and always both very bad and highly predictable.

The future looks like this. Soon enough there will be problems to solve, and someone will say, "OK, now what do we do?" With no continuous improvement methodology in place and with the kaizen team gone, there is no longer any problem-solving support or leadership in place; and since the supervisor is likely overworked, the people on the team will say, "Remember the good old days when we had a utility person; that is exactly what we need." Soon enough they revert to "the good old days," and a utility person is scheduled. They then jump in and work on any problem that arises and in so doing hide the real issues. After a steady diet of this, pretty soon the cell has regressed to just where it started—or worse. So although the cell improved to a rather healthy 23% GP, the concept of the kaizen event alone cannot create an environment where these gains are sustainable.

The kaizen event plus ongoing kaizen plus some other "stuff"

When you analyze the impact of the ongoing kaizen activity by the workforce, a totally new dynamic is achieved. First, you get the "shot-in-the-arm" effect of the kaizen event, but you also get the benefit of two massive dynamics: the benefit of sustaining the gains already achieved and the benefit of continuous improvement. The net effect is that the gains from 100 to 188 h/p/s is sustained, and on top of that the process improved to 304 h/p/s to a very healthy and sustainable 31% GP.

This clearly shows the significant advantage that the lean approach has on being competitive. Some lean tools via kaizen projects can assist you, at least in the short term. However, only when the full arsenal of lean manufacturing is unleashed are the gains massive and sustained. Hence you cannot just apply a few tools; you must implement the foundational elements, as well as the strategic and tactical techniques (see the House of Lean in Chapter 8) along with the individual behavioral skills, to get the full and sustained benefit.

The lean approach, compared with just the kaizen event, is clearly beneficial and seems so obvious. So why doesn't everyone just do this?

That is a great question to which I have given a great deal of thought, and I can only give a partial answer. Although inadequate, I am sure my answer will give you insights into how you might address the problems and improve over those who have gone before you. I have three major responses to this question:

- First and foremost, almost all management is impressed with short-term gains. Many business processes drive this thinking. There are monthly reports to management, the quarterly stock dividend, and the behavioral drivers of all drivers—the desire by management to get a big bonus and to get promoted. All these things, and probably others, drive short-term thinking. A foundational element of lean is the need to be long-term thinkers.
- Second, it is a lot of hard work. A lot. Take another look at the skills that were taught to the Theta cell team. The skills used in the management-led kaizen event were the first seven on the list. That is imposing enough. However, to implement not only the kaizen event but also the employee-led, smaller kaizens, which are the heart of the continuous improvement phase, there were over 50 line items. This alone explains why many managers love the kaizen event. You seemingly get most of your benefits for only a fraction of the effort . . . sounds like a winner. However, as we mentioned earlier, there are two problems: the benefits are short-lived, and the extra 8% (from 23% to 31%) profit will separate the winners from the almost winners. So if you want the gains, if you want all of them, get ready to put in the effort.
- Third is the need for management to change. If all that happens is that a few kaizen events are planned, then the need for management to change is very minor. After all, since they already do many projects, this is just business as usual, only now we are using a few really cool Japanese "tools." However, in a full transformation, managers and other leaders must make significant changes in their roles. Specifically, they must become "lean leaders." And the bottom line is that management *is not willing to change that much*. It is much like James Womack said, "I often say that managers will try anything easy that doesn't work before they will try anything hard that does . . ."

The magnitude of the cultural change is almost impossible to comprehend

Most managers naïvely underestimate the time, effort, and empathy it takes to create change.

When the term "lean" was coined in 1990, the consensus of what composed this wonderful system was that "lean" was a "set of tools" that could drive your business to higher levels of productivity, quality, customer service, and profitability. In the last five years, there has been a great deal of talk about lean being a culture-changing mechanism. Some experts in the field of industrial cultures, such as Deal, Kennedy, Kotter and Shein, have written extensively on culture and culture change. Now many of the lean community have taken up the mantra that to implement lean, you need to change your culture. Some say it with a deep understanding; yet others say it as an "Oh, by the way," attitude.

Unfortunately, there are too few who truly understand the depth of cultural change that is required, and there are even fewer who know how to go about executing this cultural change. In his book *Leading Change*, John Kotter says, "The combination of cultures that resist change and managers who have not been taught how to create change is lethal." Hence, even if they consciously think about the cultural change, they almost universally underestimate the magnitude and the dynamics of the required cultural change. I don't mean they underestimate the effort required by 10 or 20%, as they may do on a cost estimate. I mean they underestimate it by 100, 200, or 300%! Yes, if you ask top-level managers of successful change transformations, they will tell you it took double, triple, or quadruple the effort to reach the cultural goals they had first envisioned. After doing culture-changing initiatives for nearly 30 years as a consultant, I can say with certainty that I have not been a part of even one single culture-changing engagement where the effort and the time to make the desired changes were *overestimated*."

The design of the change does not address many key issues

When we embark on a culture-changing event, we are dealing with the greatest variable most businesses have, the human condition. However, the truth is, given a series of stimuli, even the best of consultants cannot predict with accuracy how a culture will change over any protracted period; neither can they predict how much it will change, nor how fast it will change. Then there is the problem of change interaction. Each change has a sympathetic or antagonistic effect on several other cultural variables. The net effect is that the present case is a highly mobile and highly unpredictable condition that can be analyzed after the change with accuracy, but it cannot be predicted very far in advance before the change.

Resistance to change

For most people, learning new and different things is not very natural. We all like to go along in our routines and live a predictable, stable life. There are a few real radicals who are always looking to change for change's sake, but the majority are not so inclined; we seek and appreciate stability.

Furthermore, most people will tell you that "people naturally resist change" and that "it is just part of our DNA." This is a common and almost reflexive statement people make when we speak of change.

I disagree, not completely, but I believe that is only a partial truth. In fact, I most often hear this from the very people who are trying yet being very unsuccessful in getting others to change. You see, if true, it provides a way-too-convenient excuse for their failings in getting others to change.

When this topic is broached, that is, "People naturally resist change," I will often ask listeners something like, "Can you tell me about the two biggest changes that occurred in your life?" A quick summary of the items from the people in the group will normally show that getting married and having children top almost any list. And with a little trepidation and the reality of a shotgun wedding being an answer, I will ask them, "Did you do these two activities voluntarily?" Usually an interesting discussion follows that unlocks an almost universal and obvious truth that although we are not averse to change, we almost always resist "being changed," and of course that comes with the added specification, "by others."

So to those who think resistance to change is a part of everyone's DNA, I say "phooey." Rather, what I firmly believe is that the majority will change, if you can get them to *want to* change. I do not profess that getting them to want to change is a simple feat. If I knew that formula, I would be an incredibly wealthy consultant and a national hero at some level. I am neither.

Head and heart engagement

However, there are two basic rules I follow with change management. For the clear majority of us, we first need to wrap our heads around the need for change, which means we need to get intellectual understanding; I call this head engagement. Second, we need to get people to want to change. Since this deals with acceptance at the motivational level, I call this heart engagement. For people to embark on a change, they need both a head and a heart engagement.

Unfortunately, most managers only plan as far as the head engagement, if even that. Repeatedly I have seen change initiatives undertaken where the top management does nothing more than proclaim "We are going to change" or "We need to change" and expect the rank and file to buy in with their heads and their hearts as they are expected to change what they do on a daily basis. That is where most management change efforts stop. Some, more enlightened managers will poll their people and put together what they earnestly believe is a meaningful plan for change. These efforts are often far more successful but still fall far short of success. The best treatment I have seen to guide you in cultural change is the writings of John Kotter. On our website at www.wilsonleanrefining.com is a list of his books and other fine books discussing change, change management, and cultural change.

A lean transformation with a deep cultural change

Let me relay a recent experience with a lean transformation. Six months into our transformation, which was going well, Eric, the GM, asked me why it was going so slowly. The employees had just completed their six-month employee engagement survey, and he was disappointed to find that although engagement was up 11% overall, in the Production Department they had made no progress.

Eric went on, "We have modified the floor activity, changed to cellular manufacturing, installed water striders, redesigned the work activities of the supervisors, and done dozens of things to promote more engagement; yet the floor shows no improvement in our survey data." Then he said, "After all that we have done, I expected better results," summarizing his concerns with distinct frustration.

I replied, "What you say is only partially correct, Eric. Yes, we have made major changes on the floor, but likewise you did similar things seven years earlier in your reengineering effort; and then after dropping that after two years, you started a similar activity with your Six Sigma initiative. So don't just expect everyone to blindly jump on your bandwagon."

I reminded Eric that I had previously discussed this with him, and he nodded knowingly when I told him, "The magnitude and timing of cultural change is hard to predict, and we will only get a feel for this after 12 to 18 months. As I told you before, you need to earn people's trust, and that does not come overnight. Furthermore, we have not 'changed to cellular manufacturing'; rather we have modified 6 of the 11 cells, and although the other 5 are being redesigned for one-piece flow, it is not yet a reality. And are you aware that only 4 cells are currently being serviced by a water strider? Regarding the redesign of the job of the supervisor, Javier (the lean coordinator), at my request, did an audit, and only 2 of the 5 shift supervisors are actually doing their leader standard work, although they were completely involved in their job redesign."

I went on, "First, what you see with your mind is not the reality on the floor; it is more what you hope for. You need to audit more and spend more time on the floor to stay in touch with the reality of the gemba. And second, all of this is very normal; you cannot expect people to willingly turn on a dime. Many of them have 30 years of experience and training in an earlier system; now you expect them to change at your pace. For 30 years they were taught and trained to behave in one manner, and they are reluctant to change; don't forget this is the third change initiative in the last 10 years."

That discussion did not go over well with Eric. He simply turned and walked away. Later that week we discussed it more, but I was never sure that I had convinced him that he needed to spend more time at the gemba, nor was I convinced that he understood that he was grossly underestimating the effort involved in changing a culture.

Then on my next visit a month later, I got the worst news I think I could have gotten. Eric again approached me and told me, "I was upset that the leader standard work was not being done, so I 'took a trip to the gemba' and did a little "motivating," and now if you audit, you will find all the leader standard work being done on all three shifts."

I was furious and almost lost it completely. But instead I took a deep breath and counted to 10. It helped. Very calmly I told him, "Eric, you are the GM over three plants. Each plant has a plant manager, who in turn has three production managers, each covering a single shift, and each of them has four or five production supervisors. You bypassed all those people, all this structure, to get a small group 'motivated'?" I told Eric, in no uncertain terms, that he had violated all we had tried to teach regarding system management, delegation, responsibility, engagement, and respect for people; he had modeled the wrong behavior. I reminded him, "Eric, you had made great progress when you swore off the old, heavy-handed, top-down directive style of management and had worked hard to effect a substantial change. But in this moment of crisis you had reverted to the very style of management we had worked so hard to change." He had nothing to say; he just turned and left, once again. I thought he was going to fire me.

Later that week, Eric returned, and with a certain amount of embarrassment, he admitted he had let the crisis get the best of him. It was very clear that he wanted his organization to grow and prosper, and he realized his actions created a setback. Eric sought my advice.

I so wanted to give him a lecture. Earlier I thought he had actually turned the corner in modifying his management style, only to see him regress in the face of a minor crisis. Rather, I took a deep breath and counted to 10, and the old saying came to me: "Never reply when you are angry, never promise when you are happy, and never decide when you are sad."

Since I was clearly both angry and sad, I put off Eric and told him that we could talk after lunch and that I would meet him in his office. I thought about the entire event, and I paid particular attention to how much anger, sadness, and frustration I felt. And then it dawned on me. I was not so mad at Eric as I was at myself. Just as Eric had anticipated greater changes in the plant, I had anticipated greater changes in Eric. And for so long I was able to reinforce that with the data I was gathering about Eric and his staff, just as Eric had done about his observations on the floor regarding cell completion, the introduction of the water strider, and the implementation of leader standard work. Both of us had experienced what we call "confirmation bias," and when the reality of it all hit us, we both went into denial and reverted to earlier, destructive behavior. And it was all a major symptom of underestimating the magnitude of effort it takes to truly transform a culture. Reality had not met our expectations.

After lunch, we met. I told Eric how happy I was with the progress the company had made and how proud I was to be chosen as the sensei for the lean transformation. I then reminded him, using numerous examples, how the employees were making progress and assured him that this setback, although consequential, was correctable. I shared with him my thoughts and how I too had made some of the same mistakes he had and how we both had a similar response. Finally, I reminded him that mistakes are not really mistakes; rather, they are teaching opportunities.

Then I told him that as his teacher, I had some homework for him. From our many discussions, I knew he was having some minor problems with his teenagers, Stanley and Stella. So I asked him, "Before you fully evaluate the unified effort and teamwork it takes to change a culture—which among other things is changing the behaviors of people—I want you to answer one question for me. Why is it that you cannot get your 14-year-old daughter to make her bed in the morning and get your 16-year-old son to take out the garbage?" Eric thought I had lost it. He was calm and polite, but he was convinced I was completely out to lunch. I then asked him "to humor me and get Anne, his wife, involved in the discussion." At this moment, Eric called me "certifiable," but then he took a deep breath and appeared to count to 10. After which he said, "Well, I guess we hired you and are paying you for your advice, so I probably should not ignore this homework. It'll be done when you return next month."

On my next visit, when Eric came to pick me up at the airport Monday morning, he was so excited I could not believe it. He told me that:

- Anne was the key that unlocked the entire issue.
- They had converted a whiteboard.
- And Stanley asked his advice on his physics problems.

"It was the homework," he said, "It was the homework that got me to see; well, really it was Anne." After a few minutes, I was able to get him to put his thoughts in chronological order. As it happened, one evening after dinner, he sat down to put his thoughts together about Stella and Stanley when Anne asked what he was doing. Eric explained the incident with me and his "homework" assignment. And Anne said, "Eric, you have been raving about how much work you are getting from the floor workers, and even though it is creating some unease with your supervisors, productivity is way up. Just what are you doing differently?"

I told her, "We realized that the floor workers could do much more if we just got them the information they needed and created a visual reminder system. So we did, and they responded in spades." Anne immediately walked over to the whiteboard reminder we have on the wall and wrote down the basic chores for Stella and Stan and placed a week's worth of boxes next to each chore. She then got Stella and Stan and explained to them that 'you are to do these chores and check off the boxes; Dad and I will monitor them only and no longer remind you each day. The responsibility is yours.'"

Eric continued: "Things went wonderfully for several days. Then Stella did not make her bed one morning. Her excuse was that Anne did not wake her in time. So when she got home from school, we made a change. It was now Stella's responsibility to set her alarm to wake up whenever she wanted, as long as she got her room made up and had breakfast before she left for school. The whole process of delegating the responsibility is working wonderfully."

"But," Eric told me, "then I asked Anne, 'If this is so obvious and simple now, why have we been harping on them—for like forever—to get this done?' Anne thought and then said rather introspectively, 'It was not them who needed to change so much as it was us; we were the impediment. You were the enforcer who caught them when they failed, but with your work schedule and travel, you were inconsistent, so the message was never clear. Me, down deep I wished they would always be our little kids and am having trouble with them growing up and probably do not push them very hard in the responsibility realm.'"

Eric explained that it was an amazing epiphany. He went on, "Let me tell you, not only are things getting done, but there is less stress in the house, and for the first time Stan is asking me for help with his physics. We even had a talk about a girl at school he is sweet on. It's great."

Eric explained more: "At work the parallels are similar. Once we created a system with greater clarity and clear responsibility for the workers, just like we did for Stella and Stan, they stepped up. Just like at home, the limitation was not the floor workers but the supervision and management. Once we created an environment where all the workers knew what to do and how to do it and we gave them the resources, they did it and they reveled in it. Stella and Stan, just like the floor workers, are motivated by a sense of accomplishment and other intrinsic motivators. In addition, I now have a greater appreciation for the depth and the magnitude of a cultural change. Yes, the floor workers need to change, but their change is limited by the ability of their supervisors and managers to change. The same goes for my family. And that involves not only the changes that I had to undergo as a parent in devising a better system but also the changes that Anne had to undergo in terms of letting go. Finally, I can see the need to 'go to the gemba' and be observant in a cold, dispassionate, objective, introspective way. It is important to fight through the issues of confirmation bias by continually questioning and listening. I have a new appreciation for the depth and breadth of change that is required in a cultural change. Earlier I thought I had this 'cultural stuff' pretty well mastered, and now I learned that we all need to sharpen our cultural antennae . . . and they can always be sharper."

And exactly what did Eric do with this newfound "awareness?"

After some reflection, Eric gathered his direct reports and relayed the entire story. He started with our discussions, described his reversion to "kicking some butt," and told the whole story of himself, Anne, Stella, and Stanley and how much he had learned from the "homework," both for his family and at work.

He then put a question to the entire team: "How can we do this lean transformation better?" There was a lot of discussion, but at the end, a small team was charged with reviewing and reflecting on all that the workers had done and how to do it better.

They complete their first "hansei," deep reflection

After going to the gemba and meeting as a group several times, the team members evaluated their original plan, which was to:

- Convert to manufacturing cells using the "surgeon" concept
- Install andons and other types of transparency to communicate from the cell
- Add water striders to support the cells
- Create the new job of team leader

They concluded that the plan was solid and should be continued as it was.

However, they felt they needed to reemphasize the importance for the management team to provide the proper leadership. They then decided to have "open discussions" in small groups with the leadership team, including all supervisors, on the topics of:

- Leaders as change agents and the need to initiate, observe, and follow up on actions.
- Leaders as learners, paying special attention to how all the changes are progressing to see what they could learn from what they had already done.
- Leaders as teachers, especially making sure that all fully understand the new techniques.
- Leaders as role models, especially making sure that leader standard work is done as taught with both policy deployment and personnel development in mind.
- Leaders as superior observers and supporters, especially making sure they go to the gemba to observe and listen. The trip to the gemba is not to criticize or correct, it is not to send out messages, nor is it to check status. Going to the gemba is for input, unbiased input. They agreed that they all needed to be more observant, spend more time listening and less time talking, and then reflect on what they had observed while at the gemba.

And the results of the deep reflection on the meetings

After the initial meetings by the senior management, and after the open discussions that all the executives, managers, and supervisors attended, they concluded that they were making good progress and that the review of their leadership skills was a good idea but really just provided a reminder. Finally, they concluded that they really did not need to change much except for the executives and managers to spend more time at the gemba and possibly be more patient as the cultural adjustments took hold.

The results of the lean transformation

When a company goes through a change such as outlined above, the benefits are huge. In the case of this plant, although the leaders had added more support personnel, they had made major improvements by every metric. First, quality was no longer an issue; customer returns dropped to <25 ppm. Even counting the added support people of cell leaders and water striders, productivity rose from 95 pieces/person/shift to 234 pieces/person/shift, and full-load, nonexpedited on-time delivery was up from 92 to >98%. They had not had a single environmental incident and not had a lost-time accident in 30 months. This change took 36 months, and even today they are still changing and still improving.

THE "SURGEON" CONCEPT FOR THE VALUE-ADDED WORKER

Value-added workers are people on the line doing assembly or parts manufacturing who have some of the lowest-paying jobs in the facility.

Yet these people are the only ones adding value. It is everyone else's job to support them. Like it or not, these are the folks who are executing the strategic plans so painstakingly developed by the senior management.

Hence they need to be treated as you would treat a surgeon in the operating room. All the tools and materials they need to be successful are brought to them just in time and in perfect condition. Everyone else moves around and caters to their needs. Unfortunately many firms resist this concept as it seems to turn the organizational pyramid upside down.

Thanks to Richard Dunn and Karn Gill of EXCO Engineering, who coined this term.

REGARDING THIS "TOOLS-ONLY" APPROACH, DID OTHERS HAVE BETTER ADVICE . . . EARLIER?

Yes, there were others with a much better perspective. Early in the period where America was gaining interest in the Toyota Production System, and long before *The Machine That Changed the World* was published, others were studying the Japanese methods. One early harbinger of lean was Norman Bodek, who started Productivity Press and brought the writings of Ohno, Shingo and others to the American public. Others who were writing in the early 1980s included Richard Schonberger, *Japanese Manufacturing Techniques: Nine Lessons in Simplicity,* and Robert "Doc" Hall, *Zero Inventories*. Both of these authors, in their first books, did an excellent job of describing the technical aspects of the Toyota Production System, including small lot production with frequent deliveries, quick changeovers, double shifting, production leveling, pull production systems, workload balancing, multiskilled workers, cellular manufacturing, and total productive maintenance—all focused not only on inventory reductions and a total quality system but on cost reductions as well.

Later these authors published follow-up books. Schonberger's was *World Class Manufacturing: The Lessons of Simplicity Applied*. And Doc Hall's second book was *Attaining Manufacturing Excellence: Just in Time, Total Quality, Total People Involvement*. Each author's second book followed in four years from his first.

Each of the second books of Hall and Schonberger contained a dramatically different tone. In each of their follow-up books, there was a great deal of writing dedicated to the roles of management, leadership, total facility engagement, and the human element in general. The attention to the human element was so important to Hall that it was part of the title, "Total People Involvement." In Schonberger's second book, the topics of management engagement were also manifest as well as the many topics of managing the change, handling the resistance to change, leadership, world-class management for the CEO, project champions, overuse of staff personnel, plant organizations, team building, teachers and users, and partnering. There was an entire chapter on strategy, focusing on the key role of management, as well as an entire chapter on training. All these topics were either absent from or lightly treated in his first book.

In Hall's second book, in the Preface, he states, "The first challenge is to gain an overall grasp sufficient to lead an organization through a broad scope of manufacturing change," and in the book he focuses clearly on two major topics in a lean transformation: leadership and change management. This second book also contains information on leadership thinking, problem-solving atmosphere, employment security, gaining purpose, people flexibility, attaining total people involvement, and the changed role of staff employees. In addition, Hall advances the idea that workforce participation is management participation, and there is even a chapter titled "Reforming Permanently," which is a treatment in sustaining the gains through cultural change.

After reading their initial and second books, I can draw some conclusions. First, both of these academics (Hall at Indiana University and Schonberger at University of Nebraska) had learned from their studies of and visits to Japanese companies. Second, through some mechanism they seemed to learn faster than others. Their initial books were very technical treatments of the Japanese lean systems, at that time called JIT manufacturing or JIT with TQM. Both books were very good and are still great reading. They, however, focused almost entirely on the technical topics of the TPS and hence covered the tools and the continuous improvement aspect of lean nicely. Their second booksshowed a great deal of interest in not only technical issues but also leadership, managerial, and the human topics. They had learned from their teachings and real-live field transformations that to implement a lean transformation took a deep dive into the human element as well—what Toyota calls the "respect-for-people" concept. Third, since it takes some time to write and publish a book, it appears that it took both of these practitioners less than three years to conclude, If we want to transform these companies, we need to do more than teach the tools; we need a deep dive into the human element as well.

Yet it took the lean community, as documented by James Womack, over 15 years to learn this reality. In addition, it was not until around 2008 that the Shingo Prize added Cultural Enablers to its criteria.

Hall and Schonberger, however, had awakened to the reality that implementing a lean transformation was as much about the management, leadership, and people development as it was about the tools. Unfortunately, there are still many firms that have yet to awaken to this reality. Sadly enough, neither Richard Schonberger nor Doc Hall could garner the same audience as some of the larger lean organizations like the Lean Enterprise Institute, the Shingo Institute, and the Association of Manufacturing Excellence, which, up until recently, were all largely focused on teaching and marketing the tools of lean.

The works of Schonberger and Hall were honed by experience *and* observation, not just by observation. I recommend their books to all; you will find great writing by great professionals who fully punctuate why teaching the tools is simply not enough to support a lean transformation.

So should we throw darts at those who learned more slowly?

From what I have written here and elsewhere, you might think I want to blame the other academics for something.

Not true.

Do not think for one second I hold these academics, especially those who were very active in creating the Lean Enterprise Institutes, in anything but the highest esteem. They created an awareness and catalyzed a movement that even the greats who preceded them could not accomplish. They brought lean to the public eye, and they got hundreds if not thousands of companies to stand up and take notice. You will recall from our discussion about the 2007 Industry Week/MPI Census of Manufacturers that a little less than 70% of U.S. plants were implementing lean manufacturing.

I was and still am amazed that largely through the efforts of the LEIs, these academics created orders-of-magnitude more interest than either of my two heroes, Dr. Joseph Juran and Dr. W. Edwards Deming.

Not only did they create the interest, but they have followed the lean mantra of "learn-do-reflect," and they have reflected on what has worked and what has not. Even though I wish they would have learned earlier that "Lean does not equal tools," they did learn, and they helped others change as well.

I have no issues with them.

No, my issue is with the large number of practitioners both in industry and certainly among the consultants—who believed what these academics said, lock, stock, and barrel. They acted as if, for some reason, these academics had a monopoly on the truth about lean manufacturing. And furthermore, if there was something these academics did not bring forward, it simply must not be relevant. I am disappointed in my fellow practitioners and consultants for not following the lean mantra of "learning by doing."

> **CULTURAL CHANGE LEADING INDICATOR NUMBER ONE—LEADERSHIP**
>
> The number one leading indicator of cultural change and success is leadership.

They chose to learn by copying . . . even when it did not work so well.

The difference between the academics who studied lean from the outside and the practitioners who were trying to improve the companies from the inside is that the practitioners who did the actual work also had the actual data that proved that things were not working as the academics had predicted. They had goals that were not met, they had projects that were not completed, and they had transformations that were taking way too long.

Anyone on the front line of a lean transformation knew that:

- Lean leadership played a key role.
- If you did not get the whole facility aligned and engaged, things would not go smoothly, if at all.

And anyone who worked with a lean transformation more than three years learned the hard way that:

- Kaizen events alone would not effect a lean transformation.
- Management plays a crucial role.
- If you wanted to sustain the gains achieved in kaizen events and continuous improvement, you absolutely needed to create a corporatewide cultural change.

They learned through practice that "culture trumps everything."

Unfortunately, because these practitioners did not challenge the established paradigm, the entire community did not learn as quickly as it could.

The academics are not wrong . . .they just miss some crucial stuff

I can search and search and hardly find any errors in their work. But what I do find is an incompleteness. It's an incompleteness that I found when *The Machine* was published and that I also found in the writings of Ohno, but it is an incompleteness that is slowly becoming understood.

I was a huge Deming disciple from back in the 1970s and thought at that time, and still do, that his 14 Obligations of Management are still key to unlocking the manufacturing potential in the United States. Consequently, I never wavered in my belief that every lean transformation needed to be driven by visionary leadership utilizing a totally engaged workforce. Those are the keys so often missing from our Western approach to change. We often lack the dynamic, informed leadership with the skill set outlined by Dr. Deming, and we continually underestimate the need to engage the entire workforce.

And ultimately the goal of a lean transformation is to change your culture. For a much deeper review of the topic of culture and the five leading indicators of cultural change, see Chapters 4–9 in *How to Implement Lean Manufacturing* (2nd ed.). Also, check the website of my company, Quality Consultants, at www.qc-ep.com; we specialize in cultural change, and you will see our motto: "The only effective way to improve quality is to make it part of your culture, this is our aim."

> **DR. W. EDWARDS DEMING'S POINT 1**
>
> "Create constancy of purpose for continual improvment of products and service . . . allocating resources to provide for long-range needs rather than short-term profitability, with a plan to become competitive, to stay in business, and to provide jobs."

WHAT CAN YOU EXPECT?

My message to you comes in three parts:

- First, there is nothing wrong with the philosophy and principles of lean. But the effort must have strong leadership, and you also must execute the entire lean transformation well; then you will be successful.
- Second, the method to execute a lean transformation is no secret but requires a serious effort to make a major change in your business culture.
- Third, the many lean efforts that have gone astray or failed to meet objectives should not signify a failing in lean; rather, it should be looked upon as a work in progress. You need to treat what you have done, your successes as well as your failures, as a learning experience on your way to becoming a world-class performer. Rather than quit your lean transformation, you need to build upon your early successes and learn how to sustain those gains through a serious cultural change. You need a constancy of purpose to implement lean over the long term. Given this effort, you can succeed nicely.

CHAPTER SUMMARY

There is a great deal of information, mostly anecdotal in nature, that suggests that the lean movement is not living up to expectations. The clear majority of the information suggests that when a lean initiative fails to meet expectations, it is due to the way the lean transformation is executed rather than a fundamental flaw in the philosophy of lean; with that, I concur.

One very large error that is accelerated by impatient management, looking for quick and easy gains, is to design a lean transformation based almost solely on using the tools of lean. The philosophy of Toyota, described in *The Toyota Way, 2001,* does not support this approach.

In fact, some early writers, notably Richard Schonberger and Robert "Doc" Hall in the 1980s, exposed this tools approach as insufficient to make major gains; yet their warnings went unheeded.

So when it was found that balanced pull production cells operating at takt were not, in and of themselves, sustainable in the long run, the adopters changed and recognized that we need to do this with a strong leadership and an engaged workforce, not just with a bunch of staff folks doing kaizen events. This, in turn, required a far greater degree of management involvement. Furthermore, it did not surprise me at all that the lean transformation model that was once "What we need are the tools" has morphed into "Oh, by the way, we also need engagement as well as leadership, and finally we need to change the culture."

Nonetheless, many lean transformations are still based on this tools approach. And in many cases, they achieve short-term gains, more often than not, compromised by long-term losses. And the all-too-typical response is to blame the lean concept and say it will not work.

In reality, the concept of lean is solid as a rock . . . if executed using the Toyota model, which basically means according the same emphasis to the human element as is given to the continuous improvement element. And there is nothing wrong with the lean philosophy and principles, and if executed properly, huge gains can be expected in almost any endeavor in both the short term and the long term; which should be the objective.

Many believe that lean is not a long-term solution to manufacturing excellence; to this I say "phooey." To those who say it doesn't work, I say you haven't worked long enough and hard enough. You have just gotten started and have not completed the journey. The gains are there.

Toyota started with its famous kanban system in 1953 and implemented jidoka in the auto factory in 1955. Yet it took the company over 30 more years before it had what it called the Toyota Production System. Not surprisingly, Toyota is still improving on these methods after yet another 30 years.

However, today I find companies giving up and declaring lean a failure after one or two or three years. In light of all this, to them I say, "Wake up and work harder."

I have come to understand that lean is not failing if executed properly or even approximately well. Rather than failing, your implementation is not yet mature. It is a work in progress, and declaring its demise is both premature and immature.

Part 2 Lean and Refining

CHAPTER 4

HOW MUCH HAS LEAN PENETRATED THE REFINING BUSINESS?

Aim of this chapter—*In this chapter we explore how deeply the concepts of lean manufacturing have penetrated the oil industry. To find out if any companies are employing this powerful technique and to what extent the concept of lean manufacturing is understood within the U.S. refining industry, we will cover my personal experience with Chevron, investigate what is on the blogs and the Internet, and review what 31 in-depth interviews with oil executives revealed about the penetration of lean manufacturing in the oil business. Finally, I will introduce you to the one example of a lean transformation in the oil industry that I could find. It is the story of Aera Energy, an example of lean done right.*

MY EXPERIENCE PRIOR TO 1990

Before 1990, the term "lean" had not even been coined. Hence there was no lean consulting, no books on lean, and no lean industry. There were books on the Toyota Production System and several on JIT (just-in-time). Most of the books were translations of books originally written in Japanese and came through Productivity Press. The topic of JIT was being pushed by APICS (American Productivity and Inventory Control Society), but none of this got much traction in manufacturing, or anywhere else for that matter. At that time, the new wave of consulting centered on quality. Many consultants and consulting firms were engaged with companies implementing TQM (total quality management) or quality improvement programs based on the teachings of

Dr. Deming. These consulting firms were trying to teach his 14 Obligations of Management and largely selling that as a method to support and accelerate process improvements.

Well, it so happened that earlier I had been reading about Japanese methods and was particularly impressed by the NBC whitepaper, "If Japan Can Do It, Why Can't We?" Later I read Dr. Deming's book *Out the of Crisis,* and I only needed to attend one of his seminars—and I was hooked, 100% hooked.

We started with statistical process control and small-group problem solving

Once I was hooked, I got on a reading campaign that I still maintain. Many of the books I read are focused on these Japanese methods, especially Japanese management and leadership methods. Most of my peers in the oil industry, who had been exposed to Dr. Deming, did not take him seriously at all. For example, a four-day Deming seminar I attended in San Francisco in the mid-1980s was attended by over 30 Chevron employees. By the third day, only three of us were left.

CHECK OUT THE READING LIST ON OUR WEBSITE

There is a great deal of good information in books, especially to pique your curiosity and get you into "action mode." Visit our website at www.wilsonleanrefining.com, and you will find a very large list of books on lean and a variety of other topics including specific skills such as SPC and HK planning. There are also books on other related topics such as motivation theory, general management, and listening.

Since scarce few were interested, the early days of using Deming's philosophy at Chevron were very slow. I was not sure how to start, so I started with something technical and something I already had some knowledge in: statistical process control (SPC). SPC was a practice of increasing interest in the manufacturing industry, and it also was something I could do within my own division. I needed no support or approval from anyone. I brought in Dr. Donald Wheeler to teach his weeklong SPC class. He was brilliant and a great teacher, but I could see that one dose of SPC would not get the refinery doing SPC.

I took it upon myself to become the SPC expert. Although I had taken statistics in college and Wheeler's course, I knew that would not be enough to support my staff. I later took a second and third course from Dr. Wheeler and a class in advanced statistics from Dr. David Chambers, a colleague of Dr. Wheeler's. I took a class in DOE (designs of experiments) from Lynn Torbeck. Lynn was the best person on this planet, in my opinion, to apply the principles of DOE in a real-world setting. He was all about using statistics to gain knowledge about the system so it could be improved.

I read, I studied, and I took other classes. However, while I was reading and studying, I was doing, and that taught me the most. I was always reviewing some process using SPC, and it was this "doing" that was my real teacher. After some time, I developed classes in SPC and taught SPC to the supervisors, engineers, and technicians in my division. Later I taught them measurement system analysis (MSA) and even gave classes in designs of experiments (DOE). This worked very well for three reasons:

1. In the Technical Division, our basic task was to problem-solve the many issues in the refinery. Now when they analyzed a system, the engineers and technicians almost reflexively put the data on a control chart. Those results were startling.

2. Most of the staff members were very young, so they embraced the new techniques much better than the "seasoned veterans." It was not a fight to get them to use these new tools.

3. Most importantly, they routinely gained additional knowledge and came up with superior solutions. In fact, by using SPC and statistical problem solving, we made millions of dollars for the refinery.

Because SPC helped them do a better job, the engineers and supervisors became enamored of these techniques, and some became quite skilled in the use of SPC. But even though this was exciting to me, not many outside of the Technical Division were impressed. At this time, I was the manager of the Technical Division, which was the support group for the entire refinery. The division consisted of nine groups including design engineering and drafting, process engineering, IT, environmental engineering, the laboratory, two reliability groups, and facilities planning. plus long-range and short-range oils planning. I had 9 direct reports with a total of 13 supervisory personnel and over 140 people on the payroll. It was a large and vibrant group.

We expand into team problem solving

However, we really had no other engaged employees outside of my division. Then I took a trip to Chevron's Richmond refinery and purely by chance ended up in a class on team problem solving taught by Peter Scholtes and Brian Joiner. This was a game changer.

First, we created two or three teams from the Technical Division and tested the methodology. It worked extremely well, as we solved several problems that had been nagging the refinery for years. Next and very quickly, I approached my peers, the manager of operations and the manager of maintenance. I proposed we put together some teams from multiple divisions, and after only a little selling, they agreed. This resulted in cross-functional problem solving, and the benefits from this were huge, larger than anyone imagined. And without knowing it, we started a process that facilitated engagement in the refinery. We worked on many projects and solved many longstanding and serious problems. We solved operations problems, technical problems, maintenance problems, and even some issues in both human resources and finance. (Some of these are detailed in Chapter 12.)

We start to break down some of the silos

It was amazing to see how these cross-functional teams performed. We could easily get them to do things, such as information sharing, empathetic listening, and cooperation, which in the normal day-to-day operation were simply not happening. Without trying to, we were effectively breaking down the barriers that are referred to as "silos," those huge emotional barriers that normally exist between divisions. The work ethic, the cooperation, the creativity, but most of all the trust that was present on these cross-functional teams simply was not evident in the day-to-day activities of the refinery. It was both refreshing and amazing as well as enlightening. I learned about how to attack and solve very difficult problems, many of which had persisted for years, even decades.

We persist for two more years

Over the next two years we commissioned and led over 20 teams. Most were extremely successful, some were successful, and some failed to meet expectations. We learned from all three categories and improved the process as we created new teams. What we found was that the following three questions, in order, provided the guidance that correlated best with success:

1. Did the team consist of volunteers or those who were "voluntold?" Early in our effort, my peers felt it best for them to select the team members from their divisions so we would have "good, motivated" workers on the interdivisional team. It turns out that although this was both thoughtful and well-intended, it was a huge error. Once we started using volunteers, we were universally successful and never went back to selection by the division manager or anyone else for that matter.

2. Did the team have a good leader?

3. Did the team have a good sponsor? A sponsor was a management person whose job it was to clear high-level problems.

What I learned from this

After more than two years and numerous successes and a clear model of statistical and team-based problem solving to follow, I retired from Chevron. Because of our huge successes, I figured our efforts at El Paso would persist for some time.

Dumb me! Really dumb me!

It took about 30 days, and then most everything we had started began to unravel. The team leaders were redirected to other activities. With no one to organize the effort, no new teams were formed. And the entire team effort died within a few months.

I learned, right then and there, the need for both strong sponsorship and strong leadership.

I was surprised and disappointed, and now I realized how frail changes like these had been. Immaturely, I had thought that since we could make better analyses and better recommendations, these techniques would persist; we had been hugely successful. I had completely underestimated the amount of effort and time it takes to change the way people think and act. Later I learned that changing the way people think and act is the operational definition of changing a culture—and it is not easy.

I have never forgotten the lesson of those successes, yet how quickly the entire system would regress to its former state. Based on this experience, I coined a saying I use often: "Culture trumps everything." I had totally underestimated the work and time it takes to make a real cultural change, and I have yet to find even one person who has overestimated the time and effort it takes.

Cultural change takes great effort, great focus, and tremendous persistence. This concept alone is enough to explain virtually every lean initiative that people have said was a failure.

So the bottom line of my work with lean in a refinery is this:

- It is possible to create a culture of continuous improvement and respect for people, especially one focusing on problem solving to reduce wastes using an engaged workforce in a refinery. Not only is it possible, but with the right transformation design and the right management team, it will be successful, very successful.

- We had no problem getting engagement of the rank and file. The problem in getting a fully engaged workforce was more often the resistance of the management team.

- The real problem was not the "problem" we were trying to solve; rather, it was the issue of changing the culture. The culture in our refinery in El Paso was very strong. Unfortunately, it was also very inflexible as well as inappropriate to create either a continuous improvement effort or a respect-for-people effort. The culture, controlled by the senior management, vigorously resisted any effort to actively engage the entire workforce, despite the obvious successes of our group problem-solving work. Of all the Chevron refineries where I had worked, we at El Paso had the best company-union relationship. However, there was an underlying local concern by some of the senior management, and especially by the corporate management, that this type of engagement would weaken their bargaining position relative to the contract with the union. That was never my belief at the time, and now even less so. Now I see those attitudes very clearly and for what they really are . . . a red herring to justify the management position, which is "In spite of the data in front of us, we simply do not want change."

I TRY TO GET A CLIENT IN THE REFINING BUSINESS INTERESTED IN LEAN MANUFACTURING

In 2006, one of my business objectives was to get work with a client in the refining sector. From my prior days in the oil business, I had lost most of my contacts in the refining sector. El Paso Refining (the old Texaco refinery that was across the street) had dramatically changed with mergers and a subsequent sale to Western Refining. Nonetheless, I contacted a highly placed VP for Western Refining with a solid recommendation from an intermediary. I called him to schedule an appointment. He was aware of me through the recommendation and my work with Chevron. Unfortunately, he had no knowledge of lean, and furthermore, after a brief phone discussion in which he had few if any questions other than "So what is lean?" he told me he was not interested. A dead end.

I search the Internet

I searched the Internet and found literally nothing except a couple of blog postings. I found some applications in the upstream process of oil-well production and even oil-well maintenance but literally nothing about refining.

I write my book, and I get speaking engagements

In 2009 I published my first book, *How to Implement Lean Manufacturing*. My primary purpose in writing this book was to share what I had learned; my secondary reason was to gain credibility as a consultant. I wanted to create additional credibility that I could couple with my 20 years of refining experience so I could use that in applying lean and work with refineries to exploit this wonderful methodology of lean manufacturing.

My book created many speaking opportunities. In the next two years, I spoke at several conferences. Three were the 2010 and 2011 iSixSigma Energy Conferences in Houston and the 2010 IndustryWeek's Best Plants conference in Cleveland. On these occasions my presentations focused on expanding the use of lean into refining and describing why and how well it would work. Later, collaborating with Jason Farley, we coauthored an article that is almost always number one on the search list when you search for "lean in petroleum refining" or "lean manufacturing in petroleum refining" (see the Appendix.) Surprisingly to me, this created practically zero interest from any clients. I would get an occasional call from someone whose interest could best be described as mere curiosity. I continued to write dozens of articles for *iSixSigma* magazine, *IndustryWeek* magazine, and *Energy Digest*. None of this created interest from the petroleum refineries, nor did it create much interest from the upstream business either.

I join LinkedIn and other blogs

On the topic of lean in refining, there was little going on in the Internet, so I decided maybe I could cultivate a little. I was already on LinkedIn and joined several groups including:

- LinkedIn: Energy
- Petroleum Downstream
- Oil Refining–Global Technology Forum
- Plant Reliability and Maintenance Professionals

And what I found was essentially nothing on lean. I found some discussions of process improvements and an occasional reference to Six Sigma but practically zero on lean as a concept or as a philosophy and certainly nothing about lean as a means toward continuous improvement or an actual manufacturing system . . . nothing.

I started topics on "Why is there no discussion on lean applications in petroleum refining?" It took two years until I finally got one comment. In addition, I subscribed to the *Oil Pro* magazine and forum. Again very little on refining in general, mostly upstream stuff, and I could find literally nothing on lean. There were several articles about Six Sigma and even some about Lean Sigma but nothing on lean. Well, there was one article in 2015 referencing Hess using aspects of the Toyota Production System, but a careful reading shows that although Hess was using some TPS buzzwords, what it was doing was making some good process improvements but without any cultural change, without any documented continuous improvement effort, and certainly without any real form of employee engagement. That is not lean manufacturing; that is basic process management.

I continue to get weekly summaries on LinkedIn and the blogs, but honestly, I am not expecting much.

I interviewed several refining executives, managers and others

From 1990 to the present, I have worked with several companies in the oil refining business, some that support the oil businesses and others in the chemical and continuous process industries. In these engagements, my efforts are mostly directed at solving problems, teaching problem solving, or teaching one of the lean tools such as kanban or hoshin kanri planning. However, in the case of two refining companies, I was retained for the implementation of their Lean Sigma efforts.

At each engagement, I would take some time to interview a few people, mostly senior-level executives, to gain insights into their understanding of lean manufacturing. In all, I had formal interviews with a total of 31 executives, managers, engineers and consultants. Two were VPs of refining, another was the VP of hydrocarbon processing, and there were two refinery managers. In addition, there were eleven senior managers who reported directly to the VPs or the refinery managers. Also, I interviewed three other mid-level managers, two first-line technical supervisors, and six engineers. I completed the interviews speaking with four external consultants.

Particularly revealing was an interview I had with one of the consultants just prior to writing this book. He had worked in the oil industry for over 30 years. In fact, we had hired him as an external consultant when I was with Chevron. It surprised me when I contacted him and asked, "How much do you think lean manufacturing has penetrated the oil refining business?" His reply was a question: "What is lean manufacturing?" I was flabbergasted. He is a well-respected leader in his field, has tons of hands-on experience, has written extensively, and yet had no idea what lean manufacturing was. I gave him the *Reader's Digest* version, and he quickly replied, "Can't do that in refineries." He went on, "They barely use the intellect of the engineers; computer technologies have locked people into their offices. And the culture in a refinery does not use the intellectual resources of the operator at all. No one is permitted to 'lead from the front.'" He added, "The management would never allow that." A rather damning and negative view of not just lean applications, but problem solving in general and the general use of people in refining. We spoke for quite some time, and although I found his comments negative, he readily supported them with specific examples. He made a rather compelling case.

A second refining consultant, one whom I know quite well, rose to the level of regional VP before retiring from his company and starting his consultancy after 30 years in refining. When asked about lean in oil refining, he told me quite frankly that "I see practically nothing." He went on, "And although I have only a superficial understanding of lean manufacturing, that has come about since my retirement. In my days in management, I knew nothing about lean; *there was just no interest*." I have known this man for over 30 years, and when he says he has only a "superficial understanding," he is being quite modest. But his point was that in his days of refinery management, no one was interested. Pretty hard to be more specific than saying "there was just no interest." Not some or very little, but "no interest."

I also interviewed a consultant in the field of lean manufacturing. Although he had no clients directly involved with refining, he had some experience in the continuous process industries. He had, however, actively tried to find clients in the oil business, without much success. Specifically, based on his interactions with people in the oil business, he felt they believed they had the answers on how to improve things, and although they had no lean experience and very little or no knowledge at all regarding lean transformations, they dismissed the concept of lean manufacturing totally. Again, based on his interactions, he felt they were very impatient and were looking for the proverbial silver bullet. He went on to say, "They openly admitted they did not want to change their culture."

Since any culture-changing initiative must be led by the senior management, the replies from the three VPs and two refinery managers were quite revealing. One of the VPs just plain told me he had never heard of lean manufacturing. He had, however, heard of Lean Sigma but not in any refineries. He called it "an upstream effort at continuous improvement." I asked if he had ever considered it for his refineries, and he said, "Like I say, it is an upstream thing," showing both a lack of awareness and a lack of interest. Of the two refinery managers, one professed no knowledge, and the other said he had not heard of any pure lean applications but was personally involved with a Lean Sigma effort at another location. I asked him how he would describe the "lean portion" of the Lean Sigma effort, and he quickly and simply said, "It is about waste reduction; Six Sigma is about variation reduction." He had a minor understanding, so we went on. I asked him whether he had seen a cultural change in his prior Lean Sigma activities. He did not know and said he had not thought about it as such.

All such change efforts require complete engagement from the senior leadership. Thus with this level of understanding of lean, it is not surprising to see no lean efforts in refining. When I interviewed refinery mid-level managers and those reporting directly to the refinery managers, I found similar comments to those of their refinery managers. None of these managers could articulate the essence of lean beyond the concept of "waste reduction," and no one could see much difference between a lean transformation and a Six Sigma initiative. No mention of continuous improvement, no mention of respect for people, and certainly no mention of either management and worker engagement or cultural change.

To my surprise, among the first-line technical supervisors and the engineers who tended to be younger and with much less experience, I found some encouraging awareness. For the people in this group, it was typical that they had read books on management and process improvement. Two of this group had read books by Dr. Deming and found them very applicable, but not really used in their facility. Yet three others had read my book, and on more than one occasion we had serious discussions about lean applications in refining. One young supervisor said, "In a large integrated oil company, like ours, making changes to a single refinery is very difficult if it doesn't come from the top. The centralization limits innovation." Another said, "A huge problem we have is the massive effort we have for evaluating and planning work, even routine repairs. Makes sense for large projects, but not for routine work. Our bureaucracy is stifling." Yet another said, "We have a hyperview on costs, yet we do not understand the primary value stream." One particularly insightful young supervisor said, "Refineries are really good at automating waste . . . as opposed to eliminating it." Yet another said, "SPC is not used, and true problem solving and root cause analysis is lacking. And it seems that every problem creates a solution that complicates the process, like adding extra inspection steps." These young engineers and young supervisors had read more about lean manufacturing and had a much better grasp than their more senior managers. In addition, they could see the applications of lean manufacturing in their own refinery environment.

So what about the Lean Sigma efforts? Don't they count?

No, not at all. What these people are doing is not lean manufacturing. What they are doing is not even Six Sigma. It is a watered-down version of the original Six Sigma concept. Let me give you some information from my previous book, *How to Implement Lean Manufacturing* (2nd ed.):

The concept of Six Sigma has achieved a great deal of publicity and, quite frankly, a great deal of success lately. Most of the publicity is directly or indirectly related to the earlier efforts put forth by GE, which was made very public by their now-retired CEO, Jack Welch. Six Sigma owes its roots to Motorola and the efforts of Mikel Harry. I first read about it as a design concept and tolerancing mechanism popularized in the publication "Six Sigma Producibility Analysis and Process Characterization," put out by Motorola University. The purpose of this technique was to start at the design phase and try to produce a "Six Sigma" quality product. This was defined as one with less than 3.4 defects per million opportunities. And the focus was largely on variation reduction.

Later, the concept of Six Sigma was broadened to become a problem-solving technique, and the Six Sigma curriculum has become standardized with Six Sigma Blackbelt and Greenbelt certifications now available. The concept was broadened from one of variation reduction to working on three critical aspects of manufacturing: cost, delivery, and quality with an emphasis on lead-time reduction (delivery). The three categories of Six Sigma projects were those classified as:

1. Cost . . . critical to cost, or CTC
2. Delivery . . . critical to delivery, or CTD
3. Quality . . . critical to quality, or CTQ

Since the early days at Motorola, the Six Sigma concept has grown and has had various degrees of success. Welch in his writings claimed that GE sent $10 to the bottom-line for each dollar they spent on Six Sigma efforts. Most large GE facilities had a complement of Blackbelts and Greenbelts and set up these internal consultants as cost centers. Today, to most people:

"Six Sigma is a project-based, problem solving initiative which uses basic as well as powerful statistical methods to solve business problems and drive money to the bottom-line of the business." (from www.qc-ep.com).

In fact, more recently it has morphed into almost exclusively a cost reduction program and each project is evaluated against an ROI or other financial measure of success. This being the case, Six Sigma is neither a manufacturing system nor a manufacturing philosophy at all; rather, it is a fine set of tools that can enhance problem solving in any sort of business, manufacturing or otherwise. The Six Sigma problem-solving concept is a sound one and helps make the company a more powerful money-making machine.

Summarizing, Six Sigma was created by Motorola to achieve Six Sigma levels of quality focusing on variation reduction. Through success at GE and other companies, it morphed to become a cost-reduction program—every project needs a clear ROI. Lacking a strong culture-changing mechanism, it rarely leads to sustainable change.

In its early stages, Dr. Deming would have vigorously supported it. Recall he said that management is prediction. Well, with reduced variation you can get better predictions hence better management. On the other hand, the way Six Sigma is managed today, I am certain he would not be a current supporter. Six Sigma is not a manufacturing philosophy nor is it a culture-changing mechanism. It fails to create continuous improvement and has no mechanism for total engagement of the workforce. Yet when it is contained within a lean transformation and managed for what it originally was . . . solid statistical problem solving focusing on variation reduction . . . it can be a real asset to a lean transformation.

Keep in mind our interim definition of lean:

- Lean is the creation of a culture of continuous improvement and respect for people.
- The means to lean is to problem solve your way to the ideal state through the total elimination of waste using a fully engaged workforce.

Lean Sigma programs in the refineries that I have worked with, and I know of no others that have embarked on such an effort, are just as the Six Sigma programs described above. They utilize a project-based, problem-solving approach using engineers and managers trained in Six Sigma skills and qualified as either Green Belts or Black Belts, and their job is to transfer money to the bottom line.

No longer is either quality or delivery of any significance unless it translates directly into financial gains. Furthermore, Six Sigma programs have a huge and bureaucratic structure to keep track of project status, project costs and gains, current qualification levels of belts, current status of belts, sponsors, and champions.

ERIC HOFFER ON CHANGE

"In times of change, the learners will inherit the earth, while the learned will find themselves well equipped to handle a world that no longer exists."

Where these initiatives fail to get the gains of a true lean transformation is fourfold:

1. They are not continuous improvement. They are project based, and projects have a beginning and an end. The beauty of these projects is that they are large; they need to be to justify the time and effort of the program costs. Again, refer to the history of the Theta cell; the first kaizen event is very similar to a Six Sigma project. Unlike the progress on the Theta cell, there is no methodology within the Six Sigma process to capture the gains from the over 100 kaizens that followed.

2. The people doing the changing are a select group of highly trained engineers and supervisors. The initiatives are not "using a fully engaged workforce," as there is minimal engagement of the rank and file. So they miss the continuous improvement

aspect, and in the case of the Theta cell, they would have missed the productivity improvement step from 188 pieces/person/shift to 304—a huge 60% improvement for which Six Sigma has no means to capture.

3. Although the Six Sigma methodology has a "control phase" so that the gains might be sustained, this control phase pales compared with the efforts used in a lean initiative, which include documentation, training, visual management, leader standard work, and an audit system. I am a Master Blackbelt and trainer and have trained more than 200 belts of all colors and can tell you with certainty that the control phase of the Six Sigma DMAIC (define, measure, analyze, improve, and control) is far and away the weakest in the training and even weaker in the execution of the five phases. It cannot hold a candle to the lean approach of sustaining the gains.

4. Most importantly, a Six Sigma effort is not a culture-changing mechanism. First, you are doing projects that most businesses were doing before Six Sigma and will continue to do after Six Sigma. Second, the only people that are changing are those engineers and supervisors who are involved as belts, project sponsors, or system champions. It is generally less than 5% of the workforce. Since so few people are changing, it is not possible to change the culture. And most Six Sigma efforts exacerbate an existing problem with Western business practices, as they reinforce the emphasis on short-term financial gains.

Although many Six Sigma and Lean Sigma efforts have shown some fine benefits, they are unlike a lean transformation as they:

- Cannot work toward continuous improvement
- Cannot sustain the gains
- Cannot engage the workforce
- And most importantly, cannot change the culture

Lean sigma efforts and lean transformations, although they sound very similar, are not. Do not be fooled.

WHERE HAS A REAL LEAN TRANSFORMATION OCCURRED? AERA ENERGY IN THE SOUTHERN CALIFORNIA OIL FIELDS

In my quest to find a lean transformation in a petroleum refinery, all my efforts were "dry holes." And my efforts to find a true lean transformation in the upstream business were also "dry holes." That is, until I serendipitously got reacquainted with an old college friend, Gene Voiland.

Gene and I had both graduated from Washington State University's College of Chemical Engineering in 1969. We quickly lost track of each other as we were pursuing our careers, he with Shell and I with Chevron. However, we would occasionally cross paths at a WSU football game. Gene had retired in 2008 but has stayed very active, and one of his passions is the WSU College of Chemical Engineering. In fact, he is so passionate that he and his wife, Linda, gave the college a multimillion-dollar donation. Among other recognitions the university renamed the College of Engineering to the Voiland College of Engineering and Architecture.

Gene became aware of my writings, and since he had performed a lean transformation at Aera, he asked if I might be interested in working with WSU to apply lean principles at the College of Chemical Engineering. Of course, I was interested. While we worked together on this project, I learned of his work with the lean transformation at his former employer, Aera. I was particularly taken when, while talking about lean, he said, "It dramatically changed my company; it changed me." And even more so, I was impressed with the clear and strong passion with which he made that statement; it was obvious it had come from the core of his being.

I began some discussions with Gene, and he agreed to let me interview him and use his information for this book. He later sent me information from which I could easily glean that he and his management team had truly created, managed, and sustained a lean transformation in Aera.

Just what is Aera?

Aera Energry LLC is a major California oil production and exploration company. Its primary product is crude oil shipped to nearby California refineries. It was formed in 1997, and Gene was the first CEO.

At the time of formation, Aera produced about one-third of California's crude oil production. Aera's largest producing area is the Belridge Oil Field Complex. Belridge employs around 400 people and has over 10,000 producing wells, with 7,200 pumping units producing 127,000 b/d of crude oil. The "shop floor," as Gene calls it, is 66 square miles and has not only producing wells but steam generation, electrical cogeneration, electrical substations, and transmission systems as well as massive water-treating facilities. At crude prices of $50/bbl, the crude cash flow exceeds $2 billion a year. This is not a small business at all, and Gene and his team transformed it into a lean enterprise.

Why was the transformation at Aera a lean transformation?

To answer that question, let's go back to the interim definition of lean that we discussed earlier in this chapter:

- Lean is the creation of a culture of continuous improvement and respect for people.
- The means to lean is to problem solve your way to the ideal state through the total elimination of waste using a fully engaged workforce.

Let's now review that definition, element by element, and compare it with the Aera story.

Lean is the creation of a culture

As I uncovered the history of Aera, I found that the first thing the members of the management team addressed was their culture. Once the management team arrived, they started a series of brainstorming and awareness sessions to determine the culture they needed to be successful in their new venture. Since nearly all the management team had come from one of the major oil companies, every one of them was already inculcated in the closed, strong, and inflexible cultures that are typical of the majors. Hence they had a striking number of commonalities, many of which would not be conducive to success with the new, smaller venture. Gene and his team decided, at day one, that to be successful they would need to have a different culture. They would need to be more entrepreneurial and build a culture that:

- Recognized the reality that if they did not prosper, they would not survive
- Had a bias to action rather than a bias to study
- Would allow mistakes and not be fear based
- Was based on deep levels of trust throughout
- Was highly transparent
- Recognized that the best people to redesign the processes were the people who were working on the processes
- Brought the technical and managerial support to the field rather than otherwise

From the basis of the culture they wished to cultivate, they prepared a vision and values statement and then went to work to live these documents full time—across the corporation and in good times and bad. Gene confided in me that it was not as easy as it sounds and that some of the team had a problem changing from the old "major oil company" culture and hence some managers were changed out.

The culture that Gene and his management team created, was a huge change from the major oil company culture that I saw at Chevron and others.

- In the culture of the major oil companies, a common saying was not "Let's get going," but rather "Analysis to paralysis."
- When you did act, you made sure you had numerous signatures on your documentation before you proceeded to action. My mentor called it "getting their fingerprints on your murder weapon."
- And look out if the project failed. They were always looking for scapegoats . . . and scapegoating was a highly delegable activity.
- As for the business reality that we might not survive, that was not even a thought that entered their minds; and even if the

facility did not survive, "They would find a place for me" was the prevalent thinking on survival.

- Trust was not a commodity you could find in any quantity. There was huge distrust between management and the workers. There was distrust between the functional intracompany silos. For example, the operating divisions did not trust the maintenance groups, and engineering did not trust either. But worse yet, among the management ranks, you could not trust your superiors or even your peers. Since the folks in the management ranks saw themselves as upwardly mobile and many were striving for the same positions, it was common for partisan company politics to be seen throughout.

- As for being transparent, it was not. Secrets were the means for many to maintain power, especially between the company and the union and often between management and the rank and file, as well as "secrets in your own silo" to hold intracompany power. "Transparency" was a bad word.

- Any process redesign needed lots of engineering and management review, re-review, and re-re-review. Delegation was not an exercised commodity, and "to lead from the front" was as foreign as most of the oil we ran.

- Most of the managers, who wore suits and ties, were seldom found in the field or at the job site. A little vignette to punctuate this point. When I arrived at the El Paso refinery, I was a middle manager, and we all wore at least a sport coat and a tie. I was used to going out into the plant, and on my first week I took a tour by myself. I showed up in the crude unit control room, and Froggy Barnes, who was later to become a good friend of mine, was the operator on the board. He did not know me, so I approached him, and before I could say anything he said, "OK, so what did we screwup now?" I introduced myself and said, "Nothing that I know of. Why would you ask that anyway?" His reply, "We never see any of you coats and ties unless we screw up or we have a visitor from the home office; so who's visiting?" Pretty clear picture and pretty easy to diagnose this part of the culture, huh?

You can see that the early efforts that Gene and his team expended to define their culture were both a major undertaking and a critical factor toward success.

A culture of continuous improvement

Gene and his team prepared a sound business plan, and as part of that plan, a key driver was well "on time." They calculated that for each percentage point of well on time that they improved, they could generate $20 million of annual revenue. They had created a compelling business case for change. Focusing on this metric, they began their improvement process. Again, breaking down this metric, they found that maintenance was a key factor, so they embarked on a total productive maintenance (TPM) program across the entire corporation. They brought in an external consultant and implemented TPM, using the lean model of TPM so common in the Japanese Big Three of Toyota, Honda, and Nissan.

I do not have space to show the many metrics that improved at Aera over the years, but let me show you two. Figure 4.1 illustrates the key metric, well on time. Note that the improvements were continuous and sustained over a protracted period. These were not short-term gains only.

Well "On Time"

Why this process? Higher well "on time" = higher oil production = higher revenue

Waste Reduction Areas

- **Find down well faster**
- **Get rig on down well faster**
- **Fix well faster**

Result: Well "on time" increases

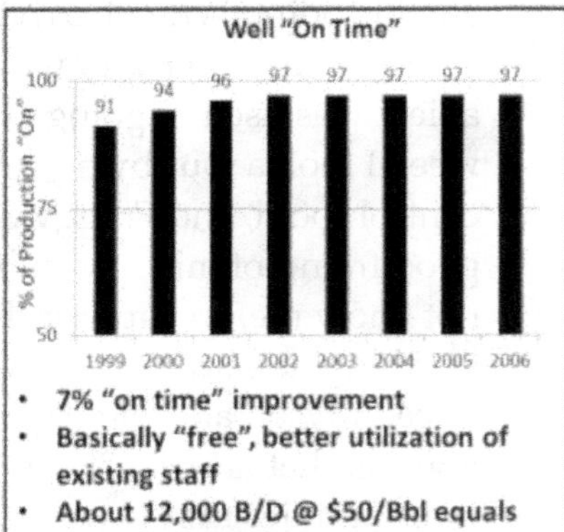

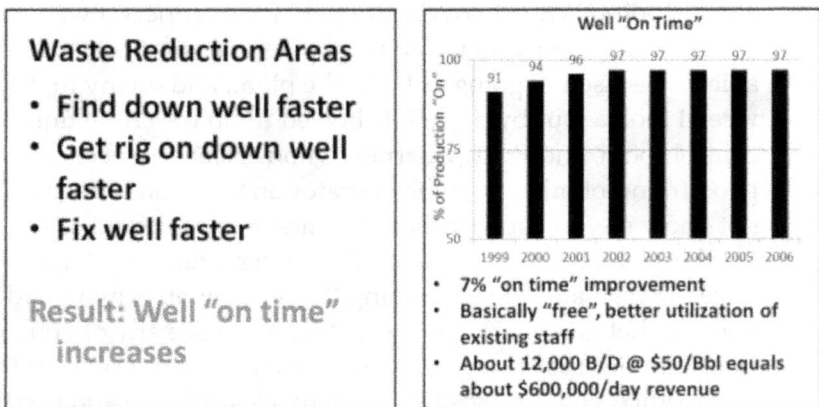

Figure 4.1 Well "on time" improvement at Aera from 1999 to 2006

Second, an often overlooked "business" metric is safety performance. In every lean transformation that I have been involved in, safety and morale have improved dramatically even though no new programs or projects were initiated; you get improved safety and morale and usually improved quality with a lean transformation. Aera's safety performance over a 14-year period, as it implemented lean, is shown in Figure 4.2.

Figure 4.2 Aera's safety performance from 1998 to 2011

A culture of respect for people

It is hard to show a metric for "respect for people," although safety performance is often a very good indicator. But it was built into their values, some of which were:

- Honesty
- Integrity
- Respect for others
- Passion to improve

But more so was the method in which people manifested their respect for others. Recall that an item that Gene and his team wished to inculcate into their culture was to recognize "that the best people to redesign the processes were the people who were working on the processes." Early on Aera designed and installed TPM, a keystone of the vaunted TPS. Gene and his team brought in a consultant to design, kick-start, and support the effort initially. The "kick-off" project was the vibration pen. This pen is simply a small vibration monitor that can be used to indicate and quantify machines in distress and shut them down before major or catastrophic events occur. These major events in an oil field not only cause excessive maintenance costs; they exacerbate downtime and often create safety problems. So, it is far better to provide surveillance and to shut down the equipment in a timely fashion. This event was designed and implemented by the rank-and-file operations and maintenance personnel. It was not a management-dictated and engineering-intensive effort; the folks in the field, with their hands on the equipment, implemented the project. And you can see some of the benefits in the well-on-time graph in Figure 4.1. But more importantly, it is a manifestation of having a solid vision and living that vision on a day-to-day basis. By implementing this project, in this fashion, the company is holding those people in high regard and letting them make an impact in the areas that affect them the most. It is showing respect in its finest form.

Their efforts also went into the area of autonomous maintenance. In the oil business, this is a touchy subject. Autonomous maintenance is a concept that transfers simple chores, usually called maintenance, to the operations personnel who surveille the equipment. Since they visit the equipment far more often than the maintenance personnel, they are closer to the work. Hence things such as a minor leak or a broken pressure indicator can be fixed right on the spot, before it becomes a big problem. And this can be done without the bureaucratic work of filling out maintenance requests, scheduling the work, scheduling the people, completing the work, and then following up to make sure it got done. For the operator, it is often easier to *just do it!* However, this blurs the lines between operations work and maintenance work, and thus the unions frequently and vigorously resist that. When Gene and his team implemented this, it went off without a hitch; they obviously had the trust of the union.

An item that surprised me was their approach toward transparency. You will recall that was a clear cultural element they wished to display. I asked Gene, "Just how transparent were you?" His answer was, "We would open the books to anyone . . . and did." Not a matter of just going to the gemba, or advising via newsletters, or even holding "town hall" meetings; they opened the books. Pretty revealing.

The means to lean is to problem solve your way to the ideal state through the total elimination of waste using a fully engaged workforce

When Gene and his team prioritized their problems, the first major issue they vigorously attacked was reliability. I showed you some of the results. That is just scratching the surface of what they did, but they got these gains by problem solving thousands of projects, large and small, using the entire workforce, not just the engineers and supervisors. Early on in their effort, they engaged Mr. Oba, a sensei at the Toyota plant in Georgetown, Kentucky, and he worked with them on an ongoing basis, making visits to the Aera facilities and giving them guidance. Everyone was taught the seven wastes, and Gene and his team created a central group to support the transformation. This group was the proper type of support. The members of the group did not design, execute, or lead the lean transformation; they were there to support.

> **LEAN—**
> **OUR LEARNING MANTRA**
>
> Lean skills are learned by doing . . . and in no other way.

Problem solving was a way of life, making "smart mistakes" was OK, everyone knew the "target conditions" to meet, the focus was on creating value through the elimination of the seven wastes, and everyone was totally engaged.

This is an example of lean done right.

It would be easy enough to turn this story into an entire book, and maybe Gene should do that. The members of the Aera team did several things exactly right to launch a lean transformation. Among the things they did very well include:

- Starting with a compelling business case.
- Educating the people in top management so they were competent in the change.

- Creating a program that starts small, proves success, and grows.
- Using a major project, the vibration pen, early on to create engagement, success, and credibility and achieve an "early win" to show progress.
- Prioritizing "trust" as a key value and working hard to develop and maintain it.
- Meeting frequently as a management team to stay focused.
- Providing leadership at all levels: checking their performance compared with the Six Skills of Lean Leadership outlined in Chapter 1. You will see with clarity, they hit on all six.
- Keeping a long-term focus.
- Creating a financial rewards system that was totally reflective of their vision and values and goals.
- Most importantly and widely understood, but practiced far less, leading the transformation from the very top. Gene, who was the CEO, and his direct reports in the C-suite designed, guided, monitored, and modified this huge change transformation. This was a management-led, not a staff-led, transformation.
- Making the change totally through the line organization. Although they had a lean support team, it was only a resource to the line organization. The support team did not initiate change. This was a transformation done by the people doing the work, consistent with the desired cultural change. This was not a change done by external consultants, staff personnel, or a select group of Black Belts and Green Belts.
- Not using a tools approach. Rather they found the key business problems and then brought in the necessary lean tools as they needed them. Their first and biggest tool was TPM.
- Simultaneously keeping an eye on the "results" of the business as well as the "processes" of the business.

- Clearly recognizing that their approach, a lean approach, was not a new thing. Rather, it was a new way to do the things they needed to do anyway.

This is a very impressive list and goes a long way to avoid the Six Roll-out Errors and Ten Lean Killers outlined in my book *How to Implement Lean Manufacturing* (2nd ed.). That notwithstanding, I would like to share four additional points I found extremely interesting and revealing as Gene and his team executed this lean transformation.

They defined the specific culture they wanted . . . right up front

When Gene and his team embarked on this new venture in 1997, their first effort was working with the senior leadership to define the specific culture the leaders wished to have. Either Gene and his team had been taught the power of culture or they just sensed it, but that is where they started. That was a stroke of brilliance that set the stage for all else to come.

They learned by doing and observing

I asked Gene, "Where did you and your team learn about lean?" Gene replied, "Several Aera managers were already very interested in the lean concepts. We already had developed a close relationship with NUMMI (the Toyota-GM joint venture in Fremont, California), and we openly shared information with them. As for me, I had already been exposed to many quality principles while with Shell. I listened well and extracted many key techniques for use at Aera."

A second item surprised me a great deal. I asked him what he had read about lean. He said, "*Lean Thinking,* by Womack and Jones, and your book." Somewhat embarrassedly, he chimed in, "But we observed a lot," as if that were a secondhand substitute for reading. Well, Gene had it right, and the clear majority of people who try to learn by reading have it wrong. Oh, don't get me wrong; there is nothing wrong with reading. However, even though I am an author, I will tell you, and have said this many times, that "to learn, you need to do." There is no substitute for that. A distant second for learning by doing is anything else experiential; observing what others are doing is a type of experience that often is much better than reading. In fact, in my opinion, we actually learn very little from reading unless we use it to improve our awareness and that causes us to do two things:

- Question what we know
- Drive us to action

Although Gene and his team did not read much, that may have been a blessing. What his entire team did was travel and observe and use the "shop floor" as an ongoing experiment to "problem solve to the ideal state." If you want to learn, you absolutely must *do*. The team did that. And it is no coincidence that the very first skill of lean leadership is "Leaders as superior observers." This is what Gene and his team did intuitively; they learned by doing, and they went to the gemba of accomplished others and observed.

They went to the gemba and worked with the rank and file—"they walked the talk"

The method that the Aera executives used to gain invaluable insight was their ability to "get outside themselves." The first major aspect of this concept was an early practice they adopted. It is what I call the "Disney concept," because I first read about it in a book about Walt Disney. He would make all his executives work two jobs at the theme park: one as a character and one as either a vendor or a helper at one of the rides. As I recall, the purpose was to allow them to recenter their thinking.

"Getting outside yourself" is a concept I teach to my clients so they may see their businesses from a different perspective. I have already recounted how the Aera executives would visit other facilities and other businesses to broaden their perspective. As a first effort and at the individual level, the Aera executives, and all managers for that matter, were required to "go to the gemba and work." Not just "go to the gemba and observe." Yes, these managers also went to the oil fields and processing plants and performed the jobs of the operators and the maintenance workers. They sampled crude and transferred products. They installed and removed blinds. They changed out and cleaned pressure gauges and even repaired rotating equipment. In my experience, there is nothing more enlightening to senior leadership than going to the gemba and working. For the typical executive who spends the majority of his time in his office or at meetings, it is easy to get a sanitized, unrealistic view of the business, missing much of the realities of actual life at the gemba. At the individual level, "going to the gemba and working" can be very eye-opening. And as you recall, one cultural imperative that Gene and his team wished to develop was to bring "the technical and managerial support to the field rather than otherwise." There is hardly any better way to truly live your vision and exhibit true lean leadership. Recall that one of the six key qualities of lean leadership is "Leaders as role models: They walk the talk."

They "checked" their performance using external agencies, and they further "got outside themselves"

Earlier I spoke of their efforts as a team to visit other facilities. Well, it is one thing to go and visit; it is yet another to critically compare your business process performance with the rigors of clean, clear data analysis. In their early days, they began a process of best-practice sharing across business units. They used an annual audit to highlight opportunities to share; this was done internally, and significant progress resulted. However, they also started using outside agencies to audit their facilities. Among these agencies were the two primary owners: Shell and ExxonMobil. In addition, they had their corporation scrutinized by the Marshall Institute, the Toyota Supplier Support Center, and finally the Foundation for Industrial Maintenance Excellence, which evaluates and administers the prestigious North American Maintenance Excellence (NAME) Awards.

You can see a significant effort to get outside yourself and broaden your perspective. Keep in mind that the first and the most important quality of lean leadership is to become a superior observer. Implicit in being an observer is that you first must look, and that implies that you look beyond the end of your nose. Otherwise you will see nothing new, nothing different, and you will miss multiple opportunities to improve.

The folks at Chevron were not good at "getting outside themselves"

Let me share an experience I had with Chevron, for the sake of comparison. As a rising young executive, I was sent to a weeklong training affectionately called "Charm School." The afternoon of the final day, the CEO joined us for open discussion. Following that, we adjourned for dinner. The CEO was Ken Derr, and at dinner I was seated next to him. I took the opportunity to ask him this question: "If you could change just one thing in Chevron, what would that be?" Without batting an eye or missing a beat, he said, "We need to quit comparing ourselves to ourselves." I was a "Chevron guy" and deeply inculcated into the Chevron culture and really did not understand his statement. I asked him to say some more about that. He went on, "This last quarter, if you read our report, you will see that corporate earnings are up 3% over the same quarter last year, and the board feels good about that. Well, maybe we should feel good; but do you know that Shell increased 4% and Exxon 7%? Don't you think that is relevant?" Well, that discussion catalyzed an awareness that later changed—entirely—the way I viewed Chevron.

THE "OSTRICH EFFECT"

The ability for businesses to "get outside themselves," at its worst, is the lack of interest by managers and others to look outside their businesses and compare themselves with others or recognized standards of performance. It is also this same reluctance to look at other locations, other businesses, and even other industries to see possible new techniques to utilize. My friend Robert Simonis calls this the "ostrich effect."

Then, by mere coincidence, when I returned to El Paso, the Solomon report arrived. (Solomon is an organization that does an industrywide review of refineries and benchmarks them on a multitude of business and operating qualities.) This was our third report, and our overall Solomon evaluation showed us performing in the third quartile. In a few areas, we were in the first quartile, but in the key factors that drove refinery economics, we were in the fourth quartile in all three studies. In earlier studies, we had spent more time trying to prove Solomon wrong than absorbing what could be learned, reminding me of what Ken Derr has said. And once again, we rejected the study and went into argumentation and denial. Once the performance of our refinery was compared with that of others and we did not look so good, we sought to argue with the data and the methodology. I am sure, somewhere, at some time, we made some changes, but the overall effort was to discount and diminish the report. We simply did not want to look outside ourselves to see areas for improvement.

Spurred by these data, I commissioned a team to complete a "Critical and Comparative Analysis of Our Distributive Marketing Area." It was a comprehensive study of the ability of Chevron's El Paso refinery to compete with the other six refineries in its distributive area. It showed El Paso was sixth on the list of seven in terms of profitability. We were very strong in producing gasoline, especially high-octane blends. In addition, we made very competitive diesel, aviation, and jet fuels.

These were part of our economic strengths, but there was a huge vulnerability with the refinery. On all products—gasolines, diesels, and jet fuels—we were not competitive at all compared with the Gulf Coast and West Coast refineries. Their refineries had more and better cracking upgrade capacity and the ability to buy "opportunity crude." We were only protected by the inability of those refineries to pipeline product into our area where we enjoyed some of the highest gasoline margins in the United States—not only in El Paso but in our New Mexico markets as well. We were very vulnerable. However, when this was presented to the senior corporate management, they almost laughed, saying, "No one is going to get permits to build a pipeline in this environment." To the contrary, we had researched it and found that much of the facilities needed could be grandfathered. Nonetheless they categorically rejected the entire idea as folly, and we moved on. This is just another indication that efforts to gather outside data and make business evaluations were discouraged and ignored, and in this case even ridiculed.

Later, after I retired from Chevron, two pipelines were built to bring product into El Paso. The first one came from the Texas Panhandle, and the second one came from the Texas Gulf Coast. From this time forward, there was severe downward pressure on product prices in this area. The bad news was Chevron suffered economically . . . however, I can now buy cheaper gasoline. I guess every cloud has a silver lining.

Aera was far more introspective . . . with obvious benefits

Earlier I mentioned that the oil industry was very insular and inward looking. When Aera was formed, the company executives recognized this liability and worked hard to make sure they did not fall into the same trap as the fully integrated oil companies; they got outside themselves. This was a very strategic move to becoming lean, and coupled with other cultural traits, it allowed Aera to prosper.

In 2002 Aera's Belridge Producing Complex won the prestigious North American Maintenance Excellence (NAME) Award for its maintenance and reliability program and results. In 2004 Aera's North Midway Sunset Unit was awarded a NAME Award. Aera is the first oil exploration and production company to win the award previously given only to manufacturing organizations. In 2011 Aera's development team was awarded the Association for Manufacturing Excellence (AME) Manufacturing Excellence Award. Aera is the first energy company to receive the award.

A lean transformation done well

So you can see that Gene and his entire team at Aera truly went through a lean transformation as they created a culture of continuous improvement and respect for people. And the means to lean is to problem solve your way to the ideal state through the total elimination of waste using a fully engaged workforce.

CHAPTER SUMMARY

The title of this chapter is "How Much Has Lean Penetrated the Refining Business?" A *lean* summary would be—damn little.

However, it is worth looking at some information. When I first embarked on my lean journey at Chevron's El Paso refinery, other than the efforts I personally catalyzed and supported, there was no management interest in lean manufacturing. After I retired, the entire effort stalled, despite some amazing results. As I later looked for clients using contacts I had, plus being active and engaged in blogs and groups on the Internet, I found nothing; and almost no one found me. Later I wrote dozens of articles that were published in several national e-zines. I wrote two books on lean manufacturing and spoke at large, well-attended national conventions as well as with smaller, narrower groups; again with very little response. Along the way, and just before writing this book, I interviewed several refinery executives and found out that the concept of lean manufacturing was largely not understood in oil refining, especially by the executive leadership. However, along the way I got several jobs in both the continuous process industries and refining, although none was a complete lean transformation. In the continuous process industries, I was asked to assist in the continuous process improvement efforts by teaching specific skills, mostly the so-called lean tools. The work in the refineries was to support Lean Sigma efforts. After reviewing the differences between a true lean transformation and the Lean Sigma that these refineries were doing, we concluded that there was a huge difference between a Lean Sigma initiative and a lean transformation. A lean transformation is a much more powerful process with true culture-changing abilities.

Finally, I introduced you to the one true lean transformation I could find in the entire oil business: the lean transformation of Aera Energy in the oil fields of California. It was a true lean transformation and a tremendous success.

CHAPTER 5

WHERE WILL LEAN FIRST APPEAR IN THE REFINING INDUSTRY?

Aim of this chapter*—In this chapter we consider if lean is likely to appear with the fully integrated oil companies (FIOCs). And if so, why? Or if not, why not? In addition, we will explore one major aspect of what is changing in the business of refining in the United States.*

LEAN HAS BEEN AROUND FOR OVER 30 YEARS, AND ALMOST NO ONE IN THE OIL BUSINESS IS PAYING ATTENTION

The concept of lean manufacturing has been highly visible for more than 30 years now. And according to the 2007 Industry Week/MPI Census of Manufacturers, "Nearly 70% of all plants in the US are employing Lean Manufacturing as an improvement methodology." Wow! Almost 70% of all U.S. manufacturers are implementing lean manufacturing.

But among the FIOCs and even the smaller refining-focused businesses, there is virtually no penetration of lean manufacturing. There is a decent volume of Six Sigma and even Lean Sigma activity in the upstream sector, but as I explained earlier, this is not to be confused with a lean transformation. In addition, in refining, whether with the larger FIOCs or the smaller refining-focused firms, there are no lean transformations under way and only a little Six Sigma and Lean Sigma activity.

Why is that?

Will lean work in a refinery environment?

To answer this, we need to answer two questions. First, is the methodology transferable to a refinery? And second, will it yield results?

As for transferability, to me that is the trivial question. The short answer is, it will transfer just fine. Lean has proved itself in a broad range of businesses for more than 40 years, so why not refining? Richard Schonberger, in his wonderful book *Japanese Manufacturing Methods: Nine Lessons in Simplicity,* said, "Management technology is a highly transportable commodity," and later, "The Japanese had little trouble learning our techniques, and we will have little trouble learning theirs." To more fully explore this topic, see Chapter 7.

As for the results question, the answer is a resounding yes. It will not only yield results; it will yield amazing results both in bottom-line contributions and in delivery and quality, as well as make major contributions to safety and environmental compliance and improvements in morale. To further explore some of these opportunities, see Chapter 12, which will give you several tangible examples of improvements, actual accomplishments, using this lean model.

You have already heard a little about the power of lean manufacturing in oil refining, so I am convinced it will work just fine. In addition, I have experience in the upstream business, and it has proved itself there. And you just read the story of Aera in Chapter 4.

So in response to the question, "Will lean work in a refinery environment?" the evidence is overwhelming and the answer is *yes*.

CHANGE MANAGEMENT

For leaders to make a change, they must get the support of their constituency, and they must carefully address three topics (at least):

1. What is wrong with the present situation?
2. What is the better place?
3. How are we going to get there?

These are the three key questions to change management.

If it is so good and it is out there, why isn't anyone buying?

That's a good question. But, don't be overburdened by the need for a logical explanation. If life were all so logical, do you think people would smoke, knowing full well there are severe and negative health consequences? Some of the reasons are as illogical as that.

However, the most revealing reasons I can find came from the interviews discussed in Chapter 4. They are:

- Many are simply unaware of lean manufacturing. Some are aware but have a wrong concept of lean manufacturing.
- Others think it just does not apply.

The common denominator of these reasons is that people's awareness of lean is very low. But I think there is a different reason altogether.

ON AWARENESS

"Men stumble over the truth from time to time, but most pick themselves up and hurry off as if nothing happened."

—*Sir Winston Churchill*

Why change?

It goes to change management. If leaders wish to change something in their organization, they first must make it clear that the present condition is no longer a good place to be. So when these managers are happy with the present state, very often they do not look for something better. I think that is the case with many oil companies.

Year after year they seem to make money . . . lots of it. So, whether the refineries are making money because of the way they are run, or in spite of the way they are run, they are profitable. And when a company is prosperous, the managers are less likely to change and certainly much less likely to take on a culture-changing issue, such as a lean transformation.

But could they not make much more money using a lean system?

Absolutely. However, that would require they go beyond question 1 of the Six Questions of Continuous Improvement regarding the present condition and explore question 2, "What is the desired condition?"

To find this better condition or place, first you must look. And I have given you data that support they are not even looking. So why don't they look, at least a little?

The FIOCs are very, very insular

Based on the penetration of lean in other industries, you must wonder how much time the FIOCs have spent looking outside their own walls and how much time they spend in introspection trying to find a better way to perform. It's hard to see them other than isolated and complacent. Hence I doubt very strongly if we will see any epiphanies soon coming from the FIOCs.

By their nature these FIOCs have cultures that are very insular, inward looking, politicized, and bureaucratic. These characteristics lead neither to reflection nor to an attitude that might support change.

However, as is always the case, when there is money to be made, there are entrepreneurs waiting in the wings. And this self-assurance by the FIOCs can be a huge source of vulnerability, vulnerability to losing the business to others who are more open-minded and creative and to those who "have a better idea."

Sounds a lot like the American auto industry a few decades back. I suggest the FIOCs take a step back and reassess—lest they lose the whole ship.

They need to awaken and learn from the auto industry or else watch history repeat itself

Most of American manufacturing has as least studied, if not undertaken, the lean transformation as a means toward continuous improvement. Yet you can see from earlier discussions, oil companies are largely unaware of lean manufacturing and the power it brings to a business. This lack of awareness, coupled with its logical conclusion, a lack of action, places them in a precarious situation.

We can draw a parallel with the American automobile industry, which in the 1950s and 1960s simply laughed at the Japanese manufacturers and derided their quality. They were in militant denial about the wave of competition they were soon to face. When the wave came in the 1960s and 1970s, they did not know how to respond. To change to a lean manufacturing system was something they actively, openly, and aggressively chose *not* to do. Yes, they had a wealth of publicly available data and chose not to use it. It was the largest case of "industrial denial" I have seen in my lifetime.

Later as it became apparent that the Japanese auto industry was getting a foothold, rather than face the facts, the American car companies chose to use government intervention, via import quotas, to keep their market share. Unfortunately for them, but fortunately for us auto buyers, that did not work, and soon enough there were full-scale Japanese automobile plants in the United States. My mentor, Toshi Amino, built the first one, Honda's plant in Marysville, Ohio, completed in 1982. In the meantime, Chrysler has gone through two doses of corporate welfare and GM has gone through one, while Toyota has taken over as the largest automobile company in the world.

It is debatable whether Toyota is number one or number two, depending on the data you use, but what is not debatable is the growth of Toyota into an industrial giant. According to *The Machine That Changed the World*, from Toyota's founding in 1937 until 1956, it had produced a total of only 2,685 automobiles. The company grew and grew, and soon enough it was seen as the quality producer of automobiles. Then in 2012, it became the first automobile company to produce 10 million cars per year. And although competition has forced the Big Three of GM, Chrysler, and Ford to become more productive, their manufacturing system is still significantly inferior to the Toyota Production System or any other lean manufacturing system, as we call it here in the United States. There is no question that the Big Three have since changed, and possibly Ford has an opportunity to compete long term with Honda, Toyota, and Nissan; I doubt GM and Chrysler have that capacity or the will. The question for the Big Three is this: "Are you willing to change internally, at a pace that exceeds the rate of external change, in your business environment?"

That is the survival question, and not all are up to the challenge.

And my question for the oil industry is, "Are you up to this challenge?" To date I have seen absolutely nothing to suggest the FIOCs are willing to change in this way. Getting an FIOC to change is like trying to dance with an elephant. Like the auto companies, the FIOCs are relying on mergers, spin-offs, and restructuring to protect themselves and achieve their financial objectives. This is the equivalent of voluntary Chapter 11 bankruptcy. In addition, they are lobbying heavily for preferential legislative treatment. This only works for so long and only locally.

When faced with worldwide competition, your greatest weapon is productivity that stems from good management and good leadership.

Since the U.S. refining industry appears to believe that it is bulletproof and insulated from international competition, it has no incentive to improve beyond staying in business, which in the United States means being competitive with your neighbor. Well, what if the FIOCs are not bulletproof? What happens when their neighbor is India or China with its dramatically different cost structure? What about Saudi

Arabia? Oh, by the way, Saudi Aramco, in June 2016, acquired sole ownership of the largest refinery in the United States, the Port Arthur 603,000-b/d refinery. What options does a major producer such as Aramco now have with major U.S. refining capacity? Do you think it can now create some synergies with its worldwide refining capacity of over 5 million b/d? It does not take a very imaginative person to see how Aramco can put severe downward pressure on other U.S. refiners. In addition, it announced its intentions to acquire more refining capacity (see "Exclusive: After Motiva Split, Saudi Aramco Aims to Buy More U.S. Refineries—Sources").

And there is a cultural conflict within an FIOC

When Quality Consultants was started in 1990, it was based on providing consulting support to businesses so they might be more competitive and profitable through cultural change. Cultural analysis, cultural design, and cultural modification were our primary products. In our work, we often refer to the works of Terrence Deal and Allan Kennedy, some early authors in the subject of business cultures. Their most notable publication is *Corporate Cultures* (1982). In it they categorize corporate cultures based on two parameters: risk and the speed of feedback on actions. Their model, in a 2 x 2 matrix format, is shown in Table 5.1.

Table 5.1 Four types of business cultures

	Perceived Risk to the Business		
		High	Low
Speed of Feedback	Fast	The "macho, tough guy culture," like oil exploration. Make it big or lose big	The "work hard, play hard culture," like sales and some manufacturing
	Slow	The "bet your company culture," like R&D	The "process culture," like accounting. Lots of rules and procedures, bureaucratic

This always intrigued me regarding the FIOCs, because the upstream business, especially exploration, by its nature, is largely a very high-risk, fast-feedback business with its "macho culture." The point is that the macho culture is a very different culture from one that is desirable in a refinery that is halfway between a "work hard, play hard" and a "process" culture.

Recall that an operational definition of culture is "just how we do things around here." So with different cultures, there will be different values, norms, and behaviors.

While in the oil business, I always wondered if the concept of a fully integrated oil company was even a good idea since it needed at least two dramatically different driving cultures, which could create grossly different behaviors and many internal conflicts. If there were competing cultures, there would be egregious and often irreconcilable differences in opinions regarding expansion and capital outlays as well as research and development.

PUTTING THIS ALL TOGETHER

Lean has been around for over 30 years, and according to survey data, over 70% of all U.S. manufacturing is using lean as a process improvement strategy.

But not the oil industry. Not one major oil company is using lean methodology. Partially this is due to complacency—after all, the oil companies have been very profitable. However, they also have an insular, inward-looking nature. Although the lean methodology has demonstrated unprecedented improvements in manufacturing as well as a wide range of nonmanufacturing firms, the oil companies have shown practically no interest in exploring lean as an operating philosophy.

Whether the company is one of the FIOCs or one of the smaller refining-focused firms, no one is using a lean methodology as its operating system. None of the firms are changing to lean.

Actually, I don't see any of them changing their operating systems at all. Whether you look at them from the outside or inspect them from the inside, they look just like they did in 1970 . . . with a touch of technology added.

Compare the changes in the automobile industry, in food processing, or in retail businesses with changes in the oil industry, and the distinctions are stark indeed. In comparison, the oil industry looks like a dinosaur carrying an iPad.

Makes one ask . . . So the *external* environment must not be changing, huh?

Jack Welch is quoted as saying, "If the rate of change on the outside exceeds the rate of change on the inside, the end is near." And the rate of change of the manufacturing systems in the refining business is not changing much at all; there is very little "change on the inside." Some might be quick to respond that this is not an issue, because the rate of change on the outside is pretty slow.

But I am not so sure. Let's look.

So what does the overall picture of U.S. oil refining look like from 2005 to the present?

The facts are shown in Table 5.2.

In 2005, the 14 largest refiners processed 14.9 million b/d of crude oil. Of this volume, the 7 FIOCs processed 10.06 million b/d, or 67%. Only 33%, 4.9 million b/d, was processed by the smaller, refining-focused companies. Then, by 2015, the 14 largest refiners processed 13.9 million b/d of crude oil. Of this, only 3.9 million b/d was processed by the FIOCs; that is only 28%. In the 10-year period from 2005 to 2015, the refining-focused groups have purchased refining capacity totaling over 10 million b/d, and they have risen from 33% and are now producing 72%—more than doubling their capacity. This is a massive swing. Of course, part of this is due to the spin-offs of the ConocoPhillips and the Marathon refineries, but that just punctuates the point that the integrated oil companies are better off not running both the upstream and the downstream business units.

Valero, strong in 2005, increased refining capacity by nearly 500,000 b/d and tops the list in 2015 at nearly 2 million b/d. And in this same period we see some small, minor players from 2005 come into the top 14. Motiva increased to fifth on the list and now processes 1.076 million b/d, more than any FIOC except ExxonMobil. PBF Energy was not on the 2005 list, but by 2011 it was thirteenth with 342,200 b/d. It had increased to 502,200 b/d in 2015 and is now at 900,000 b/d with its recent purchases of ExxonMobil's Torrance and Chalmette refineries. This likely makes PBF sixth or seventh on this list. In addition, refiners like WRB Refining and HollyFrontier have entered the top 14. On the other hand, Royal Dutch, an FIOC, which in 2005 processed 1.14 million b/d, was a major player at number five; now it is fourteenth on the list, processing only 426,400 b/d.

A Bloomberg article stated, "Oil companies from Chevron Corp. to BP PLC are selling more refineries than any time ever." It goes on, "A glut of refineries up for sale by integrated oil companies after the

Table 5.2 U.S. Refining Capacity, 2005–2015

	2005			**2011**			**2015**	
	Company	**Cap, b/d**		**Company**	**Cap, b/d**		**Company**	**Cap, b/d**
1	ConocoPhillips	2,177,900		ConocoPhillips	2,041,900		Valero Energy	1,964,300
2	ExxonMobil	1,940,000		ExxonMobil	1,951,000		ExxonMobil	1,855,700
3	Valero Energy	1,484,500		Valero Energy	1,682,300		Marathon Petroleum Corp	1,731,000
4	BP PLC	1,475,150		BP (UK)	1,368,050		Phillips 66 Company	1,611,200
5	Royal Dutch	1,143,050		Marathon Oil Corp USA	1,142,000		Motiva Enterprises	1,076,000
6	Marathon Ashland Petroleum LLC	935,000		Chevron Corp	1,027,271		Chevron Corp	948,271
7	Chevron Corp	909,000		Royal Dutch/ Shell (NL/UK)	976,650		Tesoro Corp	844,180
8	Sunoco	880,000		PDVSA Venezuela	854,050		PDV America	761,240
9	Petróleos de Venezuela SA	828,000		Koch Industries (Flint Hills)	771,578		BP PLC	649,000
10	Premcor Inc.	771,750		Sunoco	673,000		Koch Industries	560,000
11	Flint Hills Resources	745,513		Tesoro Corp	657,300		PBF Energy	502,200
12	Motiva Enterprises LLC	730,000		Saudi Aramco	376,750		WRB Refining	482,000
13	Tesoro Corp	559,200		PBF Energy	342,200		HollyFrontier Corp	467,350
14	Saudi Aramco	365,000		Holly Corp	285,350		Royal Dutch/Shell Group	426,400
	Source: "Refiners Add 2.7 Million b/d of Crude Refining Capacity in 2005," *Oil and Gas Journal*, Dec. 19, 2005.			*Source:* "Top 20 Largest Refining Companies/Refiners in the U.S. as of Jan. 1, 2011," *Petroleum Insights*, July 1, 2011.			*Source:* Energy Information Administration, "Refinery Capacity 2015," table 5..	

global recession dragged down profits are now available for 80% less than they fetched in 2006."

So there is change . . . big change. And no longer are the big guys eating up the small guys. Just the opposite! The former small guys are now not so small and are acquiring refineries—at pennies on the dollar.

There is no question that the FIOCs are shedding refining capacity.

So what are the reasons the FIOCs are shedding capacity?

In this section, we will explore this in detail. You will read several quotes giving a variety of explanations. Some are easy to see; some are pure spin; some have some substance. But when the dust clears, behind it all are three driving forces that are compelling the FIOCs to sell off this refining capacity:

- They are not making enough money in refining.
- They don't like the exaggerated public relations exposure of environmental and safety issues prevalent in the refining business.
- They have limited capital and want to invest it in more profitable operations.

We will explore each of these in more depth.

First . . . are the refineries as attractive financially as the upstream business?

Repeatedly we see data that suggest FIOCs are not making money on their refining business units; but more often we hear that refining business units are not making as much as the upstream business units. You can find several examples where FIOCs have shed refining capacity and the refining-focused companies that bought them, turned them around and made money—lots of it. It may be very hard for the FIOCs to accept that, as a group, they are not very good at running refineries; others can do it better.

Most of the earnings, year after year, come from the upstream operation. For example, from Chevron's "El Segundo Refinery Overview," net corporate income for 2013, from all sources, was $21.4 billion; yet only $2.2 billion was from refining, marketing, and transportation. To an FIOC, the money is in the upstream business units, from the acquisition and sale of crude oil. Refining, while potentially very profitable, is a secondary source of income.

The margin for the upstream portion of a barrel of crude at $70 might be $30, $40, or $50, while a typical 6% refining returns in the downstream sector, this same barrel of crude might only yield $4 to $5. Downstream operations are not nearly as profitable.

When I was first with Chevron in 1968, all the refineries were run as cost centers. The objective was to keep the oil flowing in Aramco, and all downstream operations were to push to maximum capacity and keep the oil flowing while minimizing costs. At that time, Otto Miller, the chairman of the board of Standard Oil of California (now Chevron), told us at a conference that the company could lift oil in Saudi Arabia and put it on a tanker for $0.25/barrel and sell it for $3.75—so the strategy was to "keep the oil flowing."

That was in the 1960s, so things have changed a bit. But even after Aramco was nationalized in 1974 and Chevron's earnings from Aramco shrank, all Chevron refineries were still managed as simple cost centers. At the refinery level, there was no refinery profit and loss statement, even as late as 1990. Chevron's El Paso refinery was a 78,000-b/d refinery when I retired; at $70/barrel of crude oil, that is more than a $2 billion–a–year business; and it is a small one by refinery standards. And for any multibillion-dollar-per-year business, to not have a profit and loss statement simply borders on mind-boggling. There are others that still currently operate in this manner.

Are these refineries in an FIOC financially driven?

As I said earlier, I recently completed an engagement with a refinery in an FIOC. It was a Lean Sigma implementation. It is more than interesting to note the problems we had with our Lean Sigma effort; before it was over, I called it Lean Sigma schizophrenia.

To be an effective advisor, I needed to understand the goals of the refinery, so I discussed them with the refinery manager. They were, in order of importance:

- Safety compliance
- Loss of containment
- Environmental compliance

- Mechanical availability
- Utilization
- Operating expense
- Profitability improvements

Note that profitability was the last item on the list; and the company ran this 100,000-b/d refinery following these priorities. For example, we had one Lean Sigma project with a very large, around $1.8 million per month, total earnings, which required about $500,000 of capital investment. The new facilities were easy to design, construct, and install. In addition, somewhere around $100,000 of maintenance work was also needed. Over half the gains could be achieved with no capital, and so we embarked on that aspect forthwith. Earnings flowed almost immediately, and we were counting them as everyone shared in the glow of success. Very quickly the capital was approved to spend.

But the project ran into a dead end at engineering. It seems the local engineering staff was all tied up with safety jobs. These are higher on the priority list than profitability jobs, and any engineering attention on this project was, at least one full year away. So I suggested we contract the work, but there was no one to supervise it, and to make matters worse, the refinery was in danger of exceeding its contract engineering budget. In short, the needed engineering work was not started, and likely would not start anytime soon. We effectively left $900,000 a month on the table because of a series of policies that made no sense to me. Bureaucracy and centralized control . . . you gotta love it! The small independents I have worked with would never have let that money slip through their fingers like that.

As I often say, refineries are "nonbusiness businesses." Let me give you yet another, even more egregious example of being a nonbusiness business. This refinery had lots of recoverable giveaways on both gasoline octane and vapor pressure. In addition, it had some real opportunities with diesel blending, particularly in the wintertime with its very restrictive wax cloud specification. One Lean Sigma Black Belt had completed a project to reduce vapor pressure giveaway, and we were accruing more than $1.2 million per month. I approached the operations manager with a proposal to work on additional gasoline giveaways. We still had over $600,000 per month of opportunities for vapor pressure and twice that for octane. I scoped these out and approached the operations manager to find a project champion for these projects. I was amazed when he told me to put these all on hold and start them next year. I asked why? The manager said, "They had already met the maximum value for refinery profitability improvement."

WHAT? They met the maximum for profitability improvement! Exactly what is that?

This both further amazed and puzzled me, so I asked, "Just what is the maximum profitability?" Well, it had nothing to do with refinery economics. Rather, the refinery had met its annual goal to get the maximum from its bonus calculation for the managers!

Dumb me! I had forgotten one very important adage that drives management behaviors: *"Individual goals will always trump group goals."*

Sounds like bad local economic management.

These managers, clearly, are not treating the refinery economics as if it were their own money. Eli Goldratt is credited with this maxim: *"Tell me how you measure me and I will tell you how I will behave. If you measure me in an illogical way . . . do not complain about illogical behavior."*

Before we get too critical of these managers, we need to look upward to their bosses to see just what sort of incentive programs they created.

And either consciously or unconsciously, they have created a program that *does not* prioritize economics. This goes to the root of why they are shedding some of these refineries.

I had mentioned earlier that our Lean Sigma event was schizophrenic. You see, the driving metric for the Lean Sigma initiative was "bottom-line financial contribution." Our success and failure as a Lean Sigma initiative was measured based on the true financial benefits we could accrue, and it was a program requirement that the gains for any project had to be verified by the top financial manager at the facility. Yet the local management had little motivation to pursue these financially attractive jobs.

The confusing part of this program was evident when the members of the executive staff from the home office would visit the plant. Although it was not formally on their agenda, they always took time to meet with the young dynamic leader of the Lean Sigma effort. They always asked about the Lean Sigma effort, and their questions always centered on the *financial benefit*. To the Lean Sigma team, this was very encouraging because we were gaining financial benefits much faster than the other, larger refineries. The executives from the home office were drooling over our successes. Yet the local management had delayed several projects because we had "reached our maximum for refinery profitability improvement." If this sounds like poor focus and bad alignment of goals as well as dysfunctional management sending mixed messages . . . don't be confused; it is.

What about the environmental and safety exposure?

Refineries, compared with most upstream activities, create a greater environmental and public relations risk, and hence they demand larger amounts of management attention and money spent on environmental and safety compliance.

Refining, compared with exploration and production, usually occurs near large population centers, drawing more attention to the environmental and safety issues. In large population centers, there are lots of agencies including city, county, state, and federal keeping an eye on the refineries.

In addition, there are a myriad of other issues that can easily become problematic when an oil facility is near a population center. Just a few might include odors, noise, traffic, strikes, and taxes. Out in the oil patch, more often than not, these same issues were considered a small price to pay to have the jobs created by the oil industry.

Then there is the exaggerated attention the FIOCs get. When they have an environmental issue—a refinery fire, an explosion, or a safety issue—that becomes public knowledge. Almost immediately the FIOCs find themselves on page 1 of not only the local papers, but also the biggies such as the *Wall Street Journal*. They then send out their PR experts to spin the story as best they can. But in the end, for an oil company, most publicity is bad publicity. And none of them like that. I can tell you from experience that they go to extreme efforts to avoid it. My mentor explained this to me many times and gave me many examples of how Chevron went to great extremes to "keep its logo clean."

I have no idea if the safety records or the environmental records of the FIOCs are better or worse than those of the refining-focused business. What I do know is that if ExxonMobil or Chevron has a safety or an environmental issue, it is big news. It will probably make a splash on the local as well as the national news. However, if some smaller refinery such as PBF Energy or WRB Refining has the same problem, as far as the press is concerned, it is small potatoes. Name recognition matters to the news media, and they love to take on the big guys . . . it makes great news.

The issue of limited capital

Since the refineries are earning smaller margins than the upstream business, capital investment in upstream is more attractive. And the new environmental laws and safety regulations are not only placing compliance and public relations pressures on refineries; they are causing huge capital investments as well—capital investments with no incremental earnings. That, coupled with the already lower returns, makes them less desirable locations for future capital.

Let's look at Some FIOCs that have shed capacity in the last 10 years

As I say, the corporate culture of an FIOC is not conducive to nor is it compatible with running a refining business. And I think two FIOCs have come to that conclusion and spun off their refining business totally. ConocoPhillips and Marathon, both of which were among the top 10 refiners in the United States in 2005, spun off their downstream operations. As for ConocoPhillips:

ConocoPhillips spun off its downstream manufacturing and created a new and separate corporation, Phillips 66, in 2012. (S&P Capital IQ, Stock Report, June 16, 2016, McGraw Hill Financial)

Likewise, Marathon also spun off its refining operations:

On June 30, 2011, Marathon Oil Corporation (MRO) separated its downstream business to create a new publicly traded, independent refining company, Marathon Petroleum Corp. The Refining and Marketing segment included operations of MPC's seven refineries capacity of 1,794,000 barrels per day (b/d) as of February 2016. (S&P Capital IQ, Stock Report, June 16, 2016, McGraw Hill Financial)

BP sells Texas City refinery to Marathon for pennies on the dollar

Just why would a mature, fully integrated, international oil company decide to sell a multibillion-dollar asset for so little?

The Texas City refinery was a very complex Gulf Coast refinery that BP acquired in its merger with Amoco. In 2012 BP sold it to Marathon Petroleum for:

> ". . . $598 million for the refinery, plus an estimated $1.2 billion for oil inventories already located on site at the 475,000-barrels-a-day facility. Marathon will also pay BP up to an additional $700 million over six years based on future margins and throughput of the plant." (*Wall Street Journal*, October 8, 2012)

Now $598 million, $1.2 billion, and $700 million sound like very large numbers . . . and they are for almost any business. However, let's put these numbers in perspective. First, the $598 million is the only real solid outlay that has significance. The $1.2 billion in inventories will soon be processed through the plant and will be paid for by sales in a matter of weeks as this money "in the belly of the beast" is converted into income. And the $700 million is divided over six years and is based on several contingencies. But let's say BP needs to pay $120 million each year; that makes the out-of-pocket cost for year one to be $718 million. The company never quoted the value of the refinery, but an ex-employee, who is still in the oil business, told me the refinery was valued at $12 billion. My contact said, "The $700 million over 6 years would actually be zero when all was said and done, and Marathon got a refinery for $0.05 on the dollar."

If you factor in the $700 million over 6 years, its 6 cents on the dollar. So 5 or 6 cents on the dollar. Marathon got a huge deal, and you must wonder why BP would sell its refinery so cheaply. Furthermore, my contact told me, "With BP, net profits ran in the 3% range, with 6% being a very good year and not happening very often." He went on, "Marathon paid for their investment the very first year. I think they were in the 12–14% range. Marathon did a much better job at operating the refinery." And my contact also told me that "BP had been looking to unload the refinery since the 2005 incident."

The Texas City refinery explosion occurred on March 23, 2005, when a hydrocarbon vapor cloud exploded at the ISOM isomerization process unit at BP's Texas City refinery in Texas City, Texas, killing 15 workers and injuring more than 170 others. (Wikipedia, June 22, 2016)

Maybe BP sold this refinery for reasons other than the simple economics. Do you think the bad press was a factor?

A case could be made that it had decided to get out of refining in the United States altogether and just had not announced it as such. The alternative would be to get good at refining—that creates a lot of positive cash flow for others; why not BP?

Conclusion: BP was not making much money at Texas City, and not coincidently, it had an earlier major accident there, killing 15 workers. It also needed cash to pay for its huge mess in the Gulf of Mexico. There were issues with finance, the environment, probably public relations, and capital. Sound familiar?

The sale of BP's refinery in Carson, California, to Tesoro

No sooner had BP sold this refinery than it sold its refinery in Carson, California, to Tesoro. It was a 266,000-b/d fully integrated fuels refinery and was sold for $1.075 billion, plus $1.3 billion for inventory, etc. That's a much better deal than it got for Texas City. It likely was paid 25 cents on the dollar for this refinery.

BP PLC said Monday it has agreed to sell its Carson refinery in southern California and related assets to Tesoro Corp. for $2.5 billion, as part of its $38 billion asset-disposal plan intended to repay costs stemming from the Deepwater Horizon oil rig explosion and refocus the company after the subsequent Gulf of Mexico oil spill. (Wall Street Journal website, August 13, 2012)

Now, the rationale for selling this refinery is "to repay costs stemming from the Deepwater Horizon oil rig explosion and refocus the company." Well, the "refinery speak" includes "refocusing" and the need to cover other costs. But could there be more to this than they are saying? What did BP have to say?

"With the completion of this divestment the strategic refocusing of our US fuels portfolio is essentially complete," said Iain Conn, chief executive of BP's global refining and marketing business. "BP's US fuels business is now anchored around three highly sophisticated northern refineries, which are crude feedstock-advantaged, and tied to strong marketing businesses." (BP Press Release June 3, 2013)

And the press release goes on:

"California remains an important state for us and we remain committed to supplying our customers in Northern California and the rest of the Pacific Northwest with the quality fuels they depend on," said Jeff Pitzer, BP's Northwest Fuels Value Chain president. "We've recently upgraded our Cherry Point, Washington refinery to produce cleaner-burning diesel fuel and are building a new rail terminal at the plant to take advantage of growing supplies of domestically produced crude oil."

Notice that although the press release mentioned California, the refining focus is outside the environmentally strict state of California. Sounds like BP is getting out of that area, moving north at the very least. Or could this be part of just getting out of U.S refining altogether?

Conclusion: The company is shedding capacity, in this case to create cash, but it only got cents on the dollar. Although this refinery had some environmental and safety issues, it had nothing of the magnitude of the Texas City disaster in 2005. However, there was talk of this being a "California" refinery. The entire oil business knows the environmental and safety fish bowl that operating in California has become. Who would not want to operate elsewhere?

Chevron sells its El Paso refinery

When I arrived at the El Paso refinery in 1980, there was a 27,000-b/d Texaco refinery across the street. Soon it was sold to some entrepreneurs, El Paso Refining. The entrepreneurs immediately expanded it, and by all reviews they were making money hand over fist. After two years of operation, they designed an expansion, which did not go well. Before all was said and done, they were bankrupt. The facility was then purchased by Chevron in 1993, and the two refineries were combined. This helped Chevron, as it added cracking and coking capacity. Since Chevron was long on octane strength and the El Paso Refinery was weak on octane ability, there were clear synergies. However, in 2003 Chevron sold the entire complex. The Chevron press release dated June 2, 2003, stated:

> "While the El Paso Refinery has been an important part of our system for many years, our strategic focus is shifting," said Dave Reeves, president, Chevron Products Co. "This agreement advances our corporate objective of focusing investments in areas with the greatest potential for long-term growth and shareholder return, while enabling us to continue to provide our customers with premium products through our new business relationship with Western Refining."

Whenever I hear that the "strategic focus is shifting," that usually means we just found something better to do somewhere else. And I suspect the "areas with the greatest potential" are in the upstream business units. So in refinery speak, this means "We are going to unload this capital-sucking machine that does not return much."

It, however, leaves unanswered the question of why Western Refining can turn wonderful profits year after year, using the same equipment, providing the same products, selling to the same customers, in the same marketing area. Could it just be better at it? Clearly!

Chevron sells its Philadelphia refinery

In its press release of August 4, 1994, Chevron announced the sale of its Philadelphia refinery:

> Chevron today announced it has concluded the sale of its [177,000 b/d] Philadelphia Refinery to Sun Company, Inc. The agreement became final at 12:01 a.m. today.

A bit later in the press release, it explains:

> Chevron announced in May 1993 it would sell its Philadelphia and Port Arthur, Texas, refineries as part of a strategy to focus its capital and resources in the South, Southwest and West, where the company's marketing business is strongest.

And then in the same press release, it says:

> Chevron is continuing in exclusive negotiations with Clark Refining & Marketing Inc. regarding sale of the Port Arthur refinery, and a sales contract is expected to be signed in the next four to six weeks.

Wait a minute. I am confused. Earlier Chevron said it wanted "to focus its capital and resources in the South, Southwest and West." Then it explains the sale of the refinery at Port Arthur, Texas. Port Arthur is arguably in the Southwest. And then it shuts down its refinery in El Paso. That's getting rid of about 300,000 b/d of refining capacity in an area where it wishes "to focus its capital and resources."

Well, couple that with the Philadelphia sales, add in the Burnaby refinery, the St Johns refinery, and reduced capacity at both the Richmond and the El Segundo refineries, and you can see that Chevron's perspective on refining has changed dramatically since my days in the 1990s.

Conclusion: Chevron says it wants to focus its assets in the South, Southwest, and West. I am not sure how that squares with selling the Port Arthur and El Paso refineries. Nor does it square with the shrinkage at the Richmond and El Segundo refineries. It looks simply like it wants to shed refining capacity. And I can tell you with certainty that several of these facilities were among the weaker money-making refineries in the United States; they certainly had very poor Solomon ratings. In addition, with the huge environmental issues, capital investment was large, and it too created no income; it merely allowed survival. Also at this time Chevron got its fair share of bad press due to safety and environmental issues.

YOUR ACTIONS SPEAK SO LOUDLY, I DON'T NEED TO HEAR WHAT YOU SAY

In 1984 Chevron purchased Gulf Oil. That made Chevron the largest U.S. refiner with, at that time, around 2.2 million b/d. However, by 2005, it had shed enough refining capacity to drop to number seven. on the list with 909,000 b/d. In 2001, Chevron and Texaco merged, and most of the downstream operations of Texaco were spun off as part of the merger. Later, Chevron acquired Unocal, which had earlier sold its refining capacity.

Although you will not find it in its press releases, by its actions Chevron has shouted it wide and loud that it didn't want to be a big player in the U.S. refining business—if at all.

And there are others that were sold

Some headlines:

- "Chevron Sells Pembrook Refinery to Valero"
- "Sunoco Sells Refinery to PBF Energy"
- "Chevron Inks Deal for Sale of Hawaii Refinery, Downstream Assets"
- "Delta Buys Trainer Oil Refinery in Pennsylvania"
- "Saudi Aramco Assumes Sole Ownership of Port Arthur Refinery"

And this list goes on and will likely continue to grow.

Sounds like the refining industry is changing externally

Yes it is, and the response from the FIOCs is to change internally, that is, if they wish to stay in refining. However, their response is not to become better at refining; rather their internal response to this external change is to shed this capacity.

Some, like ConocoPhillips and Marathon, simply spun off all their refining and created full, newly formed corporations. Others, like Chevron, ExxonMobil, and BP, are selling off their refinery assets based on some corporate philosophy that seems to change, depending on where and when and by whom it is discussed.

> **ON CHANGE**
>
> "In the choice between changing one's mind and proving there's no reason to do so, most people get busy on the proof."
>
> *—John Kenneth Galbraith*

Regarding refining among the FIOCs, I think their day has past. Soon enough their market share of refining will shrink to nothing. They are not structured, not organized, and certainly not performing so they can be competitive in the world of refining, and the market is reflecting this. They simply are not good enough to compete financially with the refining-focused businesses. Couple that with the environmental, safety, and public relations exposure that refineries present to the entire corporation, and add in the problem of limited capital, and the picture is clear.

FIOCs in refining are only playing a waiting game; their run has ceased.

Is shedding capacity and creating refining-focused businesses the long-term answer?

We see now that the refining-focused firms are buying up refining capacity from the FIOCs. Later in Chapter 6 I will describe how independent refiners have an advantage over the FIOCs, but that is just the start. I believe there is yet another worldwide pressure coming, and it will apply to all in refining: FIOCs and refining-focused businesses.

I have worked with independents and FIOCs alike, and I find that both feel a very strong "continental protection," in that it is now commonplace for crude oil to be transported via the water in huge volumes, while nearly all products are transported by truck, railroad, and pipeline in much smaller volumes. It is my belief that products such as diesel, jet fuel, and gasoline will be soon transported on the water, and not just by barges but by VLCCs (very large crude carriers) as well. Some say the technology is not there; tanker design makes it unlikely, and environmental regulations are too restrictive.

Maybe not today; but what if the cracking spread is significantly larger in some areas of the world? Can oil be refined more economically in China or India? What about Saudi Arabia? What prevents these countries, which have built some large refineries lately, from filling a VLCC with diesel and offloading it on the East Coast or using an old Bahamian tank field to transship it. If the refining costs in these areas are low enough, there is a driving force to do so.

Recall the U.S. automobile industry in the 1960s and 1970s. The Big Three were struggling with the reality that the Japanese could produce a car of higher quality *and* at a lower price. This combination created a huge demand for Japanese cars. The Big Three thought they could first ignore the Japanese companies; later they thought they could protect themselves via legislation, and none of that worked.

I strongly recommend, with their current level of self-assurance, that the FIOCs, as well as the refining-focused companies, need to carefully, honestly, introspectively, and dispassionately review their vulnerability and act accordingly. If not, they will become like the Big Three—precariously positioned and depending upon government intervention and corporate welfare to survive as they watch their market share shrink.

CHAPTER SUMMARY

First, lean manufacturing and its predecessor, the Toyota Production System, have been around for decades. Lean has been a mainstream topic in industry for over 30 years, and almost no one in the oil industry has made use of this proven manufacturing system. Many oil company executives have not even heard of it or are in the least bit conversant about its qualities. The FIOCs are happy with their refining manufacturing system: they are not looking around for one that might work better, and they show no signs of changing. They certainly are satisfied if not complacent. They are complacent if not arrogant.

Couple that with low profit margins, the greater safety and environmental compliance issues in a refinery, and the public relations nightmares that go hand in hand with those issues, and in a world of limited capital, it makes business sense for the FIOCs to focus on the upstream business unit—at the expense of the refineries.

They seem to be searching for an answer, although I am not sure why. Both ConocoPhillips and Marathon found an answer; they split off the refining business into its own business entity. I suspect other FIOCs will follow suit or just incrementally shed their refining capacity—like slowly and painfully pulling off an adhesive bandage—until they have disposed of it all. The alternative would be to get good at refining, at least as good as the independent refiners that recently seem to be scooping up these refineries in record numbers.

I seriously doubt they will do this. First, as I mentioned, they seem to be in no hurry to change; they would rather sell the refineries. Second, there is at least one very good business reason why they probably should not try to get competitive at refining: the culture to prosper in refining is dramatically different from the needed upstream business culture. A decent case can be made that these two cultures should not be combined.

At any rate, whether the FIOCs consciously or unconsciously shed their refining capacity, we will soon see that nearly all U.S. refining is done by the refining-focused business units. The data are clear; there is a huge change occurring right now in U.S. refining, and there are issues that need to addressed, not the least of which might be international competition.

For the question "Where will lean first appear in refining?" I seriously doubt it will occur in the any of the FIOCs. Read more about this in Chapter 6.

Chapter 6

How Can Lean Benefit the Refining-Focused Businesses?

Aim of this chapter*—Since the fully integrated oil companies (FIOCs) are shedding refining capacity and it is being picked up and profitably managed by the refining-focused businesses, there is a greater likelihood that lean will become the operating philosophy of the refining-focused businesses. This chapter will discuss some of the inherent cultural advantages that a refining-focused business has over an FIOC when it comes to refining. The chapter will also explain why it is time for the refining-focused firms to seek out a new type of competitive advantage.*

WHAT NATURAL ADVANTAGES DOES A REFINING-FOCUSED BUSINESS HAVE?

There are at least six cultural advantages that refining-focused firms have over the FIOCs when the topic is oil refining. We will explore each item in turn. They are:

- Greater management continuity
- Better short-term focus
- Better long-term focus
- Far less bureaucracy . . . easier to make decisions
- Better balance toward environmental compliance
- Less vulnerable to public opinion

Greater management continuity

I find that the management teams at the FIOCs turn over much more rapidly. This was my experience in my 20 years with Chevron, it is the experience of my colleagues still in the oil industry following my retirement, and it has been my experience as I have consulted with these firms over the last 27 years. Specifically, in an FIOC, the refinery manager and his direct reports often move on a two- to four-year basis. Since a typical refinery manager will have six to ten direct reports, that means somewhere between three and five high-level managers are turning over annually.

Once they arrive at a new assignment, the new managers then need to be indoctrinated into the new culture, get acquainted with the new processes in the refinery, meet all the people, get up to speed on the refinery's plans, and then run their division. And as it seems, no sooner do they do that, then they too are rotated out.

Just what does the rapid movement do? Since the budget cycle is annual, typically the managers will get to go through maybe three of these cycles as they try to learn the economics of the refining business. That is not too bad. However, the typical turnaround frequency for many plants is every three years, and a capital project of even modest size usually takes more than three years from concept to start-up. Since the managers are likely "rotated for their own development" to another assignment, they will not be able to go through the maintenance cycle, nor can they start and finish many projects of any appreciable size.

To further diminish the power of these people to grow and become better managers in a refinery setting, they are almost always rotated to a different location, and it is also common for them to be rotated to a completely different business unit outside of refining.

However, in the refining-focused firms I find this two- to four-year rotation cycle of management to be more like five to ten years. And it is common for those promoted to both stay at one location and spend their entire career in refining. It is a very different pattern of managerial development.

Continuity of management is a serious cultural advantage for the refining-focused businesses. As I note in *How to Implement Lean Manufacturing* (2nd ed.):

When all three change factors—internal changes, external changes, and employee changes—act on a business, I cannot give a single example of even one firm that is excellent at sustaining the changes needed to prosper. The first two enemies to sustaining the gains are not avoidable, so as a business we must manage the personnel in such a way as to minimize turnover. The issue of employee continuity is crucial to being able to sustain the gains. I know of several firms that are adequate at sustaining the gains, and they have one major item in common: a high level of employee continuity. In terms of numbers, they have developed a culture that holds on to the top management for an average of 16 years, supervisors for an average of 12 years, and hands-on workers for an average of 7 years.

This is far more in line with the refining-focused firms than the FIOCs. I find that that this longer rotation cycle is far more productive in two very important ways. First, the manager is given a chance to better learn the job and sort through what are the real needs of the facility and its people. Second, with a longer cycle there is a continuity of planning and focus that is less likely to change.

As the top management comes and goes, the manager's entire structure is wanting for direction, operating in some ambivalence and uncertainty, until the plans are made known. Also, regarding creating plans and giving direction, those managers on a three-year cycle hardly have a chance to reality-check the initial plan they just created and modify it if necessary. However, for those managers who are there, say, ten versus three years, they can make numerous iterations through their "Refinery Growth and Development Plan" and not only create better plans but learn in the process. This dynamic of developing, executing, and checking on those plans is the essence of growth in leadership.

Leadership, largely being the skill to create, sell, and execute plans, is severely stifled if the cycle is not repeated and more so if the cycle is not even completed. A good argument can be made that the rapid turnover of management in the FIOCs, while it affords greater exposure, renders learning much shallower and less effective. Bundling these two thoughts together, the longer tenures with greater continuity provide both a greater skill inventory and improved leadership skills...that is a combo that is hard to beat.

Trust is an important factor

There is yet another natural consequence of continuity of management beyond improved managerial competence and improved leadership skills. It is the critical aspect of trust. I mention it here for one very simple reason. Any business that relies on relationships to supply its service or product must rely on trust to be effective. Trust is necessary for open communications and problem solving as well as cooperation and teamwork.

Simply stated, the more that trust is eroded, the less effective and efficient the business will become.

With low levels of trust, you find enlarged bureaucracies, more procedures, more reviews and approvals, much more paperwork, and lots and lots of meetings. Low levels of trust are not just a mechanism to *create* bureaucracy. Worse yet, low levels of trust actually *catalyze* the growth of bureaucracy.

Having worked for 47 years either full time or as a consultant in over 60 companies, I cannot point to even one example where a company did not need to work on trust issues at some level or with some group. However, among the very best in this regard were two Toyota facilities where I spent significant time.

Discussing, quantifying, and improving trust is a topic most businesses avoid, partially if not totally. All too often it is too uncomfortable to discuss. It becomes very personal, very quickly, as management assesses its competence and integrity. That's tough for anyone to do. But high levels of trust are of the utmost importance to create employee engagement. With low levels of trust, engagement is simply not possible. You can get "blind compliance" or even "vicious compliance," but true employee engagement in a low-trust environment is not a reality.

The anatomy of trust

What is the anatomy of trust? Most people feel that trust is simply a matter of having very good personal traits such as openness, honesty and integrity. That is, if people are honest, open, and sincere, they are worthy of trust. Let's get a reality check on this concept. Say you have a doctor who is open, honest, and sincere. Would you then trust him to perform an appendectomy on you? What if you found out that of the last 20 appendectomies that he had performed, 15 patients had protracted hospital stays and 5 patients had major complications, including several cases of peritonitis that were almost surely caused by the surgery. If you had that information, you could conclude that this doctor, although open, honest, and sincere, simply was not very good at this procedure; he was not competent. Would you trust an incompetent doctor to operate on you? I would not. To be more accurate, we do not "trust someone;" rather, we "trust someone *to* . . ." And so I might trust this doctor to pick up my wife at the airport because he has the admirable personal qualities of honesty and sincerity. However, I would not trust him to operate on me, because he is not competent. "To trust people to…" they not only need to have fine personal traits, they must be competent and there is no substitute for this.

IS MANAGEMENT-WORKER DISTRUST NATURAL?

In the first draft of this section on trust one line read: "First there is the natural distrust between management and the workers, even if no union is involved." Robert Simonis, in his review, reminded me quite bluntly, "I do not think it is natural; it is a learned behavior. I think new employees (fresh out of college or high school) come into the workforce without this distrust."

He is correct. People may or may not initi ally trust you, but to trust people on anything substantial, they must have proved themselves to you. In some matters, there is the issue of keeping your word and honoring comments.

But these wither when compared with the number of restructurings, rightsizings, and reorganizations as well as the number of spin-offs, mergers, and acquisitions that management has orchestrated. Robert is right; this distrust is not natural. However, it has been around for so long… It just feels that way.

How is trust manifested?

For those managers who have not yet developed a significant level of skill or have not had the opportunity to demonstrate those technical skills as well as the personal traits of openness, honesty, sincerity, and integrity, they will most likely not develop high levels of trust. And like it or not, these skills and traits take time to develop and demonstrate in a refinery setting. So in a business that moves its managers on a two- to four-year frequency, it likely would be far more difficult to cultivate the levels of trust needed to be successful.

And the refining business has its trust issues, causing persistent and expensive problems. The issues of trust come in many forms:

- There is a seemingly ever-present distrust between the management and the workers, even if no union is involved. Where a union is involved, trust issues are often worse. With major trust issues, the workers are simply not engaged. They are the people with their "hands on the tiller" who have real-time opportunities to problem solve and dramatically improve the business. But their lack of engagement means the opportunities to get gains, as Dirk did in the Theta cell (Chapter 1), are often not available in a refinery setting, FIOC or otherwise.

- There are often issues of trust between the refineries and the home office staff.

- Within the refinery, there is a natural level of distrust often created by its structure. Engineering and operations do not trust maintenance, supply chain management doesn't trust operations, and human resources and maintenance do not trust anyone. When you overlay all these issues of trust, you end up with not just silos, but massive, steel-reinforced silos that create bureaucracy in spades and so much protective activity that not much gets done. Typical of these cultures with trust issues are lots of time-wasting meetings. Meetings are often poorly attended, poorly managed, and without agendas. They can best be typified by "after all was said and done, a whole lot more was said than done."

To be sure, at the root of some of the trust issues in a refinery is the nature of the business itself. A refinery can be a very dangerous place to work, for a variety of reasons. And this creates an exposure—so the operators are always wondering if the pipefitter tightened all the flanges, and the maintenance mechanics are always wondering if the equipment was properly isolated before they start work. This is natural and to be expected. However, I have not found this to be the primary reason that trust in a refinery, be it an FIOC or a refining-focused firm, is low. This first type of distrust I call "natural distrust," or simply being careful.

> **DR. DEMING ON "DRIVING OUT THE FEAR" "CLOSED CULTURES"**
>
> "Closed cultures" have poor communications channels, where open, honest, frank, and balanced discussions are not encouraged. These cultures are typified by lots of happy talk—yet problems are avoided like a cancer. In closed cultures these are the issues that the management team simply either does not see or sometimes does not want to see: either way they do not want to deal with the topic, like 'that uncomfortable elephant sitting in the middle of the room.' This is the topic that Chris Argyris (*Flawed Advice and the Management Trap*, Oxford Press, 2000) named 'the undiscussables.' In fact, as he points out, more often than not the 'undiscussables are themselves undiscussable.'
>
> Dr. Deming referred to these "closed cultures" as cultures of fear. His point no. 8 addresses this directly: "Drive out fear, so everyone may work effectively for the company."

There is another type of distrust, "toxic distrust." It comes about as a natural result of problems and the almost compulsory desire to assign blame when "things go wrong." And who gets blamed most often? You got it, the rank-and-file operator or mechanic. And this assignment of blame is wrong . . . most of the time. In *Out of the Crisis,* Dr. Deming taught us that between 90 and 95% of all problems are a matter of system breakdowns and not related to the people. So most of the time, anyone could have been there operating the equipment when the problem occurred. It was not an individual problem; rather, it was a system problem. I've heard it hundreds of times with a misrouting, an accident, a plant upset, or a fire. All too often, the questions from the higher-ups start with "Which crew?" "Which supervisor?" or "Who was the operator?" And the questions begin often before the incident is even fully understood. Let me reiterate, this is not just a problem in a refinery setting. It is a common business problem of galactic proportions. It is so large an issue that Dr. Deming made it point 8 of his 14 Obligations of Management. And furthermore, many books have been written solely on the topic of driving fear from the organization. It would be good for you to refer to Chapter 4 and the story of Aera. Pay attention to the cultural desires to foster a "bias for action" and a "tolerance for mistakes."

As far as comparing the FIOCs with the refining-focused firms, I see a "small trust advantage" being held by the refining-focused firms; however, neither group has yet mastered the management skills to get an engaged workforce. Yet the refining-focused firms have a far greater likelihood of gaining that trust, due to benefits that can be gained through greater management continuity, which is a natural part of their culture.

Better short-term focus

Ask those managers who have worked for both FIOCs and the refining-focused firms, and they will immediately tell you that the refining-focused businesses are far more bottom-line oriented. They will also tell you that decisions of the refining-focused firms are far more fact based. In addition, I have found their bonus systems, while still being broad based, are in greater harmony with the refinery goals and objectives. Furthermore, the refinery goals and objectives are not only clear and understood by all; they drive the behaviors of others much more strongly than in the FIOCs. In a refining-focused firm, there are still silos, but cooperation and teamwork are far more present as you see the facility react to a common set of goals.

Most managers in the refining-focused firms were hired away from the FIOCs, generally through the lure of much higher-paying jobs with greater responsibility. Since most of these managers came from the FIOCs, unfortunately they also bring with them artifacts of the staid and stodgy culture of the FIOCs. However, at the end of the day, both focus and alignment toward common objectives are superior in the refining-focused firms.

Better long-term focus

Since most refining-focused businesses are just that, refining focused, it is easier to keep your eye on the ball. Increasingly, they couple the refinery business with the downstream sales business, such as Valero and Tesoro have done. Nonetheless, even in this environment, it is much easier to focus on these business units, as neither one is the "hard-core macho" culture of exploration and production. Refining, marketing and sales are all classified as hybrid cultures, somewhere between a "process culture" and the "work hard, play hard culture" of Deal and Kennedy (see Chapter 5).

An example of a confusing, or maybe not-so-confusing, long-term focus

To really appreciate the long-term focus of an FIOC toward refining, consider the situation of Chevron. In 1990 it had over 2 million b/d of refining capacity. The company invested millions into its Port Arthur Refinery in 1990 to convert it to a single-train operation. Then less than three years later Chevron sold it along with its Philadelphia refinery to Clark so it could "focus its capital and resources in the South, Southwest and West." (It still baffles me that Port Arthur, Texas, is not viewed as being in the Southwest or the South.)

In that same year, 1993, Chevron purchased the refinery (the El Paso Refining refinery) directly across Trowbridge Avenue from the Chevron Refinery, when the current owners went bankrupt. However, then a few years later, they sold these upgraded facilities to Western Refining.

Taking this as a whole, it's pretty hard to square what the company says with what it does. And either there is no plan and thus no focus, or the plan changes frequently and the focus then must be very short term. Therefore, it is very difficult to see any long-term focus based on Chevron's actions.

The one thing that is clear is that its long-term behavior is to continue to shed refining capacity, both in the United States and worldwide . . . whether the company advertises it or not and maybe whether it actually has a plan for that, or not.

Far less bureaucracy . . . easier to make decisions

If you asked the same oil executive I mentioned (who had worked for both an FIOC and a refining focused firm) for the differences between the two businesses, the second thing he would likely tell you is that in the refining-focused business, he could get a lot more done, faster.

What is outrageous beyond reason is the lead time it takes for the FIOCs to make a decision. In the refining-focused businesses, the companies are far less bureaucratic. Not only do the refining-focused firms have a clear bottom-line orientation, but they also have a bias for action; on the other hand, most FIOCs are accused, and rightly so, of having a bias to study . . . and ponder. As I said earlier, a mantra I learned in my early days in the oil business was "Analysis to paralysis," and we embodied it fully.

It really became apparent to me when I was in El Paso. As a well-placed manager, I tried on numerous occasions to create some synergisms between our refinery and others in the area, most of which were independents. My earliest efforts were with the Texaco Refinery across the street. My first effort was a total failure, and the Texaco refinery manager made it clear that we were not likely to close any deals. Soon enough some entrepreneurs (El Paso Refining) bought the refinery, and I thought, "Aha, now maybe we can exploit our octane, diesel, and jet advantages."

Dumb me.

On one occasion, it happened that our neighbors had a need for high-octane blending components. We could supply what El Paso Refining needed and thereby make a small fortune at the prevailing value of an octane-barrel. I approached my boss, and he was all in. However, at Chevron, all product exchanges needed marketing review and approval from both the local management and the home office. First the marketing people went into high-resistance mode and dragged their feet. Later, when they saw what it would do to their balance sheet, they too were all in. Then came the problem—we needed to get the home office involved. We arranged a meeting in El Paso, and I recall the refining manager of El Paso Refining tell me, "I am looking to make a deal; tell them to bring their checkbook." And I did. The folks arrived from the Chevron home office and had a few questions that needed to be cleared by the higher-ups. The home office visitors could not approve this exchange, and the El Paso refining manager said, "So where's the checkbook?" The home office people had to admit they could not approve it but promised an answer by the end of the week. Well, that week came and went, and a second and a third week passed; finally in week four they had an answer. When I contacted the El Paso refinery manager, he said, "Too slow; we went another way two weeks ago. We just could not wait." That happened two more times before I finally gave up on the hope of working with the members of the Chevron hierarchy. They were just too slow.

It is ironic that when Chevron finally purchased this refinery, it cited the obvious synergisms with the two facilities as a significant asset, yet that advantage apparently was not there earlier. Go figure. The El Paso Refinery had a coker and some underutilized cracking capacity, while we had lots of crude capacity and could produce high-margin jet fuel, high-margin gasoline, and extremely high-margin, high-octane gasoline.

As frustrating as that lesson was, I already had experienced a far less personal, although larger, lesson in Chevron's slowness to act much earlier in my career. It was in the seventies, and I was in El Segundo. At that time, Arabian heavy crude was the incremental crude on the market, and it was to become even cheaper as the Saudis announced they would produce more. The problem, for most, was not the crude gravity; it was the sulfur content, far exceeding Arabian light. Consequently, the U.S. refining industry went into full gear to install incremental processing ability to handle this sulfur-laden crude. Chevron's answer was to install low sulfur fuel oil (LSFO) facilities, and I was involved in the expansion at El Segundo. A near twin expansion was under way at Chevron's Richmond refinery. In a discussion with the project manager, he told me that Chevron was the last major to get started and this forced the company to use a design and construction contractor it otherwise would not have chosen. As happened with many Chevron projects, the start-up was way behind schedule, and the entire project was severely overrun.

Finally, it started up, but by then I had left for my next assignment at Perth Amboy, New Jersey. I made a trip back to El Segundo and looked up the project manager, a friend of mine who was now the superintendent of the LSFO Division. He showed me a pile of appropriations requests for major modification to the brand-new LSFO complex. Most of the requests involved modifications to run crudes other than Arab heavy. In fact, at that time, nearly one year after start-up, he told me the division had run 38 different crudes—not one of which was Arab heavy. You see, by the time the El Segundo LSFO facilities had been completed, slowed by the initial decision to invest the capital and further exacerbated by the design and construction delays, all the incremental Arab heavy had been scooped up by our competitors, and it was no longer as cheap as expected. He was preparing a documented case study to show to senior management that Chevron needed to:

- Be faster at decision making
- Be more flexible in designs

Otherwise it could not accommodate the natural changes that occur over time. I never got to speak to him about the way his case study was received; on that I can only guess.

The slowness to act on these major issues is a key to being less competitive in the long run. However, it has a different effect on the day-to-day operations of most of the people.

In a refinery, the bureaucracy that the average engineer, maintenance supervisor, operations supervisor, or environmental advisor must overcome in his day-to-day activities is mind-boggling. It is far worse in the FIOCs compared with the refining-focused firms and is a huge impediment to actions, not to mention being a demoralizing force. And many FIOCs have responded to this bureaucracy with a structural change. They have added positions called "coordinators." It is common to find maintenance coordinators in maintenance as well as both engineering and maintenance coordinators in operations. As I unravel how these positions developed, you will see that they were caused by two things.

First, the staff groups that provide support to operations, including engineering, maintenance, planning, safety, and environmental, to name a few, are primarily Monday-to-Friday, 8-to-5 jobs. Meanwhile the operations crews work not only 24/7; they rotate shifts and rotate people from crew to crew, not to mention the typical disruptions like sick and vacation time. Hence they needed some mechanism to keep track of all the priorities and activities of these groups supplying support. That part would be simple enough to handle.

But something worse entered the picture. It was uncommon to find the supervisors in operations, maintenance and the support groups to be working on the same set of refinery objectives. In addition, poor communications, silos, and inconsistent management follow-up led to much confusion. Thus a supervisor would activate an engineer (for example) on a project only to find someone else had reassigned the person to a different job. So rather than work hard to create a common set of goals and objectives, goals that were communicated clearly coupled with consistent management follow-up, the coordinator position was created. It was the path of least resistance.

Better balance toward environmental compliance and less vulnerable to public opinion

In my days with Chevron and in my work with other FIOCs, I have found them to work very hard toward being compliant with laws and regulations, especially the environmental and safety regulations. They have entire divisions headed by a top-level manager who is often trained at the collegiate level in these fields. There are frequently separate groups for air, water, and hazardous waste management and compliance, complete with both legal and engineering support readily available. In my career with Chevron, the company was very aggressive at working to stay out of trouble with the regulatory agencies and the press. Although I cannot quote it precisely, Chevron's environmental policy contained the concept that "we could comply with all applicable laws and regulations regardless of the degree of enforcement." It was a very environmentally responsible policy, and we did just that.

While this was environmentally hyper-responsible, it was financially very, very expensive and created problems we never anticipated. One of the most egregious examples I can think of was when the RCRA (Resource Conservation and Recovery Act) regulations were enacted in the early 1980s. We aggressively interpreted the regulations and began a process of compliance at the refinery. One aspect of the regulations required that we drill exploratory wells under the refinery to find if there were underground wastes. Well, lo and behold, as we were complying, we literally struck oil. Only we did not strike crude oil; it was gasoline, diesel, and jet fuel. We found products that had been leaking from our tanks for decades. There were literally millions of gallons of product sitting atop the water table. We had to put in place an exploration and mitigation plan, and soon enough we were spending millions each year to get these hydrocarbons off the water table. At one time, we had over 60 people working on this full time not only sinking wells but putting in recovery systems and then dealing with the oil and water that were part of the recovery process. Furthermore, to mitigate future leaks we had to execute a plan to remove tanks from service and confirm the bottoms did not leak, and if they did, repair them. We spent millions on this effort as well.

In addition, we designed the equipment and developed a process to ultrasonically map the thickness of the tank floor on a 1-inch by 1-inch grid using mobile ultrasonic probes. We also created, developed, and validated a new design for an impermeable membrane with floor-leak detection capability for storage tanks. Literally we were leading the oil industry in Texas and probably most of the United States. Both items were quickly adopted by the Texas Department of Water Resources (TDWR was the state enforcement agency for the federal RCRA regulations). These were then turned into standards that the TDWR imposed on others.

This did not endear us to the rest of the refining industry in Texas. Early in the game of compliance to the RCRA regulations, it became clear that we were the leaders in compliance in all of Texas. Quite frankly, from all outside appearances we were the agent of the TDWR to create new refinery techniques that would lead to faster and better compliance. I know for a fact that the TDWR in its annual report to the EPA took credit for these "engineering advancements."

UNINTENDED CONSEQUENCES

Jay W. Forrester of MIT developed the science of system dynamics to explain system behaviors, especially social systems. In his 1991 article "System Dynamics and the Lessons of 35 Years," he writes: "People are sufficiently clear and correct about the reasons for local decision making. But, people often do not understand correctly what overall behavior will result from the complex interactions of known local actions."

The policy did not address the interaction and lack of cooperation with our competing refinery across the street. It did not address the interaction with the people at the TDWR who were used to dealing with aggressively combative firms, and it failed to address the issue that the press is far more interested in producing headlines than digging for the truth.

In our case, the environmental policy was designed to avoid negitive publicity, yet that is precisley what we got.

Meanwhile, across the street was the old Texaco refinery, now owned and operated by El Paso Refining (EPR). And life was remarkably different on its environmental front. It had the same federal and state regulations to fulfill but took a markedly different approach. While we were staffing up and working toward compliance, it was arguing with the local and state agencies on anything it could to delay the inevitable. When the oil plume beneath Chevron was found, the heat was on EPR, and it was forced to quicken its feet-dragging approach. The plume was found beneath its refinery as well, but it challenged the TDWR to prove it was EPR's. That stalled things for a while, but the TDWR blew through that red herring, and soon enough EPR was forced to undergo further exploration and plan for recovery. Well, I forget the exact number of wells EPR had drilled, but it drilled a small fraction of what we drilled. EPR appeared to drag its feet at every step; it challenged the TDWR on content, on procedure, and on interpretations of everything it could. EPR appeared to make enforcement of its facility a very expensive and tiresome effort for the TDWR.

Not only that, but this worked to its advantage. Since Chevron was so aggressive and compliant, two very undesirable things happened. First, the TDWR, when dealing with the federal agencies, could clearly say it was taking action and could point to the activities at Chevron and avoid dealing with EPR, which was clearly as guilty as Chevron, if not more so. Second, once the press got hold of this, the aggressive nature of our work made us look like the only guilty party.

For trying to be good corporate citizens—for being cooperative and for acting responsibly—we ended up spending a lot more money, sucking up hundreds if not thousands of hours of management time... and getting bad press at the same time.

For all our efforts to "keep the Chevron logo clean," we ended up creating exactly the opposite effect. We tarnished the logo in the public eye. No good deed goes unpunished.

Meanwhile El Paso Refining enjoyed the benefits of:

- Money not spent on this environmental effort that would be used elsewhere or end up in the owners' pockets.
- Hundreds of hours of management time to spend on making more money.
- Very little bad publicity.
- The same record with the TDWR as we had; it too was apparently compliant with all regulations.

Experiences like these make it clear that spending lots of money "to keep your logo clean" has its limits. If you are a smaller refining-focused firm, willing to act like El Paso Refining acted, then you have a distinct list of advantages over the larger, more visible FIOCs.

WHY WOULD A REFINING-FOCUSED BUSINESS WISH TO USE A LEAN MANUFACTURING SYSTEM?

We have outlined several of the cultural advantages that refining-focused businesses have compared with FIOCs. Because of these and possibly other advantages, the refining-focused businesses currently economically outperform the refineries of the FIOCs. But problems loom on the horizon, and simply outperforming the refineries of the FIOCs will no longer be adequate; that success has morphed into a real need for further change.

The refining-focused firms used to compete with the FIOCs when the FIOCs owned the majority of the refining market share, 67% in 2005. And since the typical refinery of the FIOC had 3–6% profits while the refining-focused firms were more like 14–16%, it was clearly easier to compete. During that time the FIOCs used their huge market share to set the market price, allowing the more efficient refining-focused firms to go along for the ride and make very good profits.

That dynamic is changing. Soon the refining capacity of the FIOCs will be so small as to be insignificant in terms of establishing refining margins, and that financial dynamic will be controlled by the refining-focused firms.

In short, the competition has just gotten tougher; the weak players are being weeded out. To maintain their profits, the refining-focused firms will have to find new ways to improve their performance to meet the new challenge of increased competition.

There is no more powerful methodology to reach this goal than lean manufacturing. A lean transformation in a refinery will allow you to achieve gains in five areas:

1. Capacity increases
2. Cost reductions
3. Availability improvements
4. Improved morale
5. Better quality, safety, and environmental performance

We will discuss these below.

Capacity increases and cost reductions often go hand in hand

As wastes are removed, it is natural for this to turn into potential capacity. If people are not waiting, they have more time to work; that is capacity. If processes are not producing off-spec product, that creates the opportunity to produce more; that is capacity. If inventory is reduced, that is turned directly into capacity. If specifications are limiting capacity and the process is not optimized, that is capacity to be exploited. Furthermore, nearly every time you make a capacity increase, you break a bottleneck. Anytime a bottleneck is broken, the huge fixed costs in a refinery are distributed over a larger volume and your "per-barrel" cost to produce is reduced. As you increase the capacity, you simultaneously reduce your per-barrel cost to produce.

Lean transformations are unprecedented in leveraging capacity, both driving operating costs down and driving the capacity of value streams up. Utilizing the tools of lean coupled with lean management, we have more than doubled the capacity of processes at many tier one auto suppliers (TOASs) while at the same time driving costs down 50%.

Refineries are much more adept at finding and removing bottlenecks than the typical TOAS, so I would not expect such large gains in a refinery. Even if the capacity improvements are not 100% in a refinery, still, significant gains can be made. In a case I discuss in detail in Chapter 12, winter-grade diesel improved fluid catalytic cracking capacity by 12% and allowed a resultant 8,000-b/d increase in crude rate at a 78,000-b/d refinery. That was done using the tools of lean and lean management. So capacity increases are available whether it is blending optimization, improving furnace operation to improve fractionation while reducing energy consumption, improving heat removal in your overhead system, increasing and/or smoothing catalyst circulation in your cracker, exerting better control over hydrogen purity, or improving cooling tower operations to prolong exchanger life and improve heat transfer. The tools of lean will allow you to exploit them all.

Based on my work in refineries, while directly managing or providing consulting support, I have seen firsthand that the opportunities are large—very large. For example, if you have a 100,000-b/d facility with operations costs in the $15/bbl range, a 20% reduction in overall costs is practical in a lean transformation. This would yield an annual savings of over $100 million a year. Or if you are looking at upgrading your value stream, we have worked with refineries to reduce specification giveaway on diesel and gasoline in excess of $30 million a year. Furthermore, we have also created greater flexibility as we did this.

With the large cash flows typical of the oil business, I think that refineries are a target-rich environment for a lean transformation.

Availability and capacity utilization improvements

In my work with refineries, I find that nothing affects operating profits more than issues with process availability and process capacity utilization.

There are a variety of definitions of these terms depending upon the industry you are in. Some of the definitions are very technical; some are simple. However, all are surprisingly similar. Here I have avoided the very technical nature of these terms and rather have focused on simple, commonsense terms. For our purposes of understanding, these definitions will be more than adequate. They are:

- Availability is defined as the fraction of time a piece of equipment or a system will operate when it is scheduled to operate. It is expressed as a percentage or as a fraction.

- The capacity utilization measures the fraction or percentage of full-run capacity that is realized in production. Capacity utilization levels give insight into the overall slack in a piece of equipment, in a plant, or in a whole refinery for that matter.

Regarding availability

The cure to low availability for a plant is normally maintenance and redundancy. Refineries have maintenance programs, and they normally have a good understanding of what shuts down their plants with the attendant huge economic consequences. My experience with refineries regarding maintenance is summarized in Chapter 7. Although there are opportunities to improve availability with a refinery TPM (total productive maintenance) program, I find that the major opportunities for financial gains lie in improving capacity utilization—which by the way, TPM will also help.

Regarding capacity utilization

Capacity utilization is frequently used as a driving metric in refineries, although as often as not, it is not used properly. To strive for high capacity utilization on the economic bottlenecks in a refinery is smart and will give you large economic gains. However, to strive for high capacity utilization on *any other* piece of equipment, plant, or stream will lead to overproduction and will be counterproductive and lead to an economic loss. Very likely, your refinery LP or planning model, particularly if it is a commercially available product such as an ERP (enterprise resource planning), probably contains a module called "Capacity Requirements Planning."

The problem!

The typical tier one auto supplier runs two 10-hour shifts, 5 days per week. A week contains 168 hours, and the TOAS typically works 100 hours. The TOAS can, at some minor premium, work an additional 68 hours to compensate for downtime or changes in demand—virtually any week it wants to. Based on a 168-hour week, the TOAS's capacity utilization is only 59.5%. Put another way, the slack time, the "unused capacity," is 68 hours per week that can be used to "catch up" should there be some downtime, for example. This "unused capacity" is available by virtue of its operating schedule.

Not so in refineries. They run 24/7 and have no slack time and no "catch-up" ability. So if your planner schedules based on 100% capacity utilization, the smallest shutdown or even slowdown will put the shipping schedule in immediate peril. A plan with 100% capacity utilization is not a viable plan.

The "solution"

To create a plan that is usable, the refinery planner "fudges" the operating rate a bit to give him some flexibility. Knowingly or not, he plans for less than 100% capacity utilization. For example, in a recent refinery engagement, after a major shutdown of the crude unit, the refinery performed a test run. It was able to comfortably run 112 mb/d of crude for over a week without incident. However, when the planner made the monthly operating schedule for the refinery, and every month thereafter, the refinery advertised the maximum crude rate to be 95 mb/d; hence it effectively had 17 mb/d of unused capacity. If it had a Capacity Requirements Planning module, it likely would show it had over 15% available capacity, 15% slack.

We see this all the time in refineries; it is standard fare.

The problem with this "solution"

So what's wrong with this? Well, that depends on your objectives. My experience with refineries is that the managers want to run them at the maximum safe rate that they can assure customer delivery. And in this case, is 95 mb/d that value?

I have seen these maximum rate numbers many times, and I ask, "What is the basis for that value?" It really comes down to a simple intuitive evaluation. And if the refinery planner has some experience, then what the refinery has done in the past will serve as a pretty good estimate of what the refinery can do in the future. But does it meet the criterion of being the maximum safe rate to meet customer deliveries?

In principle, it likely is a safe rate, and it is also likely to meet the customer deliveries. But is it the maximum? I seldom find that to be the case for two reasons:

1. It is almost always a very conservative number. You see, most often the planner gets no kudos if he increases the refinery rate a few thousand barrels per day; however, his job is really on the line if the refinery cannot deliver to the customers. Hence the rates selected tend to be on the very conservative side.

2. There is no risk-to-reward calculation made, none at all.

So what about the risk-to-reward calculations?

It is easy enough to calculate the value of producing the next 1-mb/d increment of crude; so we can calculate the rewards easily.

What about the risk for that next 1 mb/d?

I have yet to see even one refinery make this calculation, and yet it can be made and quantified. For any given crude rate, you can calculate the probability you can sustain that rate and quantify the probability of a shortfall. Furthermore, you can calculate the consequences of that failure including the probability of failing to meet the plan. Finally, you can calculate the losses, both temporary and permanent, financial and otherwise, that are associated with that shortfall.

Normally these calculations are not made — because no one knows how to make them.

In the case cited, the crude run above the 95 mb/d to the real limit is usually the most profitable crude you can run. I recently went through these numbers with the VP of finance for a refinery, as we were training on the value of breaking through a bottleneck. In her refinery, which was then making 13% on average, we would make 24% on each incremental barrel above the limit. At that time, crude was over $70/bbl, and an extra 1 mb/d of crude would earn the refinery another $6 million a year; an extra 5 mb/d would net $30 million annually.

So the question needs to be asked, "Exactly how certain are we that 95 mb/d is the *maximum* safe rate that assures delivery?" And if it is not, what is the maximum safe volume that assures delivery?

You will never know as long as you continue to use these "intuitively created maximum rates." Many of the quantifying tools of lean, especially the use of statistical analyses, can show you how to figure out what your limits really are as well as the consequences of running at this new rate. Furthermore, even if the intuitive value happens to be correct, this process of quantifying it will highlight the next bottleneck along with the benefits of breaking this bottleneck.

Both the rewards and the risks can be quantified using the many tools of the lean methodology. And I would challenge you to understand if going from 95 mb/d to 100 mb/d, with a demonstrated capacity of 112 mb/d, is a good business decision. I would suggest that with $30 million to be made, it is worth finding out. And as a corollary to that, if 95 mb/d is really the maximum safe limit to assure customer delivery, the process of determining that will not only

INTRINSIC MOTIVATORS

The power of these intrinsic motivators has been reconfirmed once again during my consultancy of 27 years. In that experience I have tried these principles and found them to be not just successful—but wildly successful as I have not only used them but taught them as well. The five intrinsic motivators are as follows:

1. A sense of meaningfulness
2. A sense of control
3. A sense of accomplishment
4. A sense of growth
5. A sense of community

find the next bottleneck; it will quantify those potential gains as well. This is all accessible to you by using the statistical tools available in the lean methodology.

Improved morale, quality, safety, and environmental performance

I have grouped these for one very good reason. During a lean transformation, we seldom need to directly focus on these issues. They are freebies that improve as a direct result of not what we do, but rather how we implement a lean transformation. And the "how" that is so important is "using a fully engaged workforce." When the the entire workforce including management and staff as well as the operators and mechanics are fully engaged, the intrinsic motivators are activated for everyone.

When a fully engaged workforce is utilized and when both the operators and the mechanics, as well as everyone else, become a part of the problem-solving and improving effort, they always gain a sense of control over the work they are doing. These team-based problem-solving efforts allow them to gain a sense of meaningfulness, a sense of accomplishment, and a sense of community as they participate with others in improving their work environments. Problem-solving efforts often stretch the knowledge base of those involved and are a natural growth experience. When this is done, their engagement increases as their information is valued and used.

But most important of all, by simply getting the workers involved and listening to their thoughts, concerns, and ideas, you will get a great response. Improved morale is a logical consequence of being listened to. If only managers everywhere, not just in refineries, could harness the power of good and empathetic listening, a whole new world would open up to them.

And help with rare events

In fact, most safety, most loss of containment, and most environmental incidences are quite rare, and "what happened when" is often only available to those on the scene at the time. I have worked in "closed cultures" (the typical refinery culture, where people are unlikely to volunteer any information), and these rare events are often hard, very hard to diagnose. However, if those same eyes and ears know they will get an empathetic listener, you are far more likely to get to the truth about the root cause of the incident. These rare events are a nightmare in a refinery, as they often have far-reaching safety, morale, cost, and public relations impacts. I know of no better way to improve the problem solving of these rare events than to engage the people who are there in a joint problem-solving effort to find the truth.

So why now?

No mystery here. Since the fraction of the lower performing refineries from the FIOCs is shrinking, this will raise the bar of competition for all refineries. The pressure to improve will be increasing as the FIOCs continue to shed their capacities, freeing their refining businesses from the bureaucracy of their detached, large, slow-to-change and cumbersome hierarchies. There will be extreme downward pressure on costs and, as Taiicho Ohno said, "the objective is cost reduction." On the one hand, reducing costs in a refinery is no problem at all. Cut some reflux here, run the catalyst a year longer, defer a shutdown or two and even contract out some of the activities. These are all proven cost reduction techniques. And if that doesn't work well enough, there are other options. A little restructuring here, some reorganization there and start holding your bills for 60 versus 30 days. This will all reduce costs... at least for today. There is no problem in reducing costs.....anyone can do that. Reduce some reflux and you'll save on energy, but lose on product upgrade. Defer a shutdown and you'll save on maintenance costs but lose on equipment availability. No, there is no problem in reducing costs. The trick is in "optimizing" costs, or in lean terms, reaching the ideal state. Not only does lean drive you to the optimum state today, it has a clear sustaining function so these gains can be realized in the long term and finally, improving and sustaining is not all that lean delivers. Beyond that, you will create a culture in which continuous improvement is a way of life. When you have an operating system that optimizes, effectively fights the battle of entropy, and if your operating systems work to continually improve the entire business, you have a competitive weapon that is unassailable.

And the first company to undertake a lean transformation will have an advantage, a huge advantage, economic and otherwise. Will that be you... or your competition?

CHAPTER SUMMARY

The refining-focused businesses have some natural cultural advantages that put them in a much better position to become lean enterprises. These include greater management continuity, a much better business focus in both the short and long term, day-to-day operations with far less bureaucracy, and less vulnerability to environmental and public opinion pressures.

Having an operating system based on the strategy, tactics, and skills that are inherent in a lean manufacturing system will create a strong competitive position.

The gains from a lean transformation will include improved capacity performance, improved cost performance, increased availability, improved morale, and improved quality, safety, and environmental performance.

CHAPTER 7

WHAT DIFFERENCES DOES PETROLEUM REFINING BRING TO LEAN?

***Aim of this chapter**—In this chapter we discuss some of the differences between the typical lean metaphor, which is a labor-intensive, tier one assembly operation in the auto industry supply chain, and a petroleum refinery. When implementing a lean transformation, there are some differences in this application in a refinery in comparison to what is in the literature. Many believe these are reasons that explain why lean will not work in an oil refinery. While there are differences, none of them precludes the ability to make a full lean transformation in either the refinery or the TOAS (tier one auto supplier). While some of the differences make a transformation in a refinery more complex and others make it more simple, we can show clearly that the disadvantages do not preclude success, nor do the advantages guarantee success.*

OUR DESTINATION IN A LEAN TRANSFORMATION

Keep in mind that our goal is to be:

- Better—that is, provide a higher-quality product.
- Faster—that is, improve on-time delivery performance at reduced lead times.
- Cheaper—that is, while improving quality and delivery, simultaneously reduce costs.

And we will do that using a manufacturing system that is lean. Once again, as you read, bear in mind our definition:

- Lean is the creation of a culture of continuous improvement and respect for people.
- The means to lean is to problem solve your way to the ideal state through the total elimination of waste using a fully engaged workforce.

THERE ARE MANY DIFFERENCES THAT CAN MAKE A REFINERY EITHER A BETTER TARGET FOR A LEAN TRANSFORMATION—OR NOT

I have divided these differences into three categories:

- The four large positive differences
- The five red herrings
- The two albatrosses

The four large positive differences

1. The economic issues of cash flow, capital intensity, earnings per employee, and operating costs

If we look at some very rough economics, even in a modest 100-mb/d refinery, at $60/bbl crude, the major raw material, crude oil, by itself costs $2.2 billion a year. To pay for this and other raw materials, operating expenses, and profits, this refinery has well over a $3 billion–a–year business in sales and probably has around $2.5 billion of capital invested. The staffing will likely be a total workforce of around 400 people; and 30%, or 120, of this workforce will be direct production personnel.

Compare this with the typical tier one auto supplier (TOAS) that might be making products such as automobile seats or fuel pumps. It more likely has $75 million a year in sales with $20 million of invested capital. Total staffing will be around 200; and 80%, or 160, of this workforce will be direct production personnel.

So if we look at invested capital per employee, the differences are stark:

- *Refinery:* $6,250,000/employee
- *TOAS:* $100,000/employee

If we look at cash flow per employee, the differences once again are stark:

- Refinery: $7,500,000/employee
- TOAS: $375,000/employee

At $60/bbl of crude, a typical profile of labor costs might be something like that shown in Table 7.1.

Table 7.1 An economic comparison, a refinery to a TOAS

Factor	Refinery	TOAS
Operating costs	20% of sales	45% of sales
Labor	25% of operating costs	60% of operating costs
	5% of sales	23% of sales

On a per capita basis, comparing the refinery model with the TOAS, we find that the refinery is 60 times more capital intensive and 20 times more cash flow intensive. In addition, the overall impact of labor costs is dramatically different. In the refinery, the labor is only around 5% of total sales, while the TOAS is 23%.

From a business perspective, these large differences cause a large difference in focus. Although these values are very dependent upon the price of crude oil, which can vary a great deal in a very short period of time, that does not alter the facts that an oil refinery, when compared with a TOAS:

- Is far more capital intensive
- Has far greater cash flow per employee
- Has a very low labor contribution to overall sales

How might this affect the lean transformation?

Applying lean principles in a refinery is very motivating with these large cash flows; the financial opportunities are huge. That gets everyone's attention and makes the projects not only priority items, but fun as well. There are also large opportunities with inventory reductions to improve flow and reduce lead times. There are major opportunities to improve safety and morale as well. However, whether our lean focus is on finances, customer satisfaction, environmental improvements, or lead times, the basic transformation is the same. We are still going to create a culture of continuous improvement and respect for people. And we are still going to problem solve our way to the ideal state focusing on waste and using a fully engaged workforce.

There will be differences in the problems selected to solve and the wastes we select to remove. Most TOASs have a cost profile that is very labor intensive; therefore a lot of profit opportunities center on productivity improvements, manpower reductions, and space reductions along with improvements to reduce transportation costs.

This will be much different in a refinery with its capital-intensive nature. More emphasis will be placed on working with the primary value stream, the oil itself. There will be opportunities to reduce lead times, make large advances in quality, and vastly improve capacity utilizations on process bottlenecks. Finally, because of its capital-intensive nature and its 24/7 operating schedule, a refinery lean transformation will place great attention on reliability improvements.

In a refinery, there is a lot of low-hanging fruit with very large financial impacts. Some of the lean projects we have executed in a refinery have project savings that would be larger than the entire annual profits of the typical TOAS; and in some cases the savings were larger than their total sales as well. These solutions have two very powerful psychological effects. Primarily, these are large opportunities, and the business will get much better, and the employees will feel good about that. Everyone likes to be part of the prosperity, growth, and success. Second, as in all change initiatives, it is good when you can show some early "wins." These early wins are easy to find and equally easy to publicize. Once they are advertised, people will get a distinct sense of satisfaction. From there, buy-in is increased, and everyone then is motivated to do even more. Once again people like to win and be part of a winner. It feeds the intrinsic motivators of meaningfulness, accomplishment, growth, and community. (For more on the intrinsic motivators, see *How to Implement Lean Manufacturing*, 2nd ed.)

2. *The operating structure*

The typical TOAS has an operating structure that is not suited to lean manufacturing. The structure looks like Figure 7.1. and its purpose is to make the ratio of direct to indirect labor look very good.

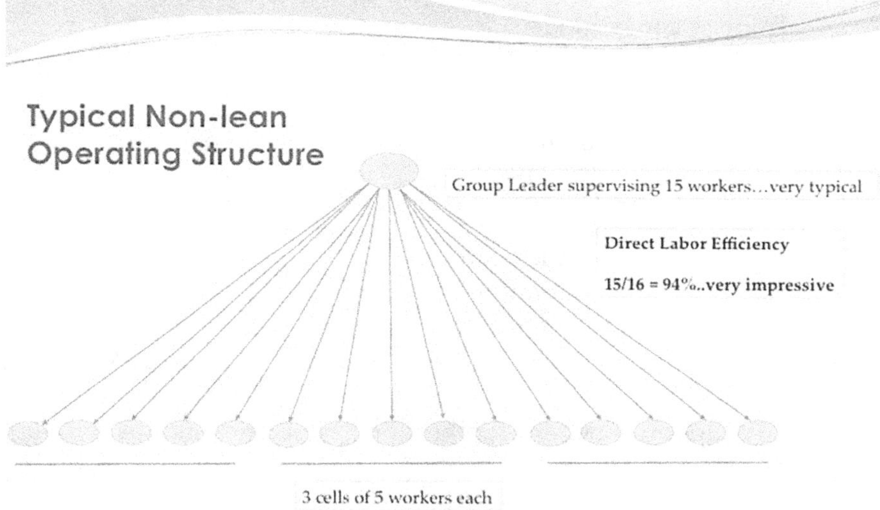
Typical Non-lean Operating Structure
Group Leader supervising 15 workers…very typical
Direct Labor Efficiency
15/16 = 94%..very impressive
3 cells of 5 workers each

Figure 7.1 TOAS typical operating structure

However, this structure will not work in a lean organization for two major reasons. First, the supervisor (group leader), who is responsible for producing the right volume of the right mix and doing so to the required quality and delivery standards, is more than fully loaded trying to supervise the 15 workers and 3 production cells. In addition to meeting the production schedule, on paper he also has the responsibilities of personnel development, especially training of his subordinates. This seldom happens, except in the most perfunctory way. Second, with employees fully loaded, as they produce to takt, there is no time for problem solving. So it is not possible to problem solve your way to the ideal state using a fully engaged workforce with this structure.

The countermeasure to this structure is the Toyota operating structure with the use of team leaders. The job of team leader is not a management position, and in those facilities with a union, it is a separate union-represented job classification. The team leader is fully qualified on all the jobs in her cell; she is a TWI-certified trainer and has been given intensive training in all aspects of her job: standard work, problem solving, and human relations training including leadership skills.

The team leader is tasked with:

- Filling in for anyone in the cell on an emergency basis
- Creating and implementing standard work
- Training cell workers
- Solving problems
- Implementing continuous improvement activities
- Doing routine duties

The primary function of the team leader is to make sure that the other workers in her cell are operating at peak efficiency. Thus she spends a great deal of time assuring that the standard work is executed properly, and if not, training the workers so they can be successful. The routine duties she might perform are not direct production duties but are mostly support functions for the work cell. Duties such as calibrating machines and filling out, analyzing, and precipitating action on production-by-hour boards are typical items. When problems arise in the cell, such as tool failures, it is the team leader's job to work with the cell first to create a temporary countermeasure so production can be sustained and then to begin the task of problem solving in real time. As she has time, she will work with the supervisor and the cell to implement the continuous improvement activities that are contained in their hoshin kanri plan (more on this in Chapter 8).

Figure 7.2 shows the Toyota operating structure. The direct labor efficiency is a not-very-impressive 79%. Yet this is a common metric used in Western manufacturing as the managers try to reduce this "noncontributing" indirect labor.

There are two flaws in this thinking. Although these flaws are lost on many, once pointed out, they provide a gap in logic that you can drive a truck through.

1. The first flaw is that the overall system optimum is not the sum of the local optima. And this effort to drive direct labor efficiency is only a local metric. The labor metric that is important is the total labor compared with the total production.

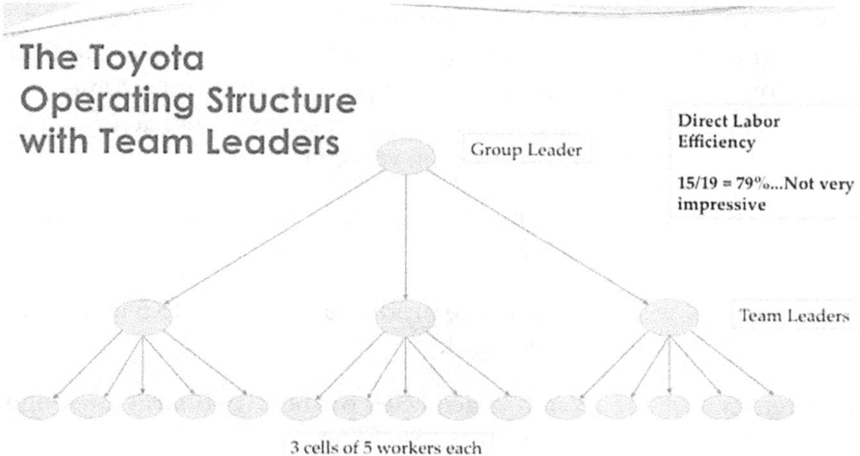

Figure 7.2 The Toyota operating sturcture

2. Second, while most managers will readily admit that the only people making the products flow are the workers with their hands on the tools, most still act as if once these workers are initially trained and the systems are installed, from then on, neither needs much support. This too is a huge flaw if you wish to be a good supplier of any product. All systems need maintenance and support.

The purpose of all indirect labor, including the C-suite, is to make the direct labor more effective and more efficient.

To drive toward maximum effectiveness and efficiency of the workforce, the employees need lots of support including responsive problem solving. To this end, the Toyota structure is almost always the proper choice.

Let me back that up with some numbers. I have analyzed hundreds of cells, production lines, shops, and control rooms. When we measure the direct labor efficiency in repetitive manufacturing, I find:

- Among those without the team leader concept, typical labor efficiencies range from 20 to 50%. The very best cells I have seen were 70% efficient.

- Among those cells with the team leader concept, typical labor efficiencies are in the 65–92% range, the very best being in the low 90s, such as Honda's engine plant in Marysville, Ohio, which produces engines year after year at 92% efficiencies. This means for each 15-person workforce:
    - The typical structure produces 15 * 0.7 = 10.5 man-days of value-added work.
    - The Toyota structure produces 15 * 0.90 = 13.5 man-days of value-added work.

Let's reality-test the overall efficiencies of these two structures and compare total cell labor with total cell efficiency:

Structure with no team leader

- Superficial direct to indirect labor efficiency: 94%
- Value-added direct labor efficiency: 10.5/16 = 65.6%

Structure with a team leader

- Superficial direct to indirect labor efficiency: 79%
- Value-added direct labor efficiency: 13.5/19 = 71%

The multiplier effect

- You get 13.5/10.5 = 29% *more work done.*
- Compounded it is 39.6% *more efficient.*

That is lean thinking versus conventional thinking. Lean thinking focuses on the system optimum at the expense of the local optima, and lean thinking dictates you pay attention to, and support, the value-added activities and those who supply it. This is why Toyota and Honda continue to prosper while GM and Chrysler struggle and are periodically on corporate welfare.

Further supporting data can be found in the story of the Theta cell in Chapter 1. Recall that the cell was struggling to get to 80 headrests/person/shift. And yet with a well-supported lean effort it improved to over 300 headrests/person/shift. If we say 300 was world class, then this cell was initially producing at less than 27% productivity compared with standard. What is even worse is that if this cell had been producing at 150 headrests/person/shift, it would have been profitable, as you recall. And then we never would have worked on it!

There is just no question that the team leader concept, or one like it that supports the direct labor worker and promotes problem solving, implementation of standard work, training, and continuous improvement, is the model to use if you wish to become a world-class performer. A colleague of mine once said, "Many non-lean companies have team leaders, but no lean company doesn't have team leaders."

So what about the typical refinery operating structure?

The Toyota model with team leaders is the operating structure used in most petroleum refineries today. The group leader, the supervisor, is normally titled a shift supervisor and is the first salaried level of management. The head operator fills the role of the team leader, an hourly, union position. So the good news is that this structure already exists.

This is a massive advantage for a refinery.

However, the bad news is that the head operator is grossly underutilized. Like the team leaders, the head operators normally are fully qualified to perform all the jobs of their workers as well as act as emergency fill-ins. In addition, they are often involved in training. There the similarities end. I see them doing scarce little problem solving, very little creating and monitoring of standard work, and no continuous improvement activities at all. And most of their routine time is spent on paperwork, of which there is plenty.

And to put this structure in place from scratch is a huge undertaking. First, you need to create the new structure and write the job descriptions defining the new roles. Next, you need to discuss all this with the union and modify the contract as appropriate. Following this is training the new team leaders as well as training the supervisors on their new roles. And then you must change all the documentation related to these new work roles including work standards and company policies.

Finally, the most difficult part, each person involved must change his or her behavior consistent with the new roles; this can be a large emotional change for several groups. For the rank-and-file workers, they uniformly like this, as they now have more support. The team leaders normally like the change, as they appreciate the new training and generally like the growth opportunity. The problems at the supervisory level are the largest and generally twofold. The supervisors frequently have trouble transitioning from being just a "work director" to being a full supervisor that not only plans the work but completes all five of the supervisory tools (see Chapter 1 and Chapter 8). When you make this change, the supervisors are the ones needing the greatest attention and the greatest support.

For example, in a recent engagement with a 150-employee facility, we have worked for almost one year to implement the team leader concept and are only 30% complete.

Simply put, having this structure already in place is more than a significant advantage; it is a huge and compelling advantage over a TOAS toward implementing a culture that is based on real-time problem solving using a fully engaged workforce.

I cannot say that loud enough nor often enough.

3. *Refineries, compared with TOASs, have a greater stock of problem solvers*

Refineries, as a group, have much more support than the typical TOAS. As you may recall, the refinery of 400 workers has around 280 support people for the production workforce of 120, around 2.3 support staff per production worker. While in our TOAS with a total employment of 200, there are 40 support people for around 160 production workers, or 0.25 support people per production worker.

Some of this is due to the nature of the work, but my observations are that they are grossly understaffed. While they have 40 support personnel, 20 are in the office providing HR and sales support and doing ordering and invoicing. There are 10 to 15 who are managers, superintendents, and supervisors. In addition, there is normally a staff of 5 people in quality, most of them doing redundant inspections and filling out paperwork. Most TOASs have a small group they call process engineers, but most of their time is spent on documentation and very little on problem solving and process improvements.

That is, unless they are a lean manufacturer. The profile of this same TOAS referenced above, if it were a lean manufacturer, would be more like 60 support staff and 80 production employees, for a total staff of 140 to get the same or even more work done. Again, refer to Chapter 1. You will recall that we added support staff and industrial engineering support. But even including that effort, the net effect was a huge productivity improvement where we increased capacity while we simultaneously reduced the number of workers required.

I have worked in not only the downstream and upstream oil business, but plastic injection molding, aluminum die casting, all kinds of foundries, electronics, electrical, lighting, healthcare, education, steel manufacturing, painting, and virtually every aspect of automobile manufacturing you can think of. Yet I have never seen as much problem solving talent in the form of engineers as I have seen in the oil business, especially the refineries. They have design engineers; they have process engineers; they have maintenance engineers; they have reliability engineers; they have materials engineers; they have electrical, electronic, and instrumentation engineers; and they have some very specific ones such as welding, structural, corrosion, and metallurgical engineers. They have lots of talented and seasoned engineers, and these engineers are the source of much of the problem solving in a refinery. That is the good news. The refineries are well structured and staffed for problem solving.

Unfortunately, the problem-solving training, guidance, and support the engineers are given is not sufficient. For example, on a recent engagement at a refinery, I was asked to assist in problem solving a situation that had persisted for some time. This group had worked days and even some nights, spending no less than 80 man-days and over 3 months on this problem, with no resolution in sight.

We convened the key personnel in a conference room and briefly introduced them to the Kepner-Tregoe (KT) methodology (see Chapter 9). Then we began the task of finding the root cause of this problem. First, we made a clear statement of the problem. After doing this, the problem had a much clearer focus, and it became apparent that 80% of the data they had so laboriously gathered and charted was superfluous. Next, we prepared the KT specification (this is a unique KT process consisting of a series of structured questions that detail various aspects that will help define what the problem is, as well as what the problem is not). By early afternoon we had found a root cause, and it was confirmed by all the data, past and present. Luckily, there was a way to quickly check the hypothesis, and that afternoon the members of the team altered the process consistent with their solutions to see if the problem would disappear. It did. We then intentionally changed the process back to the original conditions, expecting to see the problem reappear, and it did. Next, we made the problem disappear once again, confirming we could control it. Finally, we set up an operating plan to avoid the problem forever.

To be fair, we needed the data they had already collected, but the data amounted to one or maybe two days of work by one engineer to gather and present. Once the problem statement was written to Kepner-Tregoe standards, the KT specification was completed, and this was done with the assistance of a trained facilitator; the solution was straightforward.

Even with this weakness in problem-solving efficiency, the problem-solving support in the refinery is head and shoulders superior to the support in every TOAS I have ever worked with or for, as the refinery has both a good structure and a good staff. The typical TOAS works with such downward pressure on costs, it is normally "problem solving poor" and needs to hire several engineers (or problem solvers) to successfully transform the facility to a lean facility.

In retrospect, all those firms that transitioned to a lean facility will quickly tell you that the help they hired, typically industrial engineers, were the keystone that made it work.

And since "to problem solve your way to the ideal state" is a key element of a lean transformation, refineries with their strong staffs are far better positioned to affect a lean transformation than the typical firms in the assembly business. It is a significant advantage.

4. The entire foundational issue of quality is built on much better data

When Ohno embarked on his efforts to implement the Toyota Production System, he said: "After World War II, our main concern was how to produce high quality goods and we helped the cooperating firms in this area. After 1955, however, the question became how to make the exact quantity needed."

You see, by 1955, Toyota now had a very good quality system. In fact, Ohno's quality system back in 1955 is superior to that of most of the TOASs even today. He wanted to get into *quantity* control, what most people at the time called just-in-time (JIT) manufacturing. Consequently, his writings are almost devoid of the topic of quality.

Unfortunately, the refineries as well as the TOASs all have major deficiencies in their quality systems until they put in the effort to make them world class. And honestly, although I have seen many of these companies work at improving quality, I have not seen one be successful unless the undertaking was part of a lean effort.

Most refineries count on inspect, test, sort, and rework to get quality. And then they normally still have huge opportunities in product giveaway. This is much like most of the TOASs; they too rely on testing, sorting, and reworking to achieve product quality. The thought of doing it right the first time has not become a compelling driving force in either industry.

For a snapshot of the differences between quantity and quality control, see The House of Lean, in Chapter 8.

The TOASs have a documented quality system

Many TOASs have created a model of a good quality system and documented it in their quality manual. This quality manual is readily available and contains many of the lean tools. So the TOASs have a proven model to follow, if they so wish. That is the problem. Some do; most do not. Many TOASs still inspect and test and sort and rework like the systems that were found woefully deficient as early as the 1970s. Most TOASs talk a much better game than they play when it comes to quality. Most often the quality manual and its associated documents are not used much, if at all. That is, until it is time for the annual audit. Then the entire facility goes into frenetic activity to update the things they

THE INHERENT ADVANTAGE OF VARIABLES DATA OVER ATTRIBUTE DATA

Attribute data are often called "go–no-go" data or "accept-reject" data, as they have very few options. Variables data are measured data. Viscosity, weight, true boiling point, length, and voltage are all examples. Variables data come from a measurement scale that is infinitely divisible. For example, something can be 1 meter in length, or if you use a tool with greater discrimination, it may be 1.01 meters, or measured on a coordinate measuring machine it is 1.087359 meters. Attribute data give you very little information, only that you are in or out of spec. Variables data tell you that, plus exactly where you are relative to the specification limits. This allows for superior problem solving. Systems using attribute data struggle to get to quality levels of even 10,000 ppm. With variables data you can attain levels of 10 ppm and lower.

are not really doing and completing this update before the auditor arrives.

In the world of refining, each refinery does its own thing, each gets its own benefits, and the only proven and standardized quality system I know of is the quality manual that is used for government-grade products, such as jet fuels.

The TOASs have some problems not common in a refinery

When you review the typical problems encountered by a TOAS and compare them with problems in a refinery, two items stand out.

1. First, in a TOAS there are many customer returns; not so in a refinery.

2. Second, most of these products are not returned because they fail to meet the product's functionality; rather, they fail for cosmetic reasons. Take a TOAS making seats, for example. It will get rejects for "swirls" in the leather, "flash" on the plastic parts, "nonuniformity" on the stitching, and "loose threads" on the headrests. The common theme in these defects is that they failed a visual inspection. The "test" was not performed by a designed test apparatus that measured the quality characteristic; rather, the "test apparatus" was a set of eyes of an inspector, a human inspector.

   For these visual inspections, even if good operational definitions could be made to define the defects, using visual inspection by humans is the one surefire way to increase variability and reduce delivered quality.

   And with these visual inspections, when you get the response of the inspector, it is either a "go," that is, it is good to ship, or a "no-go," which means it is rejected and must be scrapped or reworked.

   These go–no-go data are called attribute data and give you very little specific information from the test. When the best data you have are attribute data, you are severely limited in your ability to improve quality.

Refineries, by comparison, do almost no visual inspections by humans in product release characteristics. The refineries use expensive, well-maintained, and well-calibrated machines that can tell you if the product is good or bad, and since the machines' output is numeric, the output also shows you exactly where the test response of this product lies on the spectrum of the specification. These data are variables data and give you a wealth of more information so you can effectively problem solve and improve.

While TOASs are stuck using visual inspection by humans to evaluate many quality specifications affecting product release, this creates an order of magnitude more variation that is largely unavoidable. Worse yet, when you get the results, they are in the form of attribute data and are not very useful for improving product quality. Refineries, on the other hand, use much better test apparatus with far less variation and they get variables data as the output.

Less variation in the test data, along with the use of variables data as output, affords a refinery a significant advantage as the people there work to problem solve their way to the ideal state.

The five, not-so-different or even "red herring" issues

There are five other topics that make for good discussion. But there really is little difference in how these topics apply to a lean transformation in a TOAS or a refinery. They are just things that need to get done.

1. But refineries already have flow

The refining industry had created a continuous flowing process almost 100 years ago, when companies converted from batch stills to fractional distillation crude units. And when it comes to the concept of flow, for a long, long time, the refineries completely outperformed the assembly operations in this concept.

It is no stretch to say that the major gains in lean manufacturing, among the TOASs and others have come about by improving flow. It was not until recently that the discrete parts and assembly industry found a way to measure flow, calling it lead time. Lower lead times meant faster and improved flow. It could be processing lead time for a single operation or for the entire assembly. Now it is commonplace to talk about manufacturing lead time or order-to-delivery lead times. Lead time as a concept was popularized, along with the technique of value stream mapping, in Rother and Shook's book *Learning to See*.

Although no one thought of it as such, since the time that businesses kept ledgers, there existed a way to calculate flow, called inventory turns. This calculation is simply the volume of that product divided by the daily rate of shipping converted to cycles per year. For example, if you have 500,000 barrels of gasoline storage capacity and ship 20,000 b/d of gasoline, on average you will turn over your gasoline inventory every 25 days, or have 365/25 = 14.6 inventory turns per year. This would be considered good in the business of refining.

However, firms that have used lean techniques now can operate at 50 to 400 inventory turns per year; that is, they have dramatically improved flow. These lean firms have reduced inventory by an order of magnitude with the resultant cost and space savings. It is no exaggeration to say two things:

1. The clear majority of the early gains in a lean transformation at a TOAS have been achieved by implementing flow (the JIT pillar; see Chapter 11). These systems that "flow" are often referred to as "pull production systems flowing at takt." When lean manufacturing first arrived in the US in the 1970s, it was then called the JIT manufacturing system, and the basic emphasis was on inventory reductions achieved through flow.

2. Refineries implemented the basics of flow around 1920 and achieved many of these gains long, long ago.

The bottom line is that this is neither an advantage nor a disadvantage toward the implementation of lean in a refinery. Since refineries had a 50-year head start on implementing flow, you will find that many of the huge gains you see advertised by the TOASs will not be the same gains we will achieve in a refinery transformation. The upshot of this is that the early emphasis of the lean transformation in a refinery will be on the foundational issues, the quality issues, rather than achieving flow.

This is certainly not a disadvantage to a refinery as it embarks on a lean transformation, nor is it a significant advantage. A refinery will just be at a "different starting position" on this issue.

2. Refineries have much more complicated and delicate equipment than a typical tier one

I have heard this stated as an impediment to a lean transformation. While the refinery has a lot more complicated equipment, I am not so sure about the delicate nature; both seem to have their fair share of analyzers and digital control systems. While the statement, even if largely true, is used as an argument to not implement lean, it is irrelevant. Though refinery equipment today may be more complex and larger, it is both well designed and well supported with trained personnel.

3. *Safety and environmental issues are several orders of magnitude larger*

Again, like the equipment complexity discussion, this is largely true—and it is just as irrelevant for the same reasons.

4. *There is too much changing and too much variety in a refinery to implement lean*

I have heard this with the point that product slate in a refinery is much larger, making it a lousy choice for a lean implementation. This is a statement made by someone who has never worked for a TOAS. My TOAS clients typically have around 300 individual products (part numbers) they manufacture. While normally around 100 of these products are 90% of their shipped volume, their product variety almost always exceeds the variety in a refinery.

As for "too much changing in a refinery," I find that to be an issue—an issue to implement, not avoid, lean as an operating philosophy.

But what kind of change is happening? For example, with a recent client that makes a complicated engine controller for a major automobile company, the product

INVENTORY AS AN ASSET—A PARADIGM THAT MUST CHANGE

There was, and still remains, a problem with equating flow with inventory turns. Since inventory shows up on the sales side of the ledger, extra inventory creates "paper profits" that make businesses and managers look better—in the short term. Hence there was little downward pressure to reduce inventories. Then when the TPS was becoming popularized in the 1970s, first by APICS (American Productivity and Inventory Control Society), inventory reductions as a profit-improvement strategy first started to get traction. For years, had we been so inclined, we could have quantified flow by inventory turns, which is commonplace today. More inventory turns means shorter lead times, which in turn means better flow. Better flow, through lower inventories, is still not the standard practice in many plants today, as it has a negative "book effect" short-term profits.

would routinely go through over 60 engineering changes per year, all requiring a change in the drawings and the bill of material as well as the process and processing equipment. This occurred as the customer and the suppliers would all be making improvements. Meanwhile, in a refinery, the specs on gasoline, diesel, and jet fuel, have changed very little in the last 10 years. These externally created changes are minor in a refinery, compared with the externally created changes encountered by a TOAS.

On the other hand, refinery processes do change rather dramatically from one day to the next. Unfortunately, most of this is caused by equipment and process reliability issues along with process instability and lack of standardization. It is very common for most refinery processes to be statistically unstable in both flow and quality, creating huge problems from day to day, shift to shift, and even hour to hour (see Chapter 10). However, these are internally created variations and are correctable.

This is *exactly* the area where a lean transformation would assist a refinery as it changed to be a better moneymaking machine, a safer and more secure workplace, and the supplier of choice. Process stability is a foundational issue in a lean transition. Or simply put, internal variation in a refinery makes it a target-rich environment for a lean transformation.

I find this issue of process variation to be an issue with *every* client I have had in my 27 years of consulting. This is nothing unique to any field of manufacturing.

5. We can't relocate equipment for lean savings, like the TOASs

There is a lot in the literature about huge savings made in the TOAS by relocating process equipment and adjusting flow paths to become more lean. This is true enough, and the wastes most often reduced are related to the movement of people and the transportation of materials. For example, in the story of the Theta cell, this accounted for a huge productivity improvement and a significant portion of the financial gains.

In the operating portion of a refinery, the equipment is largely what the lean world calls "monuments." These large, inflexible, unmovable pieces of equipment dominate a refinery. There are not large gains to be made in labor productivity, as operator staffing is so light, and so moving the equipment is unlikely to be an early focus. In addition, these monuments are connected by piping that dictates the transportation distances, this waste is not likely to be a focus either. So it is true enough—but only as it applies to the primary value streams in a refinery.

Let's look at the maintenance effort in the shops for an example. I have redesigned a few machine shops doing high-variety work, and by analyzing the movement of people and the location of equipment, we have improved labor productivity from the low 50s to the high 70s. My data on maintenance labor productivity in a refinery show that there are huge improvement opportunities here. I have seen maintenance field labor productivity numbers as low as 28% and have seen no field numbers higher than 48%, unless it was a controlled major shutdown.

Or what about all the work done by the staff? We often call this "transactional lean"—transactions such as billings or oils accounting. What improvements are available by changing the flow of the paperwork and relocating both the people and the equipment? Again, we have done this and have been able to improve labor productivity by 30 to 50%, well into the 80s. These are huge gains to be achieved in a refinery, which do not really differ from the gains to be achieved in a TOAS.

The "two albatrosses"

1. The operating schedule

An oil refinery is a 24/7 operation, whereas a TOAS typically runs 2 or 3 shifts, 5 or sometimes 6 days a week. The Toyota model, and what is used by most auto manufacturers today, consists of 2 shifts of 10 hours each, either 5 or 6 days a week. This is intentional. Since it is less capital intensive, running less than 24/7 makes financial sense. It also makes operations much easier. A key factor in lean is "flow." The concept is to keep all the products flowing during the production cycle. This allows for extremely well planned and predictable operations. It is such an important concept that Taiichi Ohno, the primary architect of the TPS, said in his book *Toyota Production System: Beyond Large-Scale Production* that "flow is the basic condition."

To maintain flow in a TOAS, you cannot stop for maintenance problems, lack of materials, or lack of people; so the system must be very robust. However, all machines will fail, people get sick, and there will be occasional stock-outs. Toyota, like many others, adopted a concept called double shifting. Simply put, at full capacity the company's plants run two 10-hour shifts, 6 days per week. In the two 2-hour windows each day, as well as the full day off, the workers complete productive maintenance and kaizen improvements, thereby avoiding equipment shutdowns. In addition, if on any shift they do not meet the production goal, they can work into these "overtime windows" to catch up. These unscheduled time blocks represent "spare capacity."

VARIATION REDUCTION

When Toyota implemented its all-out attack on inventory, it first addressed the underlying variation that made the inventory necessary. There is no hyperbole in saying that this effort to reduce variation is the genius behind process control in the TPS.

This dynamic changes rather dramatically in a refinery environment. In a refinery, while running 24/7, there is no time to catch up. There is no "spare capacity." There are no 2-hour periods between shifts, and there are no days off. If you fall behind, there is scarce little chance to catch up. In terms of uptime variation in a refinery, any variation represents a loss that is unrecoverable. Hence, firefighting, as a mode to get back to normality, is literally a way of life.

The effect of running 24/7 can be explained and quantified by the Kingman equation*:

$$Q = VUT$$

A thorough discussion of this equation is beyond the scope of this book, but I will give you a quick analysis here:

- Q stands for the inventory volume (queue) in front of the process in question.
- V is a factor representative of the variation in the process.
- U is the key variable here; it is a utilization factor for the process in question. Mathematically it is calculated as $u/(1-u)$, where little u is the mathematical ratio of the actual uptime compared with the scheduled uptime.

*The Kingman eqa tion and its implications are well explained in *Facto y Ph sics,* Wallace Hopp and Mark Spearman.

- T is the cycle time of the process in question.

In a business with a fixed or maximum inventory (Q) and a fixed cycle time (T), as the factor U increases, the factor V must decrease. In simple terms, to maintain a fixed inventory, if the utilization factor increases, there must be an equal reduction in the variation in the system. Or conversely, if all other factors in a system remain constant, if the variation is reduced, the inventory can be reduced.

Make no mistake about it—this is the primary driving force in inventory reduction: variation reduction.

For a refinery that is operating at 95% utilization, the factor for U is [0.95/(1 – 0.95)], or 19. For a TOAS operating at 20 hours per day, 5 days per week, the utilization is 100/168, or 0.595, and the utilization factor is 1.47. Mathematically this means that to operate at the same level of inventory protection, a refinery needs a huge reduction in variation, roughly 19/1.46, or 13-fold less variation compared with that of a TOAS. In the battle of variation reduction, refineries have not kept pace with the TOASs, and that explains why they have not improved their inventory turns as have the TOASs that transitioned to lean. These TOASs have focused on variation reduction as part of their operations philosophy for quite some time. (See the Appendix for Chapter 4.) It shows how the TOASs, using lean manufacturing techniques, have driven inventory turns up while simultaneously improving on quality.

Practically, this means three things. First there is a large opportunity to reduce inventories in a refinery. All inventory is a response to system variation, and variation can be reduced. Second and unfortunately, there are some factors that often control inventories other than typical process variations of flow and quality. These include some huge variations introduced and created by product leaving via large pipeline batches and by crude shipments coming in large batches as well. Third, in a refinery there is also a large amount of WIP (work in process), and much of it is stored in what is called intermediate tankage. In a typical refinery, crude and final products represent as much as 75% of all tankage by volume. However, just by count, the intermediate storage represents normally half the number of tanks, presenting a huge opportunity to improve storage flexibility and utilization.

The 24/7 operating schedule in a refinery is much more sensitive and hence far more difficult to run than a typical double-shifted TOAS.

While refineries are run at virtually 100% capacity, the typical TOAS runs at 71% maximum (120 hours scheduled/168 hours per week). This creates a significant disadvantage for a refinery compared with a typical TOAS. The TOASs are far more flexible to operate with much more "catch-up" capacity.

Fortunately, the approach of a lean transformation is designed to improve operational availability and capacity utilization, so the refinery using this approach is already prepared to solve this key vulnerability in a refinery.

More importantly, the refineries that choose the lean path will be much better positioned to compete with their peers. And for those refineries that undertake a lean transformation, they will learn how to quantitatively decide, and that means financially decide, if those last few thousand barrels per day of crude they plan to run are worth the risk. With a better understanding brought about using the statistical tool and other foundational issues, they will be better positioned to "push the envelope" while they reduce the risk.

2. *An innovation mindset*

There is a large cultural problem that would hinder a lean transformation in a refinery. The mindset in a refinery is not a kaizen mindset. Rather, the improvement paradigm is to rely very heavily on innovation for process improvement and problem resolution. Worse yet, nearly all this innovation is done by process and design engineers.

The lean model involves the use of innovation done in kaizen projects as well as small kaizens done by everyone in the facility. The concept of using kaizens—small projects done by everyone—to pick away at large problems is a cultural "no-no" in a refinery. It is seldom done.

There is nothing wrong with innovation, and innovations should be exploited; there is no conflict there. And there is nothing wrong with using engineers to drive these projects. The problem occurs when innovation is the *only* approach and engineers are the *only* ones driving them.

Let me give you an example. In one refinery, with a lot of waxy but cheap crude, the staff had a problem in meeting the diesel pour point and wax cloud specification during the winter diesel season. The problem was largely in the crude with its highly paraffinic nature. We proposed a series of small jobs to improve product testing, improve process controls, and improve blending algorithms. The gains would be in the range of $2.5–$4 million a month for the five "winter diesel" months and could easily be done in time for the next winter diesel season. However, on the books was a large capital project to replace the existing catalyst in the diesel hydrotreater with one that would crack the paraffins much better. It was expected to gain over $6 million a month, when installed. Since this new catalyst project was on the books, this refinery wasn't even interested in the smaller projects... not in the slightest. The problem with this was that there were two large opportunities missed. First, even with the new catalyst, there were gains to be made in product testing, improved process controls, and improved blending algorithms. These were kaizens that could stand on their own merits. Second, even before the necessary process and equipment changes could be made and before the new catalyst would be installed on the next major shutdown, there was going to be one and maybe even two more seasons of winter diesel. So looking at the worst case, there was $12.5 million to be made by doing these "smaller projects." Yet no one was interested. Once the new catalyst project was approved, the other projects were categorically scrubbed from the to-do list.

The paradigm is that these smaller kaizens are not worth their time; if it's not large, it's not worth doing. This is typical refinery thinking. It is also typical, as I have mentioned many times, that process changes are largely an engineering-only event, based on innovation and design changes, with little involvement by the refinery operator. Consequently, changing this mindset will be a hard one. Recall the means to lean:

- Problem solving to reach the ideal state
- Through the total elimination of waste
- Using a fully engaged workforce

This is a long way from:

- Using innovation alone
- To fix the big concerns
- Using engineers

This will be a significant paradigm to change and a significant cultural hurdle to overcome in a refinery. TOASs are much more flexible.

And the early "must do," whether a TOAS or a refinery

Getting the engagement from the unionized workers

This issue is not substantially different when comparing the refinery and it specific issues with the TOAS and its specific issues. Whether I talk to the TOAS bosses or the managers in a refinery, they each articulate the same thing:

- "These issues are overwhelming, and I hate to even think about solving them."
- "In this facility, these issues will preclude the ability to become lean."

I completely disagree. In some cases, it will be a lot of hard work and may not be accomplished overnight. But both issues are very doable, and both are crucial to success.

My experiences at Chevron

My interactions with a union have been mixed. Early in my career, I was taught that the union was a bunch of bad guys only trying to get more benefits and better pay while working less. As I thought about it later my overwhelming thought was how rational that sounded. I did not find them to be a bunch of bad guys. They were just like me; they wanted to come to work, contribute to the betterment of the company, and support their family. The biggest difference I could point to was that we dressed a little differently at work.

During my career with Chevron, I got some excellent training in dealing with unions. I had the enlightening experience of being mentored in union interactions by the master of it at Chevron, my refinery manager, Jim Keating. We had many and long discussions about interacting with the union, and Jim taught me that the best union was one that was strong, united, and representative. That is, the union's leadership should represent the membership and not be pushing its own personal agenda. Not only were Jim's teachings spot on, but they were contrary to my earlier teachings at other locations where the prevailing hope was that the unions were weak, poorly represented, and dumb, so that we could bowl them over and get what we wanted.

In addition to my one-on-one training from Jim, I attended several seminars. The most notable one was taught by Jules Justin, the absolute authority on labor-union interactions. Following his seminar, I both read and studied his two-volume set, *How to Manage with a Union*. It is a classic, and I recommend it to all who deal with unions.

While at El Paso, I was deeply involved in union activities. These included negotiating the basic work contract, meeting with the Workmen's Committee every month, arbitrating grievances, as well as having a good bit of daily interactions. For three years, I taught a class in our First Line Supervisor's School, "How to Handle the Union Steward." I enjoyed it immensely.

I came to understand and respect what Jim had taught me about unions and even more about people in general. Once I got rid of the "we-they" attitude that I had been taught early in my career and treated the union members with respect, I found all my experiences were good. That did not mean that we always agreed on everything, nor did it mean that grievances ceased. It meant that we could meet and discuss issues openly, honestly, and respectfully. We then both could leave with a feeling that we could move forward. It also meant that we had confidence in each other that we could disagree, yet we could still work together and were willing to hear each other's issues in the future. However, I must admit that was my experience at El Paso and only at El Paso. I can say with certainty that we had a better relationship with the union in El Paso than any other Chevron refinery had with the union. And based on my interviews and working with others, I am sad to say that most union interactions with refineries, not just Chevron's, are more combative and contentious, with low levels of both trust and progress, than they are cooperative and friendly.

My union experiences as a lean consultant

My experiences at dealing with a union in a lean transformation are substantial, and all have been very positive. At a recent engagement in a tier two supplier I was advised by the COO to "be very careful; they are the most belligerent in the company and will fight you all the way."

I dutifully listened and then just ignored it; I had had so much positive experience, I was not worried. After a period to see if we were sincere in our efforts, and within a few months, the union was totally engaged. So much so that the chief steward of the union was filming workstations so we could do time studies. In addition, the other stewards would also train the newly hired workers on the lean indoctrination materials, taking some load off the supervisors. And in many ways they were more engaged than the company supervisors.

The lean transformation at Aera and its interactions

Earlier I told you the story of Aera. In my interviews with Gene, I asked him about his experiences with the unions. His reply was:

"We had two. They were very supportive. Communication is the key. I spent a lot of time with them as I did with everyone. It is key to building alignment. We needed all to believe that a strong company was the key to our success and survival. There is no we-they; it is *us*. I was transparent with the unions, as I was with all. They knew the numbers, etc. It was part of our value statement that we would treat all with respect and be accountable to each other. I spent a lot of time on this and did not take it for granted."

Note his opening comment, "They," indicating both unions, "were very supportive." No mention of gaming, no mention of a combative, resistive attitude: just short and simple: "They were very supportive."

He then goes on to elaborate on how he got that support. He emphasized open, honest communications; common objectives; an attitude of inclusiveness; transparent operations; values such as respect and accountability; but most of all, lots of sincere time and effort devoted to being inclusive by him, the CEO. A very effective formula for success in any culture-changing effort. Remember it!

What's the bottom line with the union issue?

First, the need to get the hourly worker engaged is paramount. It is so significant, it is often the litmus test to making a lean transformation. Think about a refinery with 400 employees. There are typically 40 or so, including engineers, supervisors, and managers who are working toward process improvements. And they are working primarily 8 to 5, 5 days a week. The operators are there 24/7, and there are 120 of them. Just how powerful could it be to get them involved in problem solving and continuous improvement as you saw being done in the Theta cell? Imagine increasing the number of problem solvers from 40 to 160. Think how large the bottom-line impact could be. It is staggering!

In my personal experience in lean transformations, I have not found the unions to be an impediment; rather, like Gene's comment above, "They were very supportive." For example, in my experiences at El Paso, before I started my consultancy, we did a lot of cross-functional problem solving. The union workers on the teams dove in headfirst and were totally and almost instantly engaged. I believe it was because of the way we ran the problem-solving groups. We set a common goal and were open, honest, and transparent and treated everyone respectfully and as equals on these teams.

My overwhelming experience was that the unionized team members provided no resistance. Rather, they welcomed the new kind of work and responded well to being given the responsibility and the opportunity to perform. They openly remarked that it was refreshing to be involved, to be counted upon, to be respected, and to be both treated and recognized as contributing members. They openly remarked they wished we could have more of it.

The union will support it . . . but it may not come easy

My interviews with refinery personnel, those that understood the degree of engagement required by the union-represented employees, said, "This will be an issue."

I agree fully. Although I firmly believe that getting the union's engagement in a refinery is very doable, it will not come without significant effort.

Since we are looking for more employee involvement, something employees like and seek, once you convince the union you are sincere, I expect the union to be receptive. To create the trust needed so that the necessary engagement will occur, there are five issues that will need to be exhibited by the refinery management and the union management:

1. Is the refinery management willing to take the position to be supportive, with open, honest communications, and treat the union employees as equals—no more "we-they," only "us" with all union employees?

2. Is the union management willing to do the same with all company employees?

3. Will the union management and the company management be willing to give up their "individual pet issues" and bring forth the true issues that are representative of the workforce and the company?

4. Can both the union and the company overcome the past issues and the heavy emotional baggage that goes with those issues?

5. And finally, will both the union management and the refinery management be willing to commit, devoting lots of sincere time and effort to being inclusive?

So it is not a deal killer to try to get engagement in a union environment. Not at all. But if you have many years of contentious relationships, it will take a good deal of effort by both the company and the union *to decide* to make it work.

Both the union management and the company management know how to make all this work in their respective organizations; hence it is not an issue of skill. It is a simple choice. Do they or do they not want to make this work? If they do, there are a variety of ways to be successful; if they do not, nothing they do will break down the barriers and create the environment that is necessary for a fully engaged workforce to be a reality.

IS A REFINERY COMPARED WITH A TOAS A BETTER TARGET FOR A LEAN TRANSFORMATION—OR NOT?

The refinery has two major negative issues: the operating schedule and the concept of "innovation only by engineers."

The operating schedule, while it naturally places a huge emphasis on firefighting, puts a damper on problem solving. All too often, once the immediate symptoms disappear, the motivation to work harder to find the root cause is reduced. The plant is often left with a lot of countermeasures to deal with later. This is a cultural issue that will need to be addressed, and can be minimized.

"Innovation only by engineers" as a means to continuous improvement is another cultural hurdle to overcome. Although a formidable foe, this cultural paradigm will melt when everyone gets fully engaged in problem solving. You have seen the gains achieved in the Theta cell in Chapter 1, and several examples are provided in Chapter 12. Those two issues are the downside of a transformation in a refinery.

However, the upside is massive. First, the gains, particularly the financial gains, can be in the millions and can be brought in very early in the transformation. This has a powerful, positive effect that cannot be matched. Everyone likes success, and everyone wants to be associated with a winner; early gains are catalytic.

Second, having a support structure with 40 or more problem solvers with an engineering background is a substantial advantage to a facility that is working to problem solve its way to the ideal state. The typical TOAS may have one or two dedicated problem solvers.

Third, having 120 highly trained operators, in the proper structure, ready to be trained to be real-time problem solvers is an advantage of such proportions to dwarf all else. They are skilled, they are there when the problems appear, and if we train them and empower them to problem solve well, they are an asset of immense strength.

Finally, quality data available in a refinery are variables data coming from well-designed machines. They are orders of magnitude more useful than the attribute data that are a function of visual inspection by humans.

For the question "Which is a better target for a lean transformation?" presuming we have the same level of motivation and management commitment, a refinery is the clear choice, no question about it.

CHAPTER SUMMARY

While there are some differences between implementing a lean transformation in an oil refinery and a tier one auto supplier, none of these differences are so large as to preclude success, wherever it is attempted.

A lean transformation is the creation of a culture of continuous improvement and respect for people. In so doing, you will problem solve your way to the ideal state through the total elimination of waste using a totally engaged workforce. It makes no difference what industry you are working in; the approach is the same.

Refineries have two issues: the 24/7 schedule and the lack of a kaizen mindset with the full reliance on improvements being done only by engineers with an innovation mindset. That said, there are some distinct advantages that refineries have already built into their culture that will make a lean transformation easier. These include huge cash flows that make finding financial opportunities very easy and highly motivating, an operating structure that is already in place that is a necessary precursor to implementing real-time problem solving, a strong support system rich in engineering problem solvers, and a much superior quality data system.

Implementing a lean transformation, since it constitutes a large cultural change, is always a very effortful event. However, the advantages afforded by the typical refinery setting make it clear that it is easier to effect a lean transformation in a refinery than a TOAS.

PART 3 IMPLEMENTING LEAN IN THE REFINERY

CHAPTER 8

THE HOUSE OF LEAN

Aim of this chapter*—This chapter has a threefold aim. First, to introduce you to the House of Lean, shown in Figure 8.1. Second, to explain how the many elements of lean manufacturing will be applied slightly differently in a refinery environment. Third, to discuss, in more detail, the first of the foundational issues,* management. *The rest of the House of Lean will be discussed in Chapters 9, 10, and 11.*

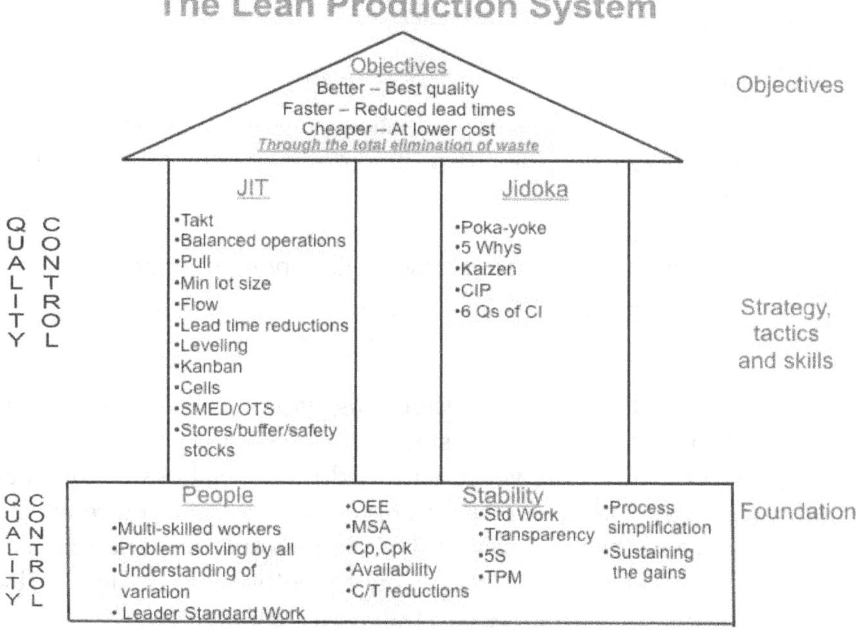

Figure 8.1 The House of Lean

THE HOUSE OF LEAN METAPHOR

Before we look at its building blocks, let's explore one very relevant aspect of the House of Lean.

This "house," compared with others, has a huge foundation. Why?

It is instructive to review some of the background of the Toyota Production System, especially what its architect, Taiichi Ohno, said, so that we can keep this in perspective. Recall the quote from his book *Toyota Production System: Beyond Large-Scale Production*. In it, Ohno says: "After World War II, our main concern was how to produce high-quality goods. After 1955, however, the question became how to make the exact quantity needed."

Consequently, what Ohno called the TPS was a production system that is a *quantity-control system,* based on a foundation of quality. Hence when you read Ohno's book or others on TPS, you will find a strong treatment of the "quantity control." Also, in the same book, Ohno wrote:

The basis of the Toyota production system is the absolute elimination of waste. The two pillars needed to support the system are:

- Just-in-time
- Autonomation, or automation with a human touch [this later was named jidoka]

So in Ohno's book there is no serious discussion of either quality or the foundational issues. He does state that by 1955 the company had a strong quality foundation; yet he simply does not elaborate much beyond that. However, when you read his books, you can find reference after reference to these topics. So it is obvious that these issues had been addressed, just much earlier. Consequently, what he does write about the TPS is built on these two pillars: JIT and jidoka.

This explains two issues:

1. It explains why so many people write about the TPS and yet do not really address the foundational issues. Possibly because Ohno did not specifically enumerate them, people may think they are somehow not needed. This is a mistake of galactic proportions. For all the time you spend on some of the sexy topics of lean such as takt, kanban, or SMED, you will need to spend 10-fold that amount of time on the foundational issues. The choice is really whether you wish to invest that time up front and give yourself a reasonable chance to be successful or whether you want to bypass those issues for now and then circle back to them later, when you find that your transformation is not proceeding as you had hoped. The choice is yours.
2. Simultaneously it explains why many people who try to implement a lean transformation using only the strategies of JIT and jidoka fail to come even close to Ohno's accomplishments. As an AlixPartner's survey found, "Only 2% of companies who responded to the survey have fully achieved their objectives and less than a quarter of all companies (24%) reported significant results."

Reread that last bulleted item. Failure to understand and aggressively act on this concept will guarantee a failure of a lean transformation. It won't hinder the transformation; it won't slow down the transformation—it will cast its failure in stone.

What does this mean to the rest of us is? We need to build that foundation of quality just as Ohno had already built it, long before he wrote his book. It is imperative to a successful transformation. That is exactly why the topics of the strong quality foundation, including enlightened management, stable processes, and people concerns are covered deeply throughout this book.

OBJECTIVES

The objectives of the lean manufacturing system are to learn how to make your products with ever-improving quality, ever-improving lead times, and ever-reducing costs.

And the means to lean will be your primary methodology, which is to:

- Problem solve your way to the ideal state
- Through the total elimination of waste
- Using a fully engaged workforce

In so doing you will create and sustain a culture of continuous improvement and respect for people.

Recall that we have more thoroughly discussed these objectives in Chapter 3, and you can see them in the top of the House of Lean.

FOUNDATIONAL ISSUES AND THE QUALITY CONTROL STRATEGY OF MANAGEMENT

The quality control foundation has three strategies. First is the concept of management; I call it enlightened management, which we will discuss in this chapter. The two other quality strategies are people and process stability; they will be discussed in Chapters 9 and 10, respectively. The quantity control strategy, tactics and skills, will also be discussed later, in Chapter 10.

Management

The success of any business effort, whether it is a simple capital project, a major plant turnaround, or a culture-changing process such as a lean transformation, is in the hands of the management team. If the management team performs well, you have an excellent chance of success. On the other hand, you will have major problems if the management team:

- Has a top-down autocratic style
- Is closemouthed and only shares information on a "need-to-know" basis
- Is adept on the computer, smooth in the business meeting and in the office, but seldom seen on the floor

- Feels it has "paid its dues" in terms of learning and now is comfortable and believes it has an excellent grasp on the technical and financial matters as well
- Feels that learning new techniques and strategies is a function of the "next level down"
- Spends little time or effort on subordinate development
- Feels that people need to be forced to work hard and can only really be motivated by fear
- Hardly knows even the names of anyone outside the office
- Rewards those who follow the company line and make the boss look good

Good management is foundational issue number one, and there is no substitute for it. I can think of no better litmus test for the success—or failure—of a lean transformation than the ability of the management team to perform skillfully, or not. And the first thing the management team must be able to do is to pass the management commitment tests (see the Appendix for Chapter 8). If the key facility managers do not score well on the management commitment tests, problems galore are right around the corner.

Leaders as champions: The Six Initial Skill Focus Areas

> **ON MANAGEMENT**
> I am not sure where I heard this quote, but I find it to be a significant truth: "It's all about the actions of management; the rest are just details."

The teaching method in a lean transformation is very different from the typical patterns found in most businesses. Rather than having large numbers attend classroom-type training, taught by a staff consultant, where "death by PowerPoint" is the training delivery method, lean training takes a dramatically different path. First, it is primarily delivered at the gemba and is one-on-one training from your direct supervisor. Second, rather than watching a boring slide presentation, you will be using real data, learning in real time as you solve real problems. Third, learning, rather than being an "event you attend," it will be a "process you continually work with," including follow up and reflecting on what you have learned. It is top down management driven, management delivered, just-in-time learning while doing. For some very specific training in which your supervisor may not be proficient, you likely will get training from either the lean support team or your sensei. The lean support team is discussed in Chapter 14.

There are Six Initial Skill Focus Areas that will be used starting at day one in the transformation. These skill areas are tools that everyone in the leadership will be using in their daily activities. To become competent at these skills, study and practice are in order.

The six areas will need individual champions from the lean leadership team (LLT) so the champions will need to become proficient in their area. They can rely on their sensei for advice and initial training and practice, but they must be reading about and actively practicing these skills in their daily activities. The six skill areas are shown in Table 8.1.

Table 8.1 The Six Initial Skill Focus Areas

Skill Area	Champion	Level of Competency*
Leadership	Refinery manager	Capable of teaching lean leadership at six months
HK planning	Refinery manager	Fully capable of teaching HK planning at six months
Leader standard work	LLT member	Fully capable of teaching to all at six months
Problem solving	LLT member	Can teach Six Qs at one month, Five Whys at three months, and KT problem solving at one year
Statistical tools	LLT member	Good understanding of data analysis and control charting in six months, other tools at one year
Meeting management and facilitation	LLT member	Fully capable to teach prior to the rollout

Note: All times are from the date of rollout.

The six champions will be the initial group of "in-house" experts for these very important skills. After 18 to 24 months, it may be appropriate to transfer the role of champions for these skills to members of the lean support team, but certainly not early in the transformation.

The crucial Initial Skill Focus Area of Learning: Leadership

In all the teaching and learning that goes on in a typical refinery, I seldom see leadership as a primary topic. This is a huge oversight. Hence leadership is the first of the Six Initial Skill Focus Areas of learning. It is the most critical, and as such, the champion will be the refinery manager. Leadership will be taught, measured, evaluated, and improved upon as a key cultural leading indicator of success.

Lean leadership by all

Leadership as distinguished from management—as a skill set

Your entire management team really needs two skill sets to be effective managers. Those are the skill set of management and a separate set of skills for leadership. The management skill set is:

- Planning and budgeting
- Staffing and organizing
- Controlling and problem solving

Through this trio of skills, the manager is expected to provide a predictable and orderly outcome of key business results such as "on time," "high quality," and "within budget."

On the other hand, the manager must also have strong leadership skills. These are the skills needed to guide the business when there are changes in the external or internal environment of the business. These skills are:

- Creating a vision and ensuring the required strategies to execute the vision are in place.
- Articulating the vision so the workforce can be aligned toward that vision.
- Acting on the vision at the exclusion of all else, thus providing motivation and inspiration to facilitate the change.

In our current and rapidly changing business environment, there is no doubt that we need more and better leadership, but that does not mean the skills of leadership will replace the skills of management. Quite the contrary—both are needed. Our businesses need to provide predictability for the business through good management yet navigate the changes needed to survive and prosper through good leadership.

Excellent management, by its very nature, is conservative; is methodically incremental; is filled with bureaucracy, reports, and follow-ups; and is focused on the short term. Thus the very best management simply cannot produce major change. Only with leadership does one get the boldness, the courage, the long-term vision, and the energy needed to activate the workforce and create the changes that are required to make a lean transformation.

John Kotter said it succinctly in his book *Leading Change*: "Management makes a system work. It helps you do what you know how to do. Leadership builds systems or transforms old ones."

What is "lean" leadership?

Lean leadership is the new style needed by management in a lean facility. It is most typified by what Douglas McGregor called Theory Y managers and by those who understand and practice the Fourteen Obligations of Management and avoid the Seven Deadly Diseases as taught by Dr. W. Edwards Deming. (For a critical and comparative analysis of lean leadership, Theory Y management, and Deming's Fourteen Obligations of Management, refer to our website, www.wilsonleanrefining.com)

The Six Skills of Lean Leadership, from my book *How to Implement Lean Manufacturing* (2nd ed.), are:

1. **Leaders as superior observers:** They go to the action—they call it the gemba—to observe not only the machines and the products but also to spend significant time with the employees. They strive to be aware of not only the products and the processes but more importantly the people. They also are in contact with their customers. A much-overlooked leadership skill they have in abundance is the ability to be an empathetic listener.

2. **Leaders as learners:** They do not assume they know it all. Rather, they go to the floor to learn. They are in "lifelong learning

mode." They are masters of the scientific method. They learn by observing and doing, but most importantly they are superior at asking questions. They learn by questioning.

3. **Leaders as change agents:** They plan, they articulate, sell their plans, and act on their plans. They are not risk averse, yet they are not cavalier. They do not like to, but are not afraid to make mistakes.

4. **Leaders as teachers:** They are lifelong teachers. When something goes wrong, their first thought is not "Who fouled up?" but "Why did it fail?" and "How can I use this as a teaching opportunity?" They teach using questioning rather than just instructing.

5. **Leaders as role models:** They walk the talk. They are lean competent. They know what to do and they know how to do all of lean that is specific to their current function in the organization. There is no substitute for this.

6. **Leaders as supporters:** They recognize they mainly get work done through others, so they have mastered the skills of "servant leadership."

Leadership "by all"

The concept of leadership "by all" is foreign to many managers, as they feel that leadership is a skill to be used primarily by the top management personnel. Quite the contrary, these are the skills that production managers use when they want to change their work procedures and make their people more productive. These are the skills that supervisors use when they want to change and improve the standard work or the skill levels of their workers. And they are the same skills that floor workers use when they want to change the workflow via a kaizen.

When I teach these skills at that level, I explain that to lead you need to:

- Have a plan
- Articulate the plan
- Act on the plan

This simple formula resonates with the production manager, the production supervisor, and the worker alike and at the same time shows how leadership can be exhibited by all.

Everyone can do this. Everyone needs to do this.

Cultural leadership

What is culture, and why is it important?

In *How to Implement Lean Manufacturing* (2nd ed.), I defined culture as "the combined actions, thoughts, values, beliefs, artifacts, and language of any group of people." In simple terms, culture is "just how we do things around here."

ON CULTURE
Culture trumps everything.

Culture is a powerful and dynamic force that guides the behavior of all the people in an organization. A quote attributed to Peter Drucker is "Culture eats strategy for breakfast;" while this is true, I find it to be an understatement.

Cultures can be healthy or not. The determining factor is, does the culture cause the business to prosper in the long and in the short term? Healthy cultures have three qualities: strength, flexibility, and appropriateness. We can investigate these qualities one by one.

Some cultures are strong; some are weak. In a strong culture when you are given a set of circumstances, the behavior of the people in the culture is highly predictable. Strong cultures have a significant advantage in that they dictate the acceptable modes of behavior and allow people to proceed with greater confidence and independence in their daily activities. Weak cultures have few redeeming values and create confusion and uncertainty and will always undermine the potential of any business. While weak cultures in a competitive environment almost surely guarantee business failure; cultures that are strong do not guarantee success—far from it. Strong cultures, by their very nature, are slow to change. Hence, in an intensely competitive and rapidly changing marketplace, this slowness to change will be a distinct liability. Unfortunately, the management of a strong culture all too often has a propensity to focus inwardly and not pay attention to the changing external business environment. A business with a very strong culture, especially one that has shown long-term success, can easily become inwardly focused, politicized, and rife with bureaucracy, making it even more difficult to change.

This points out the need for flexibility. As the business environment changes, the business culture needs to adapt. So how do businesses change, yet remain strong, so they can be successful? They need to be strong and consistent in both their values and their vision and yet flexible in "how" they execute that vision, using their values to guide their actions.

Finally, there is the topic of appropriateness of a culture. No one culture is best for all. The key cultural elements will be different from one business to the other. For example, a key cultural driver is leadership. Compare a football team with an oil refinery. On game day, the quarterback or the offensive coordinator is calling the plays. In so doing he is literally dictating the strategy. It is top-down, dictatorial leadership; it is "my way or the highway" leadership. It is anything but high-involvement management with a "let's get the input of those affected before we act" style of leadership. However, for a football team on game day, that is the correct style of leadership. On the other hand, is that the right style of leadership for your business? Is that what you want and need in a refinery? I think not. For a refinery trying to survive and prosper, it is best to be "situational leaders" with a high, a very high, dose of "high-involvement leadership."

SITUATIONAL LEADERSHIP

The basic principle of situational leadership is that the leader must assess the "maturity level" of each follower, relative to the "situation," and adjust his or her leadership style accordingly. This maturity level is a function of the "skill and the will" of the individual and therefore is not independent of the operating circumstances—it is "situational." Hersey developed a continuum to follow as leaders are able to teach and achieve higher levels of maturity in their followers. He enumerated four "styles" of leadership to use:

1. Directing and telling
2. Coaching and selling
3. Supporting and participating
4. Delegating

His book *The Situational Leader*, is a great read, and I recommend it to all.

Just what is cultural leadership?

Leadership is the art of creating a plan, articulating the plan so all can understand it, and acting on the plan at the exclusion of all else. Cultural leadership is evaluating, discovering, and implementing the type of culture your business needs. It is largely described in the triumvirate of the mission, vision, and values statements. Creating and advertising this triumvirate is the first half of the means to articulate the specific culture you wish to have.

The second part of articulating the culture is not a set of speeches, meetings, and PowerPoints. Rather, it is the behavior of the very top managers. Are they "walking the talk?" It is like the adage "Your actions speak so loudly, I do not need to hear what you have to say." It is "leaders as role models" and "leaders as teachers," two of the six elements of lean leadership.

So how can we change a culture?

There are three basic drivers of cultural makeup:

1. *The business environment itself.* For example, there are the very rapidly changing businesses in the field of technology. They will require one type of culture to survive and prosper. Then there is the very slow changing culture of education or refining. These are two very different business environments, and each environment will uniquely have a huge impact on the needed culture.

2. *The history of the business, the company, and this specific facility.* Nothing perpetuates a culture more than its history. If people can look around and see what worked yesterday and the day before and the day before that . . . then, they can do the same thing today. Literally they can see "just how we do things around here."

3. *The very few people at the very top, normally the C-suite.* Their values and their beliefs drive their actions, and once people see what the top folks do, they mimic that behavior. The behavior of the leaders, and what they will accept, sets the norm. So in terms of changing a culture, no single facility or company is likely to have an impact on the overall environment for that business (e.g., automobiles, hospitals, retail clothing, or fast food). It is hard for any single business to change the overall business environment; more likely the overall business environment will dictate that you change. History is history; it will not change either. So the only remaining driver of cultural creation in a business is the few top people, and their impact cannot be underestimated.

It is therefore up to the top leadership to consciously, continuously, and consistently manage the culture to assure it stays strong, flexible, and appropriate. There are few things that are more important to any business. And for the field of refining, where the products are commodities in a mature market, there is nothing more important. It requires:

- Clear plans made and articulated
- Consistent actions and "messages" from top management
- A means to reward the correct behaviors
- A means to routinely "check" on the cultural health
- Subsequent action plans to change when change is necessary and to not change when change is not necessary

Cultural evaluation and change is a topic too large to cover in this work; however, on our website, www.wilsonleanrefining.com, you will find more information including forms to evaluate your culture as well as an extensive reading list.

Hoshin kanri planning

Hoshin kanri (HK) planning became popular in Japan in the 1960s when it was noticed that it was frequently the policy creation and deployment model of those who had won the Deming Prize. From there it was popularized in Japan, and it was not until the late 1980s and early 1990s that books were published in the United States that described this powerful technique.

HK planning is an extremely powerful planning and policy deployment methodology. It is highly effective in addressing the six major problems that cause most planning systems to be less than effective in obtaining facility improvements. Those six major issues are:

1. *Poor focus.* Work is not properly prioritized, and hence the team may not be working on the most important activities to assure facility success.
2. *Poor horizontal or vertical alignment.* Groups are not cohesively working together to accomplish the goals that are important to the facility.

3. *A focus only on results.* The means to accomplish results are often overlooked.
4. *Insufficient levels of commitment.* Most planning systems are top down autocratic with no feedback systems to reflect "doability" and "buy-in."
5. *Lack of integration.* The goals are a set of "additional things" that need to be done.
6. *Inflexibility.* Once the annual goals are defined, they are "cast in stone."

The HK process consists of four major steps:

1. Vision
2. Policy development
3. Policy deployment
4. Policy control

There are a number of very good books on HK planning, but for the first few planning cycles, I recommend you stick to the basic 10-step HK model shown in Table 8.2. Ninety percent of the benefit comes from understanding and executing the 4 basic concepts and following the 10-step process here. More on that later.

Why you should not do full-blown HK planning—not just yet, anyway

I will be the first to tell you that full-blown HK planning is worth every penny of its effort and more. However, there are more basic things that need to be learned initially, and these basics will allow you to gain huge advances in your planning system today.

To acquaint the entire team with HK planning skills, I expect the senior management to read at least one book on HK planning in the first six months; the lean leadership team should select one from the reading list at www.wilsonleanrefining.com. My personal preference

Table 8.2 Hoshin kanri summarized: 10-step model

PDCA (Plan, Do, Check, Act)	Step	Activity	Purpose
Plan	1	Develop the mission, vision, company values, company motto, etc.	To define "who we are," how we are unique—our niche
	2	Develop key strategies (steps 3–8)	To survive and prosper in your niche. Should review what is changing in the external environment including the competition, customer needs, and regulatory changes; have key strategy "owners" to integrate laterally; use tools such as SWOT analysis
	3	Collect and analyze data	To evaluate how you compare with last year's results, with next year's demands; how you compare with competition; how you meet your customer's needs; and how you meet all external demands
	4	Plan targets and means	To determine the performance levels you need to reach and how you will measure them. Use the top-to-bottom method and "catchball" to integrate vertically and laterally. This integrates targets with the means. Catchball is also used in steps 5–8
	5	Set control items	To have specific measurables that will show progress or lack of progress
Do	6	Deploy the policy checks	To identify what is to be controlled and by whom, when to achieve results, and what is to be inspected by whom when evaluating the means. The policy checks are made hourly, daily, weekly, monthly, quarterly, etc.
	7	Deploy the control items	To identify what is to be monitored and checked, who is doing the monitoring and checking, and how frequently, essentially answering the planning question, "Who is to do what by when?" All control items need to have both normal results and abnormal results defined in quantitative terms
	8	Implement the policy	In addition to routine monitoring and evaluation, to develop projects, implement training, change existing practices, etc.
Check	9	Check results	To check annually, monthly, weekly, daily, hourly . . . all within the policy as it is deployed
Act	10	Evaluate status	To act on the "abnormal" results, to review the "normal" results for applicability and improvement opportunities, to determine what is the next step, and to return to the plan step if needed

is Bob King's book *Hoshin Planning: The Developmental Approach*. The classic is Yoji Akao's book *Hoshin Kanri: Policy Deployment for Successful TQM*. And there are many others.

However, the problem with these books is that they are written for people who already understand and have lean skills, many of which are common among Japanese managers. The books are filled with concepts like A3 documentation and project tracking, target deployment, flag method of controls, and complicated matrix analyses including quantified and complex "target-to-means" matrices, as well as the famous X matrix to simultaneously evaluate the critical key areas to the control points, to the control measures, and to the impacts. These tools, particularly the matrices, are very valuable and should be learned in the fullness of time. Unfortunately, they are complicated, and a good bit of background and practice in the basics of HK planning will make them much easier to use in the future.

Should you decide to do full-blown HK right from the start, either you will be unsuccessful from the start, or you will require a great deal of hands-on support from a consultant, which means you will be unsuccessful in the longer term because you will not learn. The way to learn HK planning is the "lean way" to learn anything; it is "learn-do-reflect." This means that in areas that are new to you, it is best to start small and get into action mode as fast as you can. Then "check" both the results and the process and "act" on what you have learned. Hence by repeated passes through the PDCA cycle, like this, you will "learn your way" to greater competency. And that is why you should start with the simplified version supplied here.

Does that mean you should do it cold turkey? Certainly not. Very likely you will need someone to assist you at the start. Your sensei should be able to fill the role of an HK facilitator. He should be there to "facilitate," and that means just that, "make it easier." Other than introduce, review, advise, and question, the facilitator should do little else. Under no circumstances should he have his hands on anything. If his fingerprints show up on your forms, he is not doing his job; if his fingerprints show up on any audits, he is not doing his job. He should introduce you to the materials that may be helpful to your program. He should then review your plans and review what you are doing and advise you if you are off course. And if you are straying, then he needs to question you so you can learn your way to get back on course. This is the role of the facilitator.

The impact of HK planning is extremely deep and wide. It is the tool to create not only continuous improvement but respect for people as well. HK planning is the second of the Six Initial Skill Focus Areas that need to be deployed at day one in the lean transformation. As you saw in Table 8.1, each Initial Skill Focus Area will have a champion. For a single refinery transformation, the HK planning champion will be the refinery manager. For a multi-refinery transformation, the HK planning champion will be the VP of refining. The role of the champion for HK planning is a little different from that of the other five champions of the Initial Skill Focus Areas. The HK champion is to become both an expert user and an expert teacher of this competency.

How to make HK planning work

There are some key techniques you can use to make HK planning work.

The first technique concerns the mission, vision, and values statements. Developing these statements is a very difficult process. For one thing, there will be differing opinions on the mission, which is who we are and what we do and sometimes how we do it. Likewise, there will be problems with the vision, which is more about where we would like to go. Finally, there is the values statement, which is what we think is important to us. First, I suggest you start with the values statement, as it is much broader and causes a new kind of focus. The key to all three of these documents is good facilitation. Since the champion for meeting facilitation is part of the group creating these key documents, you can ask your sensei to facilitate these meetings.

THE DEMING PRIZE

The Deming Prize, named after Dr. W. Edwards Deming, is the Japanese quality award given to the top companies that compete for this vaunted award. Until the late 1980s it was only given to companies located in Japan. It is very prestigious and provides a significant marketing advantage to those that earn it.

Catchball is another key technique. Most planning models collect and analyze data as well as create key strategies (steps 2 and 3 in Table 8.2), much like the HK method. However, most of these models are very weak on deployment. The deployment strategy in HK planning is a major strength of the method, and it uses the process of catchball.

In catchball, a goal is established by the folks at one level, answering the question "What is to be done?" Then those at the next level down are asked, "How can we accomplish this?" A dialogue ensues and possibly the target can be raised, or the next level down says, "We are not equipped to do that; we need a new rotoruback. Without a new rotoruback, we can only get 80% of what you want." You then decide you cannot accomplish 100% of what was requested by the management team unless you add a new rotoruback to the plan. Then you take the problem back up the organization, where it is discussed, and either a new goal is set, or alternatively the decision is made to invest in a new rotoruback. You do this communicating of "what is to be done" and "how we can accomplish this" down and up the organizational ladder until you get to the bottom of the organization chart. Most people hear this and think it is a negotiation. It is not a negotiation. Rather, it is a method to get the "what's," which are the goals, aligned with the "hows," which are the means to execute the goals. It is creating an alignment of the "results" with the "means" to achieve those results. In so doing, you get buy-in of the strongest form. This is done not only up and down the organization but across the organizational structure as well.

A third technique in HK planning it connecting the annual goals to daily activities. Most planning models, e.g., MBO (management by objectives), have annual and maybe even monthly reviews. HK planning is much different in that the annual goals are broken down into control items, control checks, and action items needed to accomplish the goals (steps 5–8 in the 10-step model in Table 8.2). These control items, checks, and action items are reviewed on an annual as well as a monthly and daily basis, and some control items are even reviewed hourly. The upshot of this is that the annual goals are inextricably tied to the daily activities of everyone. Unlike MBO, they are not a list of "extra activities" you accomplish in your extra time; HK planning guides your daily, weekly, monthly, and annual activities.

As our next technique, there is a formal "check" *o*n your key strategies not only annually and monthly, but daily and hourly as well (step 9).

Finally, to complete the PDCA cycle, once the "check" is made, action is precipitated when it is appropriate (step 10). This is dramatically different from MBO, where "checks" and "resultant actions" come on a monthly or quarterly basis, at best.

Leader standard work

The purpose of leader standard work (LSW) is twofold. First, it is a policy deployment technique. Second, it is used for personnel development. It is created for all leaders, from the corporate CEO down to each team leader. It is a documented format that is constructed with structured work such as routine reports, routine checks, and spot checks of metrics along with routine audits and, most importantly, periodic structured contact with each of your direct reports. There are also unstructured tasks, including responding to problem solving and personnel matters and possibly an unscheduled customer visit. The portion of structured time may be as much as 60 percent for a team leader, 50 percent for the supervisor, 40 percent for the production manager, only 25 percent for the plant manager, and even less for the CEO. LSW normally covers a full week's work and is updated daily and reviewed weekly.

THE POWER OF CATCHBALL

In the language of HK planning, step 4, "Plan targets and means" (Table 8.2), catchball is done to a depth and intensity that is unique to the HK planning method. In my experience, it gives your entire organization a new level of clarity, focus, and alignment. Catchball is by far the most powerful technique of HK planning.

All LSW includes a feedback loop with your respective supervisor. This then creates an opportunity for continual review of the work being done and the degree to which it is being done. This is an excellent forum for both job redesign and performance feedback. For this reason, LSW must be dynamic in nature. It is an excellent replacement for the annual appraisal system.

The danger of LSW is that managers lose sight of its objectives, and it becomes mismanaged without the dynamics, the feedback, and the proper focus. It then becomes just highly structured micromanagement—and is worse than a total waste.

LSW is the third of Six Initial Skill Focus Areas. It is the key tool used by the supervisor and manager to stay abreast of and follow up on policy deployment. It also provides a routine and structured feedback from supervisor to subordinate. In so doing it answers the questions:

- How well is the policy being executed?
- How well are my subordinates performing?

You will find several examples of LSW on our website at www.wilsonleanrefining.com. Just for kicks, we have added some bad examples as well. And don't worry; they are marked as such.

CHAPTER SUMMARY

Lean transformations may or may not be successful. One way to assure they are not successful is to pay little attention to the foundational issues. Since Ohno paid little attention to this in his writings, many feel justified in ignoring the foundational issues.

What many do not realize is that while Ohno's writings don't detail the foundational issues, it is clear he spent a great deal of time on them. You cannot read more than one or two pages of his books before you will run smack into one or more of the foundational issues. While he did not write about them, he certainly practiced them.

Paramount among those foundational issues is the topic of management. With good management, your transformation will likely succeed; without it, you will fail.

We introduced you to a new style of leadership, lean leadership, and highlighted how you can leverage your facility by utilizing "leadership by all" as a key strategy to success. The other three skills of management included hoshin kanri planning, cultural leadership, and leader standard work.

CHAPTER 9

LEAN MANUFACTURING FOUNDATIONAL ISSUE—PEOPLE

Aim of this chapter*—This chapter introduces you to a new style of people management and explains, in some detail, an activity every person in the transformation must embrace: real-time problem solving. In addition we discuss a huge untapped resource toward making problem solving a task for everyone.*

PEOPLE . . .

In my work with Toyota and other lean firms, I find they treat people vastly differently than almost all nonlean firms. First and foremost, although almost all firms will tell you that "people are our most important asset," all too frequently, that is not how they act. In an economic pinch, people are the first thing jettisoned to make an immediate impact on the bottom line. Rather than being an asset, people are treated as a variable expense to be minimized.

In my experience with both Honda and Toyota, they treat their employees as assets. They are a fixed cost with an employment-for-life agreement. It is no mystery why the Toyota plant in Georgetown, Kentucky, and the Honda plant in Marysville, Ohio, after more than 30 years, still do not have unions.

Growth and development plans

People and the proper handling of people—including training, career planning, and the commitment to a job—are forever at the heart of the TPS. The culture of Toyota is built on the people, and the company makes few compromises in this area.

In our lean transformation and to support the respect for people concept, all supervisors will utilize the following Five Supervisory Tools. These five tools are used for both the hourly and the salaried staffs:

1. *Five-year business growth and staffing plan*. This tool is uesd by members of the management team to review the facility's strategic plan over the next five years, asking themselves, "Just how will we staff that in the future? What will our structure be, and what kind of skills will we need?"

2. *Job succession planning*. Each supervisor looks at job-specific attrition that may be caused by retirements, promotions, and even those who might quit. We ask, "If Margarita left her job today, who is a capable replacement?" For each job, we create a list of potential candidates with the categories (1) Ready now, (2) Ready in one year or less with more job-specific skill development, and (3) Ready with more job-specific skill development; greater than one year of development is required.

3. *Individual performance plan*. Each supervisor with his or her subordinates, one on one, addresses growth and developmental needs in the person's current job. The discussion might start, "Boss, in about a year you will evaluate my performance, and by that time I want to be the very best employee you can imagine. To get that evaluation from you, what must I do in the next year?" It is incumbent upon the supervisor to discuss the growth needs of the individual and address what both the supervisor and the subordinate will do to create the employee's growth and development.

4. *Individual five-year growth and development plan*. This is just what the title implies. And in this challenge the supervisors address how they can assist each subordinate to improve. This information is often integrated with job succession planning. Both the supervisor and the supervised must contribute. The supervisor must answer, "What can or should the supervisor do?" In addition, the individual is challenged so he or she develops at the maximum rate and the tougher question is, "What can or should the individual do?"

5. *Leader standard work and daily feedback systems*. These were discussed in Chapter 8.

There is more information on these plans, including numerous forms, on our website, www.wilsonleanrefining.com.

The five-year business growth and staffing plan and the job succession planning are tools the supervisors and management team will use for long-range manpower planning. The individual performance plan and the individual five-year growth and development plan are discussion tools focused on employee development.

These plans interact, and when the supervisors and managers discuss the five-year growth and development plan with their subordinates, they will get information and useful input to the job succession planning done by the management team. Leader standard work has been previously discussed and is a very strong tool for individual personnel development.

WHAT IS YOUR APPROACH TO PEOPLE?

I often ask managers, "Are you attempting to engage your workforce in a joint effort in the search for excellence, or do you intend to treat them as a variable expense to be minimized?"

Do you "hire to fire," or do you "hire to retire?"

The discussion that follows is always revealing.

Problem solving by all

Problem solving by all has been a hallmark of the TPS since its inception. Everyone is engaged. It is not just an activity performed by engineers and supervisors; it is done by *everyone*—top to bottom, right to left. There are no problem-solving spectators.

There is a different attitude about problems in the TPS. In a typical Western plant, problems are a nuisance and even a sign of failure of management, engineering, or even the worker himself. Hence problems become a thing to hide and shrink away from. No one wants to accept the resultant blame handed out, and many problems go unresolved even though they are obvious to many. However, within the TPS, problems are viewed as a weakness in the *system* and an opportunity to improve the system and make it more robust. Guilt and finger-pointing are avoided, and problems are addressed and solved.

The primary purpose of the lean tools is to make problems visible. Whether it is 5S, TPM, poka yokes, andons, pull systems, kanban, or standard work, all have an element of transparency to them so the "normal state" can be readily distinguished from the "abnormal state." When you can make this distinction, and only when you can make this distinction, can you begin to problem solve. This is a secret about the lean tools. Most people see them as countermeasures only. Quite to the contrary, they are first and foremost a means to allow us to see problems, a concept we call "transparency."

THE NORMAL STATE, THE ABNORMAL STATE, AND TRANSPARENCY

The function of transparency is to see if something is normal or abnormal. 5S, andons, heijunka boards, production-by-hour boards, alarm panels, and red/yellow/green indicators are all examples of transparency.

Good transparency will allow you to determine the status of your system. We call this the "scoreboard" concept. Transparency is a key technique in process management and facilitates rapid-response problem solving. If the process is "normal," the system is performing as designed and needs no actions. If it is "abnormal," something has changed and deserves immediate attention.

Problem solving is the fourth of the Six Initial Skill Focus Areas (introduced in Table 8.1 in Chapter 8); the champion for this will have a lot to do and will have a huge impact on transformation success.

A significant advantage in refining . . . but not exploited

Refinery hourly workers have a distinct advantage over the same hourly workers in a tier one auto supplier (TOAS). Refinery workers are highly trained, especially the operators. In a typical TOAS, it is possible to hire workers on Monday, and by the end of the day Tuesday, they have completed their orientations and are on the line performing some task. And by the end of the week they are producing defect-free work, at cell design rate, ready to learn their second job. Likely by the end of the month, they have mastered that second job and are cross-trained on a third or fourth job. This is not unusual for a TOAS with a decent training program.

However, for refinery operators, at the end of month one, they have not even completed their basic operator training and likely have not even been assigned to a plant; they're still spending most of their time in the classroom. When they get to the plant, it will take them a week or more just to acquaint themselves with the layout of the process equipment and the lines. It will likely take them the better part of a year to learn their first job, complete all the situationals, and take the necessary tests.

So from hiring to competency on their first job, it takes the cell workers 40 hours on the job; it takes the refinery operators 2,000 hours on the job. This is both amazing and reasonable. The depth and breadth of the knowledge the typical refinery operator must learn is huge. Holding this knowledge is an advantage, and that is the good news.

But there is a whole bunch of bad news. In most plants, there is very little expectation for these highly trained, very skilled operators to become involved in any problem solving beyond what is called "firefighting." Certainly, they are not even expected to participate in any continuous improvement activities. As my consultant friend said, when we discussed getting line workers involved in continuous improvement activities as part of a lean transformation, he quickly and tersely replied: "Can't do that in refineries." He went on, "They barely use the intellect of the engineers; computer technologies have locked people into their offices. And the culture in a refinery does not use the intellectual resources of the operator at all. No one is permitted to 'lead from the front' . . . the management would never allow that." A rather damning and negative view of not just problem solving but the use of people in refining.

In that typical 100-mb/d refinery I referred to earlier in the book, with a total staff of 400 people, recall that there may be 40 problem solvers. This plant likely will have 120 operators. Imagine the power of unleashing another 120 true problem solvers who could not only solve problems, but do so in real time.

It is a huge unused resource, and a significant potential advantage that remains untapped. Imagine the possibilities if you could unleash this power. You could improve yields, reduce operating costs, improve energy efficiency, and improve quality while improving morale and reducing your exposure to safety and environmental incidents.

I cannot express to you how much an advantage this is over the typical TOAS. First, in the refinery, the people have far greater training. Second, the typical operating organization in a refinery is already structured for problem solving. Recall also the activities in the Theta cell (Chapter 1). Most of these workers were minimum-wage workers, but when unleashed to do problem solving, they improved the productivity another 145% over what the managers and engineers had already accomplished.

Now unleash this beast of real-time problem solving on a refinery, who have been given around 2000 versus 40 hours of training and now increase the number of problem solvers by a factor of 3, coupled with the multibillion-dollar cash flow—the opportunities are mind-boggling.

What is a problem?

For this, I rely on the methodology of the Kepner-Tregoe (KT) approach. In their methodology, there are three issues of concern:

1. If the issue is not a current concern, then it is a future concern. To resolve this, you use potential problem analysis (PPA).
2. If the issue is a current concern, but the root cause does not need to be found to resolve your concern, then you simply have a decision to make, and so you use decision analysis (DA).
3. If, however, the issue is a current concern and there is a root cause that must be understood and removed to resolve your concern, then you use problem analysis (PA).

This logic is depicted in Figure 9.1.

You will notice from the graphic in Figure 9.1 that once problem analysis is used and a problem is solved, normally a decision must be made, and so immediately decision analysis is employed. Finally, any decision will require actions, and for any action there are not only undesirable consequences but also system interactions, and so potential problem analysis is activated as well. It is a very thorough and effective methodology. It has been a methodology I have used successfully since I read Kepner and Tregoe's *The Rational Manager* in the early 1970s. On no occasion has it failed to yield results whether we were solving problems, making decisions, or looking at future threats.

So what's a concern?

However, the KT methodology starts with a "concern." Just what is a concern? This is a topic well worth discussing.

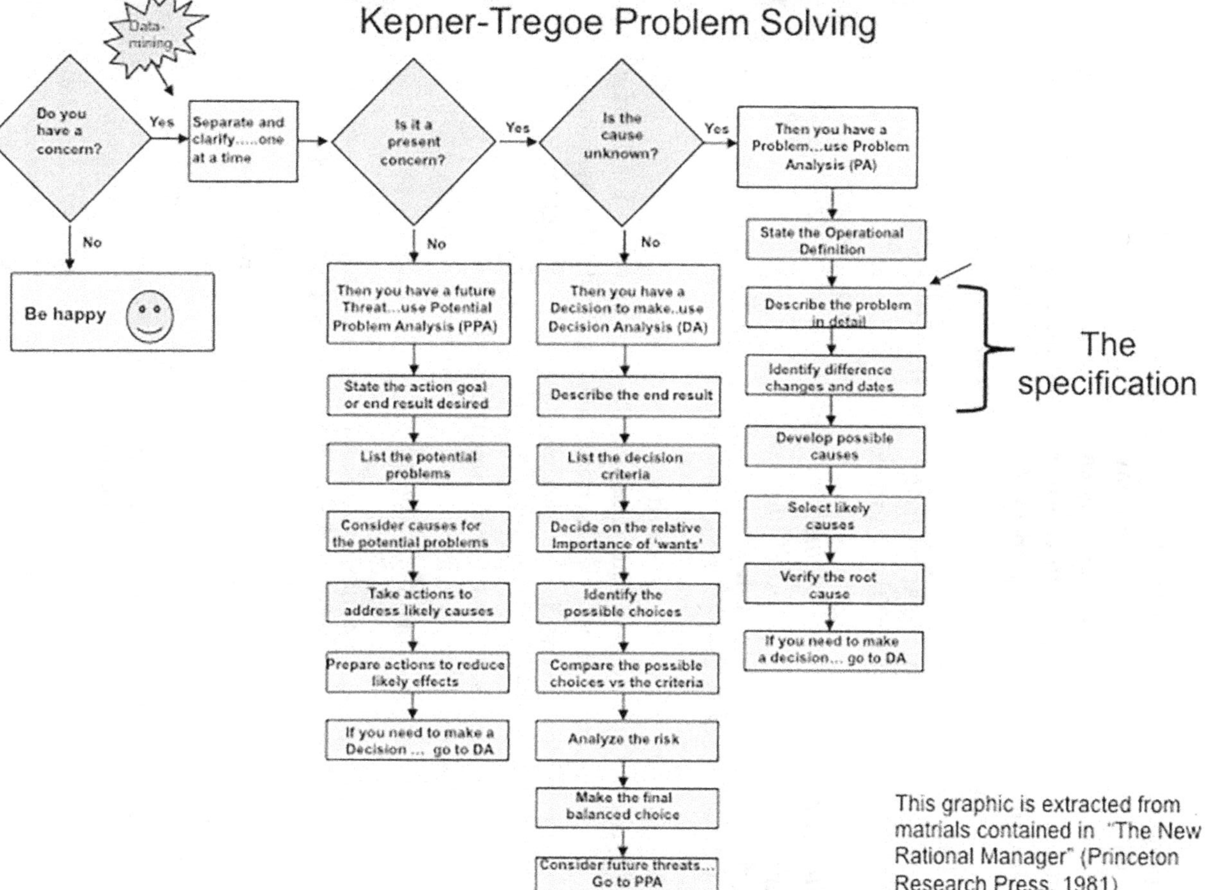
Kepner-Tregoe Problem Solving
Data-mining
Do you have a concern?
Yes
No
Be happy
Separate and clarify.....one at a time
Is it a present concern?
Yes
No
Then you have a future Threat...use Potential Problem Analysis (PPA)
State the action goal or end result desired
List the potential problems
Consider causes for the potential problems
Take actions to address likely causes
Prepare actions to reduce likely effects
If you need to make a Decision ... go to DA
Is the cause unknown?
Yes
No
Then you have a Decision to make..use Decision Analysis (DA)
Describe the end result
List the decision criteria
Decide on the relative Importance of 'wants'
Identify the possible choices
Compare the possible choices vs the criteria
Analyze the risk
Make the final balanced choice
Consider future threats... Go to PPA
Then you have a Problem...use Problem Analysis (PA)
State the Operational Definition
Describe the problem in detail
Identify difference changes and dates
Develop possible causes
Select likely causes
Verify the root cause
If you need to make a decision... go to DA
The specification
This graphic is extracted from matrials contained in "The New Rational Manager" (Princeton Research Press, 1981)

Figure 9.1 KT logic—Problems, Decisions, and Potential Problems

Concerns come from the intersection of awareness and values. Don't look and you will find nothing . . . nothing right and nothing wrong. Or have no values and nothing is bad. If you have no standards, you have nothing to distinguish good from bad. And if you can live with uncertainty, you will not have any concerns.

So how do we unravel that in a lean transformation? Simple. First, all lean tools are designed to cause information to surface. They are all designed to increase awareness. And second, the basis of all work is standards; these are our values. There is always a desired situation.

The concern is therefore the difference between "what is" (the current condition, the information that just surfaced) and "what should be" the desired condition (the standard).

What are the basic "problems" with problem solving?

In my experience, I have found there are five areas of problem solving that are seldom adequately addressed; each will undermine your ability to "problem solve to the ideal state." They are:

1. *Bad data.* Most businesses have plenty of data; in fact, they are drowning in data. Unfortunately, these same firms are starving for information, which is actionable data. That is because much of the data are unclear, and some data are poorly gathered; but most often, it is unclear why we have the data at all. When you have bad data, the current condition—the "what is" state—is not known with the clarity needed. And if the present condition is not defined, then there can be no problem.

2. *Standards that are unclear or missing.* This is a problem of epidemic proportions in virtually all businesses. Most have a good complement of product specifications and standards. What is largely missing are process standards. In a refinery, every "final and run-down" stream will have a specification. Unfortunately, the processes needed to meet the run-down specification are poorly defined, and often the specifications for these processes are nonexistent. Hence everyone "does the best they can," which is the operational definition of "no standard." Without standards, there can be no "desired state," and with no desired state in place, there can be no difference between "what is" and "what should be;" hence once again, we will not have a problem.

3. *Inadequate problem definition.* Once people have decided they have a "concern," they begin the process to define their problem, in the form of a problem statement. Clear, accurate and concise problem statements are not common. This is then a frequent impediment to problem resolution. I cannot overstate the importance of having a clear, accurate problem statement. Problems are most often loosely defined and could often be described as "some type of current nuisance."

4. *Not using a disciplined, stepwise approach.* This and the inadequate problem definition are the two top causes of weak and inadequate problem solving. In a refinery, as in industry in general, there is way too much "firefighting" disguised as real problem solving. There is way too much "Ready, fire, aim, and then fire again," as well. They are both commonplace practices and unfortunately and inappropriately are routinely accepted by the management team. The KT methodology is a countermeasure to both issues. It is unparalleled in forcing the problem solving through a rigorous "problem statement" step and the use of the KT specification, where structured questions then unlock a far more thorough understanding of the problem definition. If you have tough problems to solve, I know of no other methodologies, including DMAIC, the Eight Disciplines, OODA, TRIZ, PDCA, or even the Five Whys, that can outperform the KT methodology. Rushing too quickly and too often to action—when only part of the problem is understood—precipitates action without answering all the items in the problem specification. Until all the issues can be answered, there is a huge likelihood that the solution to the real problem is not understood.

5. *Not thinking ahead.* For every countermeasure initiated, there will be at least three issues to address:

    - The natural consequences of the countermeasure that may include other problems, smaller in impact, one hopes
    - The unintended consequences of the countermeasure
    - System interactions with the countermeasure

All three are covered by the PPA of the KT methodology.

You will note that one of the five items, is *not* finding solutions. My experience has been that once the problem is adequately defined, by using a KT specification, for example, and a disciplined problem-solving approach, the solution is typically obvious. That fact surprises most people.

What are the key problem-solving tools that will be used in a refinery transformation?

The first and most important tool is the Six Questions of Continuous Improvement. This tool should be taught to all employees and used daily. As a refresher, I am repeating the questions here:

1. What is the present condition?
2. What is the desired condition?
3. What is preventing us from reaching the desired condition?
4. What is something we can do, right now, to get closer to the desired condition?
5. When we do this, what should we expect?

   a. What will happen?

   b. How much of it will happen?

   c. When will it happen?
6. What have we learned?

The second problem-solving tool comprises the statistical and graphical tools of which there are literally hundreds. However, most of the work will be done by the five statistical tools of MSA (measurement system analysis), SPC (statistical process control, mostly control charting), DOE (designs of experiments), C&R (correlation and regression), and hypothesis testing. Likewise, there are numerous graphical tools. The most commonly used will be the flowcharting tools including the basic flowchart and swim-lane charts, critical path scheduling, the cause-effect diagram (a.k.a. Ishakawa or fishbone), the Gantt chart, and the affinity diagram. Statistical and graphical techniques are the fifth of the Six Initial Skill Focus Areas.

The third problem-solving tool, the KT methodology, should be taught to a broad range of employees. All engineers, managers, and those assigned specifically to problem solve should have KT training. In time, a lean support team (not to be confused with the lean leadership team; see Chapter 14) will be staffed, and all these members must be KT proficient. In addition, you should have at least one KT-certified trainer for the facility.

Fourth, the Five Whys will be used to dive deeper into problem solving and get beyond the superficial solutions and find the root cause. On the surface, it is very simple; however, it takes a great deal of system knowledge to do it effectively. To do it properly, you must have a thorough knowledge of the process and a thorough knowledge of the cause-effect relationships throughout the process. For a problem of any complexity, only a small population typically holds this knowledge.

CONCERNS, PROBLEMS, DECISIONS, AND POTENTIAL PROBLEMS

What if your doctor finds you have high blood pressure? She tells you the consequences, and you say, "Doesn't sound so bad to me," if so, you don't have a concern, a problem, or any decision to make. You are done. Alternatively you decide you might want to live a little longer. Your doctor says you need to go on a diet, or you could take some medication.

The root cause of the problem is the weight, so you could implement corrective action, by dieting. But in the spirit of taking the path of least resistance, you decide to take the pills. Your response is to make a decision rather than attacking the root cause of the problem. Finally, the doctor says, "And by the way, that medication may have a strong negative effect on your libido," at which time you frantically vault into potential problem analysis.

Fifth, many problems will be solved in a group format. Group facilitation should be taught to all those doing group problem solving and certainly include the senior management. This can be done locally by using Peter Sholte and Brian Joiner's *Team Handbook* and creating a local class. It is also worthwhile to send someone from the lean support team to a training in group facilitation. As part of group facilitation, create a class on meeting management, and teach it to all. This is an absolute must! Every company I have worked with needs work on meeting management. It is routinely done poorly. Meeting facilitation is the sixth Initial Skill Focus Area.

Multiskilled workers

In a TOAS, multiskilled workers are a necessity. It is common for an operator to be current on as many as 20 job stations at any time. To this extent, this is not practical in a refinery. But it is still advantageous to have employees current on three or four jobs in the same control room. It is uncommon that maintenance workers are cross-crafted. Even if the union contract allowed this practice, as it did at Chevron's El Paso refinery, I have yet to see the management exploit it well. Machinists tend to do only machinist work, welders almost always weld, and electricians almost always do electrical work. Cross-crafting can be a clear advantage if it were exploited, but few do.

In a lean system, there is also the concept of autonomous maintenance. As discussed in the story of Aera Energy, this is the concept that operators do small, minor work normally assigned to the maintenance function, such as fixing minor leaks, cleaning equipment, and changing pressure gauges. With all the bureaucracy inherent in refinery maintenance organizations, including filling out forms, planning, organizing, and coordinating, which must be done before any "doing" occurs, this is a concept well past its time in a refinery.

THINK SMALL, THINK FAST, AND THINK LOTS

There are Eight Issues of Change and Uncertainty we will consider always in problem solving:

1. Objectives are loosely defined, frequently change, and often are incompatible.
2. The environment we deal with contains unknown variables and irreconcilable uncertainties.
3. We can never have "all the facts."
4. Problems come from interactive systems that are seldom linear.
5. Any change we make will cause the system to change. All problem solving is dynamic with a moving present state and a moving terminal state.
6. At times the response we get will not be a result of our actions or intents but rather the perception of our motives.
7. For any given stimulus we apply, we will get many responses.
8. We will not let these unknowns and uncertainties cripple our problem–solving efforts.

Because of these issues of change and uncertainty, our problem–solving mantra is "Think small, think fast, and think lots—lots of cycles through the rapid–response PDCA process."

CHAPTER SUMMARY

The entire lean manufacturing concept is based on the principle that people are an asset. Aware and responsible management will treat the worker as a fixed asset, rather than a variable cost to be minimized. The management will aggressively implement the Five Supervisory Tools to fully develop all its people.

WHAT'S A "ROOT" CAUSE

Problem solving requires you to remove the root cause of the undesirable effects. And getting to the "true" root cause can be almost impossible, if not tedious. You can always ask another "why?"

What we must do is find an actionable cause and install a countermeasure that reasonable and informed people would agree will eliminate the undesirable effects in the long term. That "cause" that was mitigated or eliminated was the "root cause."

A key lean concept is that all are engaged, and this includes engagement in problem solving. We have outlined for you the basics of problem solving, focusing on the very strong KT methodology. We also summarized the "problems" of problem solving and what prevents us from finding and removing root causes and we discussed the Six Questions of Continuous Improvement.

Finally, we addressed a huge and untapped resource to aid in real-time problem solving: teaching and empowering operators so they may become real-time problem solvers.

[illegible]TER SUMMARY

> **WHAT'S A ROOT CAUSE**
>
> Problem solving [illegible] the cause of the [illegible] [illegible] you always ask [illegible] [illegible]

[illegible] are engaged [illegible] problem solving [illegible] have outlined for you [illegible] solving [illegible] [illegible] the [illegible] [illegible]

Chapter 10

Lean Manufacturing Foundational Issue—Process Stability

***Aim of this chapter**—Making processes stable is vital. In this chapter, we explain why it is so important and how it is measured and controlled.*

PROCESS STABILITY

Process stability is the concept that the process will perform within predictable and low levels of variation. This is a critical area for any refinery.

Most refineries, because they have literally thousands of control loops, think they have stable and consistent process control. Nothing could be further from the truth.

Virtually no processes, in a refinery or anywhere else for that matter, when first analyzed, are stable. I have analyzed thousands of processes, and only a handful, much less than 1%, are statistically stable upon their initial analysis. This variation and instability is felt in all aspects of the refining process, creating huge quality, safety, morale, environmental, and financial opportunities. All the empowered, motivated, incentivized, self-directed, team-based, responsible people cannot compensate for a process that is not statistically stable. In addition, no amount of reinvention, reengineering, reorganization, or restructuring can compensate for the inherent weaknesses in your processes. These processes must be analyzed and improved, one at a time, with vigor, and then a whole new world of options becomes available. Making processes stable is the first and most important process area of focus.

Stability is determined by placing the characteristic of interest on a standard control chart. If the process is "in control" (a very specific term in statistical process control described later in this chapter), then the process is statistically stable. In simple terms, stable processes are predictable.

Data Control

I find that almost all businesses, and refineries are no exception, do not do a good job of following Kaoru Ishakawa's three simple rules for data management. They are discussed in his book, now a classic, *Guide to Quality Control*. They are:

1. Know the purpose of the data.
2. Collect the data efficiently.
3. Take action on the data.

Refineries, with their big data systems, are extremely adept at collecting data efficiently. The operating data are then available in real time to not only the refinery operations controls but the managers and engineers as well. And compared with other businesses, refineries normally have the most important data recorded in some format. In addition, as discussed in Chapter 7, most data in a refinery are variables data compared with the commonly used attribute data in a TOAS. These are strengths of data management in refineries. They efficiently collect the most important data, make them available in real time to a broad range of people, and make extensive use of variables data.

> **PARADIGM CHANGE IN DATA USAGE**
>
> I find that most businesses are awash in data. Likewise, I find that the only data that are uniformly good are those parameters used for the monthly financials. This paradigm must change. That is simply not good enough if you wish to engage in process improvement. To improve your processes, you will need good data.
>
> In a lean system, data always have two uses. First is for whatever the data were chosen to monitor, whether it be financials, process controls, or product releases. Second is for problem solving. If your data are not good enough for problem solving—then they are not good enough.

However, like most businesses, refineries have several weaknesses in their data management. First, they are drowning in data yet starving for information. They frequently do not comply with Ishakawa's first criterion, "Know the purpose of the data." Regarding the third criterion, "Take action on the data," often you will find an important flow rate recorded 1,000 times per second. The refineries are not recording this quantity of data because they should; they do this because they can. This quantity of data is not helpful. Second, repeatedly you can find huge volumes of operating data, either computer gathered or hand generated, only to be stored away without even a casual glance.

Refineries, like most other businesses, have a long way to go to develop their data systems so they can effectively use the data in real-time problem solving, not just for compiling the monthly financial and management reports.

Data need to be used

One thing regarding data that I have found is that unless they are used for their intended purpose, they will not be very good when you do decide to use them. On many lean transformations, we started on our journey to "problem solve to the ideal state" and have had to back up a little to correct the data. The data were very bad. We found inaccurate data, incomplete data, misfiled data, and a litany of other problems. It is like the carving knife you only use once a year on Thanksgiving; it is invariably dull because it is not used often. Data will act the same way; if not used for their intended purpose, they will deteriorate. So when you undertake this new direction and start rigorously problem solving, you too will find that much of your data will need a great deal of maintenance. Leave some time in your plans to improve your data. You will need it; the data will need it.

Variation reduction and understanding variation

This is a topic almost skipped in Ohno's book. Yet this topic is the very essence of problem solving, process improvement, and inventory reduction, to name just a few. So why is it missing from Ohno's writings? Well, after some thought, I've concluded that he had both a deep understanding of and an ability to manage variation reduction to such a level that it was simply obvious; it was second nature to him. And Ohno—if he had a weakness—sometimes he does not state the obvious.

Do not slight this topic. It is at the heart of your company's survival, and nearly all the lean tools require an understanding of reduction in variation to work properly.

MSA

Measurement system analysis (MSA) is the statistical calculation of the variation in the measurement system. Normally there are three parameters evaluated (though some standards also require a linearity and stability analysis be done):

1. *Accuracy,* the closeness to a "true value;" to assure accuracy, you do calibrations.
2. *Repeatability,* the variation obtained when you repeatedly test a sample using the same person, the same environment, the same machine, and the same method.
3. *Reproducibility,* the variation obtained when you have different people running the same sample and using the same environment, the same machine, and the same method; that is called person-to-person reproducibility. You can also have environment-to-environment reproducibility, method-to-method reproducibility, and machine-to-machine reproducibility.

The analysis is usually done by designing a test where the same samples are tested repeatedly under controlled conditions. Using these test values, statistical calculations are made to determine how much variation is inherent in the measurement system. This amount of variation will naturally occur in all the "measured product."

A statistical evaluation is then made to determine how much this variation will affect the process control ability or the product release specification. ASTM has some acceptance standards, but the AIAG (Automotive Industry Action Group) standards have much wider industry applications and acceptance.

Frequently, process performance can be improved simply by reducing the variation in the measurement system. In the refinery, especially in the final product analyses, reduction of variation in the measurement system can provide huge financial and quality returns.

Since many products have federal specifications and there is always testing variation, often laboratories will create a "specification offset" or "guard band" to make sure they do not ship off-test product. I have reviewed several refineries and found that since they really do not fully understand their own repeatability and reproducibility, they are squandering a great deal of money they need not give away. We have found this in gasoline end point, gasoline vapor pressure, and both diesel wax cloud and pour point. In addition, we have found hundreds of opportunities with in-plant data as well.

In your laboratory documentation, there are ASTM values given for accuracy, repeatability, and reproducibility for each test. I have yet to see any refinery laboratory that did not meet these values, and did so quite easily, even engine octane values. These values are only some guideline, and since they were not calculated using your products, using your equipment, run by your people, using your methods, in your environment . . . they are not *your* accuracy, *your* reproducibility, and *your* repeatability values. When you apply the tools of MSA, you will be able to do much better on not only accuracy but also repeatability and reproducibility . . . and you can make money while you improve your measurement system.

SPC, Cp, and Cpk

Statistical process control (SPC), and especially control charting, is a set of powerful tools to assist in process analysis and problem solving. Statistical tools are so important to your transformation, they are one of the Six Initial Skill Focus Areas. The standard approach to analyzing your process is threefold: evaluating the variation in the measurement system; making your process statistically stable (predictable); and then problem solving until your process can achieve a high level of "goodness" as quantified by the process capability indices Cp and Cpk. Also know that there is a "motherload" to be found by using SPC in a lean environment, described in the Appendix for Chapter 10.

Evaluate the variation in the measurement system

Perform an MSA as described earlier. This is needed to quantify the variation associated with the measurement process itself. Frequently this variation is significant and needs to be addressed before process stability and process performance can be analyzed. In many processes, significant process improvement can be made by improving the measurement system alone.

A good measurement system analysis should precede any other process improvement activities. This is so important that in the standard requirements of the automobile industry as mandated by the

AIAG standards, an MSA must be done and the measurement system must meet AIAG standards before a capability study can be valid.

Make your process statistically stable

Place the data on a standard control chart, normally evaluated to +/– 3-sigma variation levels, and problem solve until the process exhibits statistical stability as measured on the control chart. That means there are no assignable causes of variation. When the data are plotted on the control chart, there will be no points outside the 3-sigma limits. (There are other control chart rules, and a good book on SPC will describe them; check www.wilsonleanrefining.com our website for a long list of good books on SPC.)

Once statistical stability has been achieved, we have completed the first half of the capability study. We have demonstrated that the process is free of assignable causes. It is now being acted upon only by random causes of variation. The key factor is that the process is now predictable.

THREE TYPES OF PROBLEMS

In a lean manufacturing system, there are three types of problems, affectionately called Type 1, Type 2, and Type 3.

A Type 1 problem is the most egregious; that is when you have no standard, or in oil industry terms, there is no specification. And as Ohno said, "Without a standard, there can be no improvement."

Type 2 problems are the most common; those are the problems that arise when a standard is not met.

A Type 3 problem results when the standard is not optimal.

Problem-solve until your process can achieve a high level of "goodness" as quantified by Cp and Cpk

Once stability (predictability) has been achieved, we can now evaluate the process for its "goodness." Goodness is a measure of how well the process meets its specification. Specifications come in many types. For example, in the machine shop we have a whole series of symmetrical, bilateral specifications as we grind pump shaft sleeves to +/– 0.0005 in., for example. We also have one-sided specs, as in gasoline end point, which is 437°F maximum, or gasoline octane with an (R+M)/2 value of 88 minimum.

As I noted earlier, nearly all final product streams have hard specifications, while many of the run-down streams are not so rigorously regulated. To evaluate the goodness of the process, we use the "process capability index" (this is the standard industry term). To add some confusion here, there are two formule for this term: Cp and Cpk. The derivation and formulas for these terms are readily available from any book on SPC, so I will not go into them any further here. But I wish to make two points: (1) both indices compare the actual process performance with the specification, and (2) neither Cp nor Cpk has any meaning if the process does not first exhibit process stability—that is, process predictability.

A word about specifications. Many streams in a refinery are not rigidly controlled. Normally this a large mistake. Just because product qualities can be diluted does not mean they should be. Take, for example, the topic of controlling vapor pressure of gasoline blending components. It is not uncommon to adjust the local VP requirements on some processing stream, say reformate, to save on energy at the reformer, since the workers are back-blending butane at the blender anyway. But along with the butane in this reformate stream, there often is a small amount of propane. At least it appears small. However, when you have propane in your gasoline, each barrel of propane backs out more than four barrels of butane as you try to optimize your vapor pressure specification. Not only does this shrink your mogas pool, but it also restricts the upgrade you can obtain by buying butane at $25/bbl less than mogas and blending it directly. In this case, at the reformer the process is saving a few dollars on energy, while at the gasoline blender it is squandering millions of dollars per month on lost upgrade as it simultaneously shrinks the size of your mogas pool. This once again points out that the system optimum is often not the sum of the local optima. The answer to this, which allows

GAINS ACHIEVED BY CONTROLLING COMPONENT RUN-DOWN STREAMS

By using statistical process control with real-time problem solving on these component run-down streams, not only does it lead to large economic gains, but the amount of intermediate storage can be vastly reduced as you reduce the variation on these streams.

You see, with lean you can have your cake and eat it too!

you to maximize the profits for the refinery, is to:

- Think in terms of the overall systems, not just the local conditions.
- Create meaningful internal specs on all value streams to optimize the system.
- Teach the operators how to maintain the specs; train them to problem solve and empower them to do so.
- Problem solve any problems, and do so 24/7.

The Six Questions of Continuous Improvement

The Six Questions of Continuous Improvement (introduced in Chapter 1) have an interesting history. I first learned about them when I took a course in practical problem solving that was patterned after a Toyota course of the same name. The instructors taught us the Five Questions of Problem Solving, which consisted of the first four questions listed below plus a fifth question, "When can we go see?" Since they are used for both decision making and problem solving, I renamed them the "Five Questions of Continuous Improvement." Later I reworded and expanded question 5 to include the three aspects now listed so it would better address the concept of hypothesis testing, which is integral to operating a lean facility. Also, we always practiced the mantra "learn-do-reflect," and since this always followed the activities of the kaizen, I simply incorporated them into the final question, and now I teach the Six Questions of Continuous Improvement.

They constitute the first and the most important practice that must be incorporated in any lean transformation. This will be the first skill your champion of problem solving will teach among the Six Initial Skill Focus Areas. Everyone will use the six questions, from those in the C-suite to the operators and mechanics in the field, and they will use them from this day forward.

Recall that the questions are:

1. What is the present condition?
2. What is the desired future condition?

3. What is preventing us from reaching the desired condition?

4. What is something we can do, right now, to get closer to the desired condition?

5. When we do this, what should we expect?

    a. What will happen?

    b. How much of it will happen?

    c. When will it happen so we can "go see?"

6. What have we learned?

Standard Work

As defined by Ohno, standard work is documented by its three elements:

1. The cycle time

2. The work sequence

3. The standard inventory

In his book *The Toyota Production System Beyond Large Scale Production,* Ohno wrote, "I want to discuss the standard work sheet as a means of visual control, which is how the Toyota production system is managed." Note that standard work was so important to Ohno, he said, " . . . how the Toyota production system is managed." He did not say, "How the TPS was run" but how it was managed. He was very specific.

Standard work is one of the first things I review when I am visiting a facility and the leaders want an evaluation of "just how lean are we?" I ask, "Do you have standard work? How is it prepared? And is it being followed?" I look at the revision date on some of the standard work sheets (SWSs), and if the latest date is more than six weeks old, I immediately become suspicious and dig deeper. Standard work in a lean facility is always improving; it is always changing, as the cycle to continuous improvements dictates this.

A colleague of mine likes to say, "The heart of lean is standard work." And Ohno himself is quoted as saying, "Without a standard there can be no improvement." It is not important how you say it, but for the TPS, the concepts behind the standard work sheet, and hence standard work, are foundational.

What are some of the benefits of the standard work sheet?

There are several benefits of the SWS that are not immediately obvious to those who have not worked in a lean facility.

1. The SWS is painstakingly constructed by the workers who are performing the task. They address all the safety and quality issues to make a good product or subassembly as well as incorporate the most effective and efficient techniques. The process requires input from the line workers and team leader and finally from the appropriate staff. It is reviewed and approved by the supervisor prior to the initiation of any training. This is a dramatic improvement over those procedures written by engineers and supervisors who do not actually have their hands on the parts.

2. An aspect not missed by anyone is that the SWS is the primary rate determination technique. In the labor-intensive TOAS, establishing and balancing the rate of all the workers is the primary method to control the production rate.

3. An often-overlooked aspect of the SWS is that not only is it a work direction document, but because of its extensive input, it is the "primary defect prevention" technique; it is the primary variation reduction tool in the TPS. Based on studies I have done and the experience of others, we have concluded that over 90% of all the problems encountered in a TOAS system are a direct result of not having or not following standard work.

4. Since the workers make the SWSs themselves, there is 100% buy-in, and it is routinely followed.

5. It is a "standard;" therefore it is the basis of all improvement activity.

How does the standard work sheet work in a refinery?

This is one of the tools of the TPS that does not transfer directly to a refinery—and for two very good reasons.

1. First, even though a technique of lean manufacturing is to get the process to flow, the flow is different than it is in a refinery. A TOAS normally has discrete parts, and there is a defined, visible batch size of one, which is ratcheting itself forward in the process. In a refinery the "parts" are not discrete; they are flowing smoothly as an unidentifiable, amorphous mass to the next step. It is this "ratcheting forward in a batch size of one" that literally determines the rate of production, and this step is defined by the SWS.

2. Second, and this point is critical, the typical TOAS is a labor-intensive assembly process, as is the Toyota Production System itself (see Table 7.1 in Chapter 7), and it is the SWS that defines these activities. The flow rate in a refinery is not determined by controlling the actions of the people; it is done by controlling the equipment, largely by using automatic control loops. A refinery is an equipment-intensive production system, and virtually all the rate limitations and production rates are a function of the equipment.

So why not throw out the standard work sheet?

Whoa dynamite, not so quickly! The refinery has a document that provides many of the elements of the standard work sheet. It is now called the operation procedure (OP). In my experiences in a refinery, I find that OPs, when taught and followed properly, provide the benefits listed earlier—except for item 2 on page 270. However, I often find they are made by people other than the operators doing the work—so the benefits are only partially achieved. If, on the other hand, the process as outlined above is scrupulously followed, then not only are the most efficient and most effective techniques found, but:

- They nicely address safety and quality.
- They provide a way to perform the tasks reducing process variation.

- They get virtually 100% buy-in.
- They provide a mechanism for process improvement, if that is a cultural norm.

So although Toyota style standard work sheets are not readily transferable to the refinery environment, the standard operation procedure, if it is prepared by a fully engaged workforce, will provide many of the same benefits.

When I discuss OPs, I am not limiting them to normal production operations. These OPs cover start-ups, shutdowns, and various permutations of normal operations. They also include procedures to check alarms as well as emergency shutdown devices.

The one place that does use standard work, just as Ohno designed it, is the refinery laboratory. For each test, there is a defined sequence of work, a specified time to do the test, and a precise list of the inventory needed. Note that this work in a refinery laboratory is labor intensive and the rate of work output from the laboratory is labor dependent. It is no mystery why "TPS-style" standard work can be used here.

Availability

It is the concept that the production process shall be capable to produce, when it is scheduled to do so. High process availability is a necessary characteristic of a process ready to be leaned out. In a refinery, low availability is most commonly associated with equipment downtime.

Availability is not utilization. Utilization is the fraction (or percentage) of full load capacity that is used under planned operations for any process, plant, or piece of equipment. Some refineries, to exploit their high capital investments, use "% utilization" as a metric. If this metric is used on any stream that is not a refinery bottleneck this will be problematic.

For example, in your refinery there are several plants handling light ends including butane and lighter products. Even a simple fuels refinery will normally have one or two crudes units, a fluid catalytic cracker, a reformer, and an alkylation plant all producing a variety of propane- and butane-rich streams. And often these streams pass from one plant to another, sometimes with serious economic consequences.

Most refineries have at least three of these recycle traps. There are the light ends as mentioned above, a light gasoline or pen-hex stream that can wander around, and a few streams that could be described as light diesels, all looking for a home. Any plant that measures equipment utilization as a key metric is in danger of this "recycle to look better" issue. It is only rarely that % utilization is a useful lean metric, and that occurs at bottlenecks only. Often it is inappropriately used as some local metric that leads to system suboptimization and to overproduction, the worst of the seven wastes.

TPM

Total productive (not preventive) maintenance, or TPM, is a revolutionary approach to the management of machinery. It consists of activities that are designed to prevent breakdowns, minimize equipment adjustments that cause lost production, and make the machinery safer, more easily operated, and run in a cost-effective manner. This will be an integral element of your lean effort. It is particularly important since the refinery is designed to run 24/7. (See the discussion in Chapter 7 on inventory, the *VUT* equation, and the utilization factor.)

TPM is therefore a powerful tool to improve overall performance of the plant. It has the following five pillars:

1. Improvement activities, designed to reduce the six equipment-related losses:

    - Breakdown losses
    - Setup and adjustment losses
    - Minor stoppage losses
    - Speed losses
    - Quality defects and rework
    - Start-up yield losses

2. Autonomous maintenance, which is an effort to have many simple, routine activities performed by the operator.

3. A planned maintenance system based on failure history. This is not timed maintenance; instead it is based on historical evidence.

4. Training of operators and maintenance personnel.

5. A system for early equipment maintenance.

I have several comments about maintenance in a refinery

First, the six equipment-related losses and systems losses are not well quantified, and yet maintenance costs are known to the penny. I find that maintenance costs are both overcontrolled and undermanaged within a refinery.

Second, when I observe the management of the maintenance function in a refinery, I normally see a large system intent on centrally planning distant work. This is a disaster in the making.

This massive "planning machine" is replete with gargantuan, centrally controlled, computer-driven, and bureaucratic structures swimming in rules, meetings, coordinators, and paperwork. In their efforts to become efficient, they plan themselves into oblivion and, in the end, are still highly inefficient, as evidenced by the low-productivity values.

- This is the typical outcome for any system that tries to centrally control an environment that is both distant and changing, which is exactly the refinery environment.

- Often there are coordinators in operations to work with maintenance personnel, and there are coordinators in maintenance to work with operations personnel. Whenever I see coordinators—the stench of recycle, inefficiency, and unhappiness is usually nearby. Coordinators by the drove are a sign of poor job focus and poor job alignment, usually exacerbated by inadequate or inappropriate staffing.

- This large, structured central-planning system is typically overmanaged and underperforming.

Third, rarely do I see any metrics that address effectiveness and job success. Positive outcome information is seldom found. Why is that?

Well, for one thing, maintenance functions normally pride themselves in being efficient and being underbudget, rather than measuring any metrics of effectiveness. And for another, as soon as a negative outcome surfaces, the finger-pointing begins. It is obvious that this "artful dodging" has been well practiced by both operations and maintenance personnel; these "assigning-the-blame-to-others" skills have been raised to an art form.

> **WHEN IS CENTRALIZED CONTROL A GOOD IDEA?**
>
> The military has studied this and summarizes it this way: centralized control is generally good when you are on offense, but local control is better on defense. The same holds true in a football game. The offense knows what it plans to do, and the defense must react. It is this "reacting" that makes centralized control impractical. If the circumstances on the battlefield or the football field are rapidly changing, the chain of command must be shortened, as the necessary time to respond is shortened.
>
> The maintenance situation in the plant, where problems are changing every shift if not every hour, cannot wait for orders from a maintenance staff that plans every week or even every day. Centralized maintenance is just too slow. This must change for a refinery to become a world-class performer.
>
> It is true: no longer are the large overwhelming the small; rather, the fast are overwhelming the slow.

Fourth, since the primary signs of maintenance "efficiency" are the ability to meet or beat the budget and compliance to a preplanned schedule, they are seldom good at problem solving. Especially when working with machinery such as pumps, turbines, and motors, normally the maintenance effort is simple parts replacement, rather than analysis and root cause problem solving. In a lean transformation, this cultural paradigm would certainly need to be changed.

Fifth, since the largest maintenance cost is labor, every effort should be made to keep the mechanic productive. I have seen several productivity studies done on refinery maintenance workers, and the highest I have seen, with an actual "hands-on tools" productivity metric, was 48%. Often the productivity numbers are in the 30s. These low efficiencies not only inflate the cost to produce but prolong the time equipment is out of service, decreasing availability and capacity utilization. Unfortunately, rather than focusing on making the mechanics productive, a great deal of time and effort is focused on keeping the costs of the support services of parts and tools delivery down. Hence mechanics often wait for not only tools and parts but also instructions and permits. The result is one of the largest examples of creating a local optimum—while sacrificing the system optimum.

Sixth, most refining companies have only scraped the surface of the science of equipment and process reliability. Seldom do I find a refinery that calculates the key reliability metrics such as failure rate, MTTF (mean time to failure), MTBF (mean time between failures), and MTBM (mean time between maintenance). And I cannot give you even one example of any manager in any refinery anywhere that can define reliability, except for maybe the reliability manager—that is, if the refinery has one. (Just for reference: reliability is the probability that an item will perform a required function under stated conditions for a period of time.) Since the 24/7 operating schedule of a refinery puts immense pressure on reliability, and that reliability is the very heart of this refinery's ability to make money, I find it amazing that more refiners have not explored the science of reliability more fully. It is an orchard full of ready-to-pick fruit.

Seventh, I find refinery maintenance personnel have many reasons to be an unhappy lot. I seldom find anyone who is happy with the output of the maintenance worker and normally not even the mechanic himself. The mechanic, because he is not well supported and spends most of his time waiting, which no one likes, frequently has reasons to be unhappy. His sense of accomplishment is systematically and predictably stolen from him; and worse, he sees no relief on the horizon. The operators are unhappy because they are not getting their equipment repaired in what they perceive to be a reasonable length of time. Since the operators then convey this to the maintenance supervisor, he is not happy; and furthermore he is also frustrated as he is powerless to change this. Next only to the management-union interface, the interaction between operations and maintenance is the most contentious part of human relations in a refinery.

Finally, refinery maintenance, with its large impact on both operating costs and process reliability, hence process and financial yields, deserves far more attention than it has been given. What I have seen everywhere is a one-sided effort to minimize the costs. Minimizing the costs may be the most expensive things we can do; rather, the effort should be to optimize the overall system.

I can recall a discussion that my mentor, an operations superintendent, had with the refinery maintenance manager when the maintenance workers were reluctant to work overtime on one of the large critical pumps. With a distinct and direct approach, he said, "With unreliable equipment, I can waste money faster than you can spend it."

He was correct.

And I have yet to see what I would call an "optimized" maintenance effort in any of the many continuous processing plants I have worked with—not one. The opportunities are very attractive and can be largely gained by a well-designed TPM program.

5S

5S is a set of techniques, all beginning with the letter "S." They are used to improve workplace practices that facilitate visual control, once again to make sure that abnormalities can be made obvious. The 5S principles in Japanese and English are:

1. Seiri—Sort
2. Seiton—Set to order
3. Seiso—Shine
4. Seiketsu—Standardize
5. Shitsuke—Sustain

Unfortunately, a 5S initiative is usually seen as a "housekeeping" effort and is usually done one time. But it is far more than this; it is, like most other lean techniques, both a form of transparency and also a countermeasure leading to increased productivity. The effectiveness of 5S, done well, goes well beyond that. The Marshall Institute (maintenance and reliability consultants) describes 5S and TPM as two sides of the same coin (reliability). The institute's mantra is:

- We clean to inspect.
- We inspect to detect.
- We detect to correct.
- We correct to perfect.

Transparency

Transparency, as introduced in Chapter 2, is the concept that the performance of the process or the entire line can be "seen" simply by being on the floor. A manager should be able to discern if the process performance is normal or not, in two seconds or less. We call this the scoreboard effect. That is, when you enter a football stadium, a baseball park, or a soccer stadium, you can tell in two seconds or less if your team is winning or not. In that same two-second interval, you can also assess if the team has much chance of winning or of giving up the lead if it has it. Likewise, if some process parameter is abnormal, with good transparency we can also readily discern this in two seconds. Transparency is not normally a set of charts, tables, and graphs; rather, it is a set of visual controls such as andons, red-yellow-green indicators, heijunka boards, and space markings that make the performance of the process transparent.

Over time I have come to appreciate the manifest power of "transparency done well." Good transparency goes well beyond a floor-level process understanding and creates some amazing results. It shares all the information and allows decisions to be made at much lower levels. It also opens up the culture and defangs the powerbrokers who feed on secrets. Transparency, done well, is very empowering to the entire workforce.

Transparency is manifest in our process improvement mantra:

- Create the standard—make it visual.
- Train the standard—make it visual.
- Execute the standard—make it visual.
- Reflect and improve.

Process simplifications

Process simplification is a basic concept but is frequently overlooked. It is the idea of eliminating and simplifying steps in the production process. This is one of the most powerful variation reduction techniques you can employ.

OEE

OEE (overall equipment effectiveness) is the primary measure of production effectiveness in a TOAS. It is explained here only for reference; it has very limited use in a refinery. Running the planned crude run rate and meeting the product delivery schedule are the key metrics in a refinery.

In a TOAS, OEE can be used for value stream or individual workstation performance evaluation. It is the product of these three important operational parameters:

1. Equipment availability
2. Quality yield
3. Cycle-time performance

C/T reductions

Cycle-time reductions are mentioned only because they are frequently very important to a TOAS. Not so in a refinery. Most plants are run to a rate, normally measured in barrels per day. The analogue to C/T reductions in a refinery is to break plant bottlenecks. Any time you can problem-solve and remove a plant limit, these are the low-hanging fruit of a lean implementation—that is, if the capacity can be creatively utilized. Any time a plant capacity limit can be broken, the resultant extra production is the lowest-cost product you can make. Basically, you are transforming the cost of raw materials into the value of the finished product.

Sustaining the gains

Sustaining gains is the concept that once a process improvement is achieved, the next step is to standardize it. Thus, we want to institutionalize the gains so they will be there forever. We then want to build on these gains. It is curious that almost everyone knows this, but almost no one does it, not even modestly. In my work with over 200 companies, I can't give you one example of any company that does this well.

Furthermore, this is a much larger weakness in a petroleum refinery than in the other 200 companies I have worked with. What I have found in refineries is that as soon as some new technique is implemented, no matter how large or small and no matter how effective or not, its lifespan may be no longer than the time the manager who advocated it stays in that area. This is a critical weakness that could be solved by implementing the "Prescription" described in *How to Implement Lean Manufacturing* (2nd ed.) and summarized below:

A Five Step Prescription exists on how to sustain the gains. It is not amazing, but once implemented, it is always effective. This prescription includes:

1. **Good work procedures.**
2. **Sound training in the work procedures.**
3. **Simple visual management of the process.**
4. **Hourly and daily process checks by leadership and management, LSW (leader standard work).**
5. **Routine audits by management.**

CHAPTER SUMMARY

The topic of process stability is the topic of success. Working with all your processes to make them statistically stable, that is predictable; and capable of meeting specification is a large and absolutely necessary step. Most of our production, safety, quality, and morale problems are a direct result of processes are not being statistically capable. This chapter highlighted the major tools, along with problem solving, you will use to create predictable processes that will meet your customer's needs.

This will allow you to make your facility a better moneymaking machine, a more secure workplace for your employees, and the supplier of choice to your customers.

And you can do this in no other way.

CHAPTER 11

LEAN MANUFACTURING: THE QUANTITY CONTROL TOOLS

Aim of this chapter—We now turn our attention to the quantity control tools. This chapter explores what they really are and explains why they are often misunderstood as a metaphor for lean. Finally, the chapter shows how they can be coupled with the foundational issues and used in a refinery.

THE QUANTITY CONTROL TOOLS

What are the tools?

In his article *From Lean Tools to Lean Management,* James Womack referred to the period from 1990 to 2006 as the "Tool Age of the lean movement." That may be what some called it, but there never was a "tool age." Rather, some people mistakenly believed that if they could master the tools of lean, they could fully execute a lean transformation. It was a grossly incorrect interpretation of what was at the heart of the Toyota Production System. The metaphor was "To do lean = to do the tools of lean." Using this logic, many tried and a scarce few succeeded. That is understandable. You see, there is a huge flaw in the logic, which we discussed in Chapter 3 in the section "Many Still Think of Lean as an Application of the 'Tools of Lean.'"

However, among those tools (or countermeasures, if you prefer) that were discussed, there were some new ideas that Toyota brought to the assembly line, such as kanban and SMED. In addition, Toyota also brought some very clever applications of some not-so-new ideas. If you scrutinize the list of tools that are attributed to Toyota, you will find that many of them were created in the United States or elsewhere. "Takt" is a German concept, "pull" is a concept Ohno found at a U.S. grocery store, "balanced operations" and "flow" are old industrial engineering tools that some say were developed by Ford as it tried to reduce lead time on the assembly line back in the early 1900s, and the list goes on. The "tools," those countermeasures to problems, although they are very powerful, were not the genius of the Toyota Production System; far from it.

Knowing the tools does not mean knowing lean

Rather, the genius of the TPS was how the foundational elements—the tools, the values, and the human issues—were *all* very carefully managed and woven into an integrated culture based on continuous improvement and respect for people. It was the entire integrated, living, breathing, ever-improving system . . . the integrated effort, not just the tools.

Hence learning the "tools of lean" may be a necessary condition to truly understand lean; still it is by no means a sufficient condition.

For those who have studied lean, and this means studying Ohno, you will find in his seminal book, *The Toyota Production System*, this statement:

"There is no magical method. Rather, a total management system is needed that develops human ability to its fullest capacity to best enhance creativity and fruitfulness, to utilize facilities and machines well and to eliminate all waste."

In refining, it is largely about the "foundational issues" . . . they already have "flow"

In the case of an oil refinery, the clear majority of the gains will be achieved by implementing the quality control, the foundational issues. And there is a very good reason for this.

The TPS is largely a system of discrete parts assembly. This is unlike a refinery, which is dealing with the continuous processing of large volumes of nondiscrete liquids.

When Ohno began his quest along with the Toyoda family, they soon realized the benefits to be achieved by eliminating waste and improving production lead times (or inventory turns, as we now measure them). They also realized that most of the manufacturing lifecycle for any discrete part was spent sitting still and waiting for something. It might be in a machine queue waiting for its turn so it could be worked on and then advanced in the assembly process. Or it might be a piece in a large batch, first waiting for its turn in the batch, and then, after being worked on, again waiting for the rest of the pieces to be worked on before the batch could be transferred to the next process. Furthermore, they timed and quantified the lead time and found that the part was being processed far less than 1% of the time. More than 99% of the time, this part was just sitting still, waiting; it was some form of inventory awaiting processing.

This led them to "get the part moving," and to do so, they had to figure out how to remove the primary waste, the waste of waiting. They initiated a full-blown effort at inventory removal. As they worked at this issue of waiting that affected parts and assemblies, Ohno made his now famous statement: "Flow is the basic condition."

To create flow, many countermeasures were implemented. Among these were:

- Takt
- Balanced operations
- Pull production systems
- Minimum lot size
- Leveling production
- Kanban
- Cellular production
- SMED
- Three-part inventory management

Taken together, the countermeasures on this list make up the majority of what Ohno called his "just-in-time" (JIT) system. It is JIT, because through the use of these tools, they are trying:

- To get exactly the right amount of material
- To exactly the right location
- At exactly the right time

Hence JIT is a *quantity control system,* and in 1970s and 1980s, the TPS was known as JIT manufacturing.

It was that, for sure, but it was much more. In fact, to make a JIT system work, all the *quality control,* hence the foundational issues must be well addressed, or the *quantity control* countermeasures simply will not work at all.

And since Ohno was working hard to create "flow," he would have been well advised, at that time, to check with some of his counterparts in the continuous process industries, refining being one. The refineries had largely mastered flow 50 years earlier. Consequently, since so much of the TPS is focused on creating flow, it seems that the refining industry has passed it by.

Since the refinery already has good flow, can we just ignore the quantity control countermeasures?

Not so quick. Please refer to the Appendix for Chapter 4, "An Excerpt from 'Lean Manufacturing in the Oil Refinery.'" You will see that by the metrics of "quality" and "inventory turns," the oil industry had been superior to the vaunted TPS for quite some time. However, since the TPS is based on continuous improvement, sometime in the early 1990s, the lean firms had improved and then had better quality and inventory performance than the refining industry. And they continued to improve, further opening the performance gap. From the 1990s on, they could get their discrete parts to flow faster in conveyors and cells than the refining industry companies could get their products to flow in closed piping systems. Seems hard to believe, but true it is.

So what's the bottom line here?

A lean transformation in a refinery will more intensively use the foundational issues. Likely over 95% of the effort will be on the foundational issues and only 5% on the quantity control topics. Whereas a lean transformation in the discrete parts industry will spend around 70% of its time on the foundational issues and only 30% on the flow-creating JIT tools.

DON'T LOSE SIGHT OF THE QUALITY ISSUES

Without exception, I find two issues with all my clients. First, they want to get into the "quantity control" techniques before they create a sound foundation of "quality control." And second, they grossly overestimate the effectiveness of their current quality systems. Conservatively, I estimate that most of my clients spend 70-80% of their time working on the quality control topics, even when they are trying to focus on the quantity control issues. They not only have the existing deficiencies in their quality system, but as they implement countermeasures, these "tools of lean" which are designed to create transparency cause other quality problems to surface.

QUANTITY CONTROL AND JIT

Takt

Takt is the design process cycle time to match the customer's demand, normalized to your production schedule. It is a key metric in the discrete parts industry to synchronize supply to the customer and avoid overproduction, the worst and often the largest of the seven wastes. In the processing portion of a refinery, it has limited value. It is mentioned here for reference only.

Balanced operations

Balanced operations are a simple industrial engineering technique to have all processing steps operating with the same cycle time. In a refinery, it is the practice of getting the various plants to operate in synchronization so that intermediate products need not be stored. Control schemes in many plants do not allow this, and so product is sent to storage and then returned to the plant later. Unbalanced operations often create a need for heat removal and reheating, creating energy consumption issues. In a refinery, there is a large opportunity to optimize this balance, thereby saving on operating expense and freeing up tankage for alternate usages.

Pull production systems

Pull production systems are designed to minimize overproduction, the most serious of the wastes. Pull systems have two characteristics: (1) they have a maximum inventory volume, and (2) production is initiated only when some inventory has been consumed. The opposite of a "pull" system is a "push" system, which is a scheduled system and not based on consumption but based on forecasts. Nearly all refinery systems have maximum inventories, and yet virtually all streams are scheduled. There are very few pull systems in the processing portion of a refinery. However, in the support functions of engineering, purchasing and invoicing, to name only a few, there are a wealth of opportunities to implement pull. These support functions can use a technique currently called "transactional lean" to improve the flow and reduce the wastes.

Minimum lot size

One-piece flow and minimum lot size are a means to reduce lead times in a TOAS. By reducing production lot sizes and transfer lot sizes, the product travels faster.

In a refinery, as processes batches are made, such as setting up for a pipeline shipment, the product is sitting still. The more time a product sits still, the more tankage you will need. For example, let's say you ship batches of 20,000 barrels of gasoline on one of your product pipelines three times a week. You will need 20,000 barrels of storage, plus some smaller amount to hold the run-down product while your tank is transferring to the pipeline. On the other hand, if you instead use a 10,000-barrel tank and transfer six times per week, you need half the storage required. Forgetting that you need to meet pipeline standards (and I have not missed that point), the principle is, if you reduce the transfer batch, you will reduce the required storage. This is precisely why most plants wish to operate in lockstep rate with the plants that feed them. In a refinery, that is the discrete parts equivalent to one-piece flow.

Batching in a refinery is such an accepted practice, and generally people believe "the larger, the better." Batch sizes on pipelines are very inflexible; VLCCs unload crude, and you need some large tanks to hold their cargo; at some level, it is unavoidable. Minimum lot size in refineries is a foreign concept, but there still are many opportunities to reduce batch sizes and hence reduce the needed tankage, However, there is a major effort at paradigm busting that will be required to capture these opportunities. But there are many opportunities here as well. Read about the story of reducing intermediate tankage on our website at www.wilsonleanrefining.com.

Flow

The concept in the TOAS is that parts and subassemblies do not stop except to be processed, and then only for value-added work. It is more of a concept to be attained than a reality. It is the primary tool used to reduce production lead time. The typical technique is the same as in a refinery, which is to design the process so that there is minimum inventory in the system and the plant flow rates are synchronized as close as practical. Refineries got very good at flow in the early 1900s and hence were much more efficient than the discrete parts industry. That is no longer true. Lean plants in the discrete parts industry often have more than 300 inventory turns, while refineries struggle to get 10 to 50 turns.

Leveling production

The aim of leveling is to maintain a consistent nonvariant rate of production over time for any given process. Refineries work very hard to make this work. First, they schedule plants so they are in lockstep. Second, they install feed control systems that control the flow and excess to tankage on pressure control so that the flow rate is stable. Then, as tankage levels slowly rise or fall, feed rates are adjusted. There are many opportunities to improve these systems and the issue is usually process control in the plant. Because the processes in a plant have so much variation, excess tankage is needed. You will see this in the example of reducing intermediate storage on our website, www.wilsonleanrefining.com.

Kanban

Kanban is the revolutionary practice of using cards, for example, to smooth flow and create pull in a lean system. It is also a continuous improvement tool. It is used by a wide variety of TOASs.

The system works as follows:

- The cards represent and account for all the inventories in the system.
- By controlling the number of kanban cards, we control the inventory (maximum inventory concept).
- Each unit of inventory, normally a box or a pallet, has a unique card attached.
- The kanban cards are used to trigger replenishment.
- Once the product is removed, the kanban card is sent to the front of the process to signal more production.
- Hence only what was withdrawn is replenished (replace what was "pulled" not what was scheduled).

Since virtually all refinery streams are scheduled, they are by definition push systems. Kanban is a pull system and is not intended for scheduled systems. Kanban can work for a variety of non-process systems. Two very good applications of kanban are to manage maintenance tools and storehouse consumables.

Cellular production

Cells are a key element of a lean system for discrete parts. In cells, work areas are arranged so the processing steps are immediately adjacent to one another. This lets parts be processed in near-continuous flow either in very small batches or in a one-piece flow. This, in turn, allows minimization of the wastes of transportation, waiting, and inventory—in this case, WIP. Although cells are not a good application in the main oil processes of a refinery, there are many applications in maintenance and elsewhere in the support functions. For example, in the machine shop where many pumps are sent for maintenance, a simple cell could be developed to process small pumps. We have also used cells extensively in accounting for invoicing, bill processing, and oils accounting.

SMED

SMED stands for *single minute exchange of dies*. SMED technology is a science developed by Shigeo Shingo and is designed to reduce changeover times. In a TOAS, the problem is simple. Any machine that has long changeover times must have excess capacity to account for the downtime of the changeover. Furthermore, to supply the rest of the downstream process during the changeover, a large batch must be stored up. Any effort to reduce the changeover times also reduces these two forms of waste: excess capitalization and overproduction. In a refinery, there are only a few opportunities to use SMED the same way it is used in a TOAS. First, few refinery systems make changeovers the way a TOAS does, where the process is shut down when the changeover is made. For example, when there is a changeover in crude types, this is done by a "changeover on the fly," with no shutdown. Shutdowns in the refining business are avoided due to the huge losses created by a shutdown and start-up. However, there are numerous examples where the SMED technology is very helpful in optimizing performance in a refinery, for instance, in maintenance planning for equipment cleanup and equipment repair. The SMED process is very useful to reduce the downtime as much as possible.

The basic procedure of SMED is a straightforward, three-stage process:

1. Separate the internal setup from the external setup.
2. Convert the internal setup to the external setup.
3. Streamline all aspects of the setup operation.

The best tool is the simple Gantt chart, showing all the steps in the changeover. Gather the knowledgeable people on the work, and then list all the steps. Categorize the steps as "internal setup," "external setup," or "internal but can be external;" also list the conditions to make it an external setup. From here you delete any unnecessary steps and simplify. Next, you convert as much internal setup into external setup so it can be done with the equipment running. With only internal work left, the final technique is generally to create as many parallel paths as possible.

Not only are there opportunities in equipment cleanup and equipment maintenance, but there are opportunities in equipment start-up.

Finally, I would like to share a nice example of SMED applications. In a medium-sized refinery, a young engineer, looking to save time on start-up, was reviewing the critical path for a crude unit. It seems it took over 10 shifts to install the packing in the bottom portion of the vacuum tower, and the bottleneck was the ability to get the ceramic packing saddles to the manway so they could be loaded into the column. Because of reach and weight concerns, it took a very large crane to get the saddles to the manway elevations. There was no open ground to set up a second crane, so he came up with an elegant solution based on SMED principles. He converted the "internal" time of getting the saddles to the manway, the critical path, to "external" time. He did this by building a massive structure using conventional staging and had all the saddles prestaged, ready to be loaded into the column. Then it was a simple process to install a conveyor to move the saddles to the manway. The workers loaded the saddles in three shifts, saving more than seven shifts on the critical path. This allowed them to run an additional 90,000 barrels of crude, and at $16/bbl upgrade, it was worth a gain of well over a million dollars, more than offsetting the $250,000 for the staging. A nice piece of imaginative SMED on a plant start-up.

For further study, I suggest you go directly to the author of the tactic, Shigeo Shingo. He has written two major books. His landmark book on the topic is *A Revolution in Manufacturing: The SMED System*. In his other book, *A Study of the Toyota Production System*, he expanded and refined his three stages into eight techniques. It is good reading for the lean professional.

Three-part inventory system

Cycle, buffer, and safety stocks is the threefold approach to statistical inventory management used in lean manufacturing. Any inventory beyond what is required to protect the supply to the customer, which is the next operation, is unnecessary waste. Cycle stock is the volume of inventory built up between customer pick ups. Buffer stock, on the other hand, is the inventory kept on hand to cover the variations associated with causes *external* to the plant, including demand changes and such items as transportation variations. Safety stock is that inventory kept on hand to cover the variations *internal* to the plant, including line stoppages, raw material stockouts, and anything else internal that hampers the ability to deliver the customer's demands.

QUANTITY CONTROL AND JIDOKA

Jidoka is a revolutionary 100% inspection technique developed by Toyota

Ohno wrote a great deal about jidoka, especially in cultural terms. He spoke of it as the interworking of humans and machines, which allows the machines to do the repetitive simple checking and let people do the higher-value work, like problem solving. Ohno called it "autonomation," and he spoke of it in terms of "respect for humanity."

Prior to the TPS, a great deal of manpower was exhausted using people to watch machines perform. It was Ohno's belief that machines existed to work for people, not the other way around. So workers standing watch over machines was a "disrespectful" use of manpower.

Jidoka uses such techniques as poka yoke (error proofing), which will prevent defects from advancing in the system by isolating bad materials and/or implementing line shutdowns. It is also a continuous improvement tool, because as soon as a defect is found, immediate problem solving is initiated, which is designed to find and remove the root cause of the problem. This powerful concept was first implemented by the Toyoda family in its weaving business in 1902. It was used to trigger automatic shutdowns of a loom when a thread snapped. Since then, jidoka has been continually evolving to higher levels of sensitivity. It is truly a revolutionary concept.

In addition, there is the question of where to do the jidoka. All too often the concept of jidoka is used on the final products. There is nothing wrong with this, as it prevents bad product from getting to the customer. However, for any manufacturing system with any maturity at all, the emphasis should be on process controls rather than product monitoring. Product monitoring protects the customer, but process controls not only protect the customer, but drive down both production costs and lead time. The Toyoda family applying jidoka to its looms is an example of process control.

Poka yoke

Poka yoke is a series of techniques, limited only by the worker's, supervisor's, and engineer's imagination. The purpose of poka yokes is to achieve error proofing of a process activity to make the process more robust. Poka yokes are also used in the inspection process to achieve 100% inspection. There are two types of inspection poka yokes: those that control, that is, shut down, and those that warn the operator, requiring operator intervention.

Poka yokes have wide applicability in the refinery, in operations as well as maintenance. Error proofing of "valving sequences" was a technique we used often. In fact, in very critical applications we used electrical interlocks.

Five Whys

This seemingly simple technique is not as simple as it first appears. First, it requires intimate process knowledge to understand the clear cause-effect relationships. Second, for any given cause, there may be multiple effects. It takes good facilitation to make a good Five Whys analysis. There is also more information on this technique in Chapter 9.

CIP

CIP stands for continuous improvement process or philosophy. Many can talk about it, but few can show a process flowchart for their CIP, and even fewer can adequately measure it. An example of a continuous improvement flowchart is shown in Figure 11.1.

Kaizen

Kaizen is the concept of improving a process by a series of small continuous steps, most often done by frontline workers. Often these improvements are small and hard to measure; however, the accumulated effect is significant. Over the years, kaizen has evolved to mean improvement.

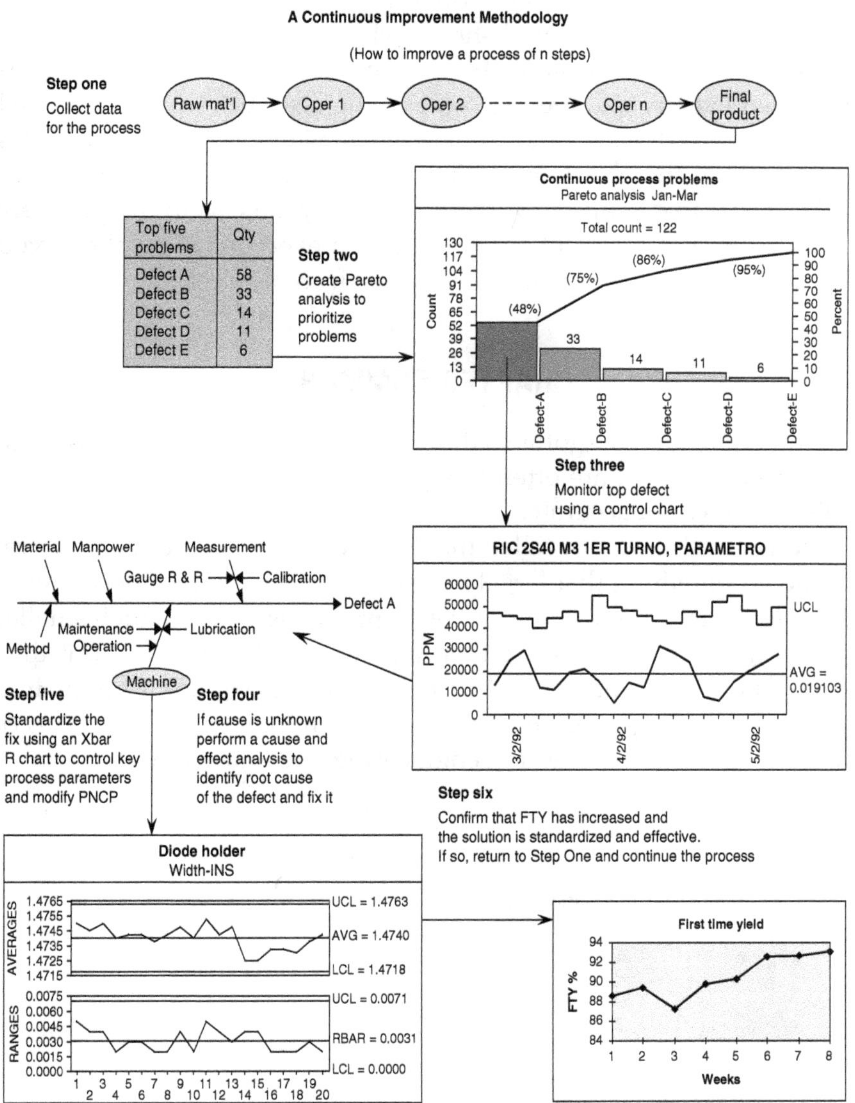

Figure 11.1 Continuous improvement flow chart (From **How to Implement Lean Manufacturing,** *2nd ed., Lonnie Wilson)*

The term "kaizen" has come to have a dual meaning. The original meaning was as stated above. It did not really refer to an activity so much as a mindset, as described in Chapter 1 in the story of the Theta cell. However, when it was imported to the United States, many companies performed large workshops (which the Japanese would more likely call kaikaku, or large change) and called these events kaizens or kaizen events, like the first part of the activities in the Theta cell improvements. It is an Americanization of a Japanese concept and has lost a little in the translation. What you call it is not so important as the fact that the frontline workers and everyone else in the plant are involved in improvement activities. When everyone is involved, you will be doing kaizen and also having a kaizen mindset.

CHAPTER SUMMARY

The tools of lean have gotten a thorough introduction into mainstream manufacturing and are often thought of as what is the heart of lean. This is simply not accurate.

Many of these tools, called the JIT pillar, really are designed to create flow, something that largely exists in a petroleum refinery. Hence these tools, these countermeasures to problems, play a much smaller role in the gains to be made than the "foundational tools." But they should not be forgotten, as they have some strong applications in a refinery including tankage utilization and in the support functions of accounting, maintenance, and engineering, to name a few.

Chapter 12

How Lean Has Worked in the Oil Business

Aim of this chapter*—You read about Aera Energy in Chapter 4, where a total company was transformed. In this chapter, I will give you some other examples of lean as it has been applied in an oil refinery. You will see how lean is being done in operations, in maintenance, and even in the support functions; yes, we used lean to improve the office work of accounting and the administrative support group, as well. In each case, you will learn how a group of engaged employees improved a situation by problem solving and removing wastes.*

WHAT'S OUR PATH TO SUCCESS?

Keep in mind that the "means to lean" is to:

- Problem solve your way to the ideal state
- Through the total elimination of waste
- Using a fully engaged workforce

In the ensuing examples, look carefully and you will see each of these three elements as the teams executed their projects.

The adjustment issue . . . or work smart as well as hard

A key question that every operator faces every shift is, "When should I adjust my process?" At El Paso, we had not taught our operators that. Normally operators were given plus-or-minus specs some distance from the mean and were told to remain in that range. This was fundamentally wrong, as almost always the plus-or-minus range was purely subjective. To fully appreciate this dynamic, you need to take a course in SPC (statistical process control), and a member of the LLT (lean leadership team) will do just that in your transformation. I am going to show you a process that you have in your refinery, where the "adjustment" rules and the instructions given to the operator made the process worse. The harder they worked . . . the worse it got!

I went to the gemba, the control house, got some cooling tower water pH data from the run sheets, and entered the data on a control chart. The process was in distress; in statistics we would say it is "out of control." I spoke to Froggy, who was on shift, and asked him how often he had to adjust the acid controller. His reply was, "Almost every sample, probably three or four times a shift." I asked him what the adjustment instruction was. He showed me the note from the supplier: "If, since the last sample, the pH has changed significantly, make an equal and opposite adjustment." I spoke to the chemical supplier who gave the operators the sample schedule and the specifications, which were 7.3 +/– 0.3 pH unit. I asked him, "What is a significant pH change?" He could not quantify it. I also asked why we sampled every two hours. His reply was that we were out of spec too often, so they closed the sample frequency so the operators could respond more quickly. Sounds logical, huh?

TO ADJUST
OR
TO NOT ADJUST

Every time an operator takes a sample, he needs to assess, "Do I adjust this process or not?" If you do not have SPC-based limits, very likely you may be undercorrecting the process or overcorrecting the process; you have no way to know. In my experience, the biggest mistake we make is to overcorrect processes, much like the operators were doing on this cooling tower.

Unless you have action limits based on statistical process control, you will get into the same problem that we had at this cooling tower—operators working very hard with little hope of success.

Next, I did a process potential study—a technique to understand what the variation would be like if the process were actually stable. (This is discussed on our website at www.wilsonleanrefining.com, and covered in detail in our SPC training.) We concluded that the tower, in its present state of control, could not maintain this +/– 0.3 unit range. However, it should be able to operate in the range of 7.3 +/– 0.6 pH unit. Then, with George, our cooling tower and boiler feedwater tech, we went to the control room. We spoke with Froggy, told him what we wanted to do, and gave him a control chart with limits set at 7.3 +/– 0.6 with the following instructions:

- Take a reading every two hours.
- Plot the data on the control chart.
- If the data fall between the lines, do nothing.
- If the data are outside the limits, make a small adjustment and document it on the chart.
- If you have concerns, call me or George.

Froggy looked over the instructions and said, "Coach, this is way outside our limits now." I told Froggy that was correct, but even then, you are outside your limits over 60% of the time, so we need a new approach. He agreed, and we got under way.

Every two hours I got a call from Froggy. He reported that everything was good until that evening when the pH dropped to 6.8; Froggy wanted to cut back the acid. I asked him, "What's the adjustment rule?" and he said, "OK, I get it."

Well, the pH turned around, and I got many calls from several operators. I took more than one trip to the control room to discuss this with the operators. We went through a short transition period while everyone learned, but for the next 14 days we did not need to make a single adjustment until the valve stem on the acid controller froze and the pH spiked. Froggy got very involved with the acid controller repair and got the mechanics to redesign it so the operators could check the shaft seal and monitor it for leaks. We then incorporated this design change into all the acid controllers in the plant.

After the acid control valve was fixed, we reset our system to sample every four hours and set adjustment limits at 7.3 +/– 0.3, and we stayed in control and in specification. The control chart after a few days looked like the one shown in Figure 12.1.

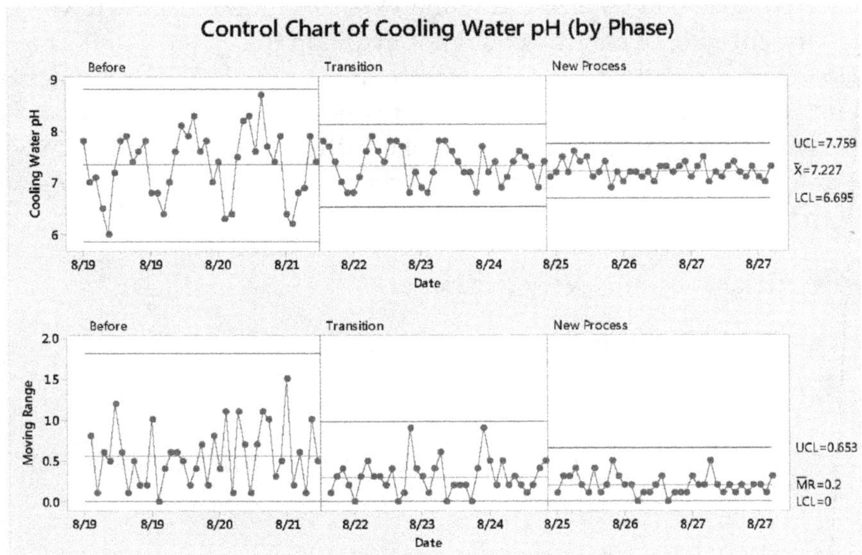

Figure 12.1 Cooling Tower Water pH

The process continued to improve. Within a month, iron levels in the cooling water had dropped precipitously, and copper counts were below the detectable limits. On the next crude unit shutdown, when we normally retube two to five cooling water heat exchangers, we had to partially retube only one. All this was done by going to the gemba to get the real story and utilizing the full strength of the workforce.

In this case, the operators had been working very hard to keep the process in control, but no one had given them the "tools" they needed to avoid overcorrection. No one in the facility had studied the process variation. No one could explain the natural process variation, and by giving the operators such general guidance, they were adjusting way too often. This aggravated the process rather than helping it. Sampling every two hours was not a problem; the problem was that they were adjusting a process that did not need to be adjusted.

In the end we could conclude that the largest source of the variation was not the process itself; rather, it was the instructions we had given to the operators to "control" the process. And unfortunately, their instruction on how to control the process—adjust every two hours—*made it worse*. We changed all that.

So by working smarter, the operators could work less and get astounding results as well. You see, you can have your cake and eat it too. That is, it is possible with lean manufacturing.

The "silo" issue

In this 100-mb/d refinery, product giveaway presented a serious profit opportunity. I had done the same thing in El Paso, and as we looked for high-payout projects, I did some data mining on RVP (Reid vapor pressure) giveaway. I was sure we had a winner here. We selected a Black Belt, Javier, for the project, and I was his mentor and the manager of operations was the project sponsor.

> **DR. DEMING ON "SILOS"**
>
> In Dr. Deming's 14 Obligations of Management, point no. 9 is: "Break down barriers between departments. People in research, design, sales, and production must work as a team to foresee problems of production and in use, that may be encountered with product or service. ("Deming Route to Quality and Productivity," 1986).

In the refinery, the long-range planner prepared the month-to-month refinery plan. The refinery short-term planner then planned on a day-to-day basis. Next, the refinery blender would calculate the blends and send that information to operations to execute. We selected all three to be on our team. In all, we had a team of nine: two from operations, one lab tech, a process engineer, the blender, the refinery short-term planner, the long-range planner, and of course Javier and me.

At our first meeting, after the first steps of creating meeting rules and getting acquainted, Javier explained what we hoped to accomplish. Javier gave a little information, and the place almost exploded. He showed that we had $15 million per year we could capture; the total RVP giveaway was on the order of $20 million per year. The manager of operations came unglued; he said that's not what he saw monthly on the giveaway report. The furor subsided, and we decided that Javier would get together with the blender and both planning groups to see if they could resolve this. They did, and it was surprisingly straightforward. The giveaway report presented to management is depicted in Figure 12.2.

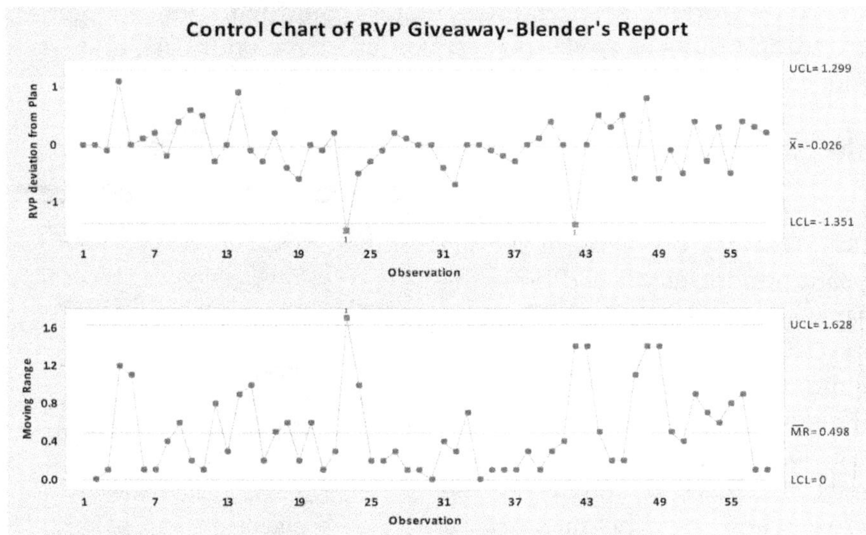

Figure 12.2 Blender's gasoline RVP giveaway

As you can see, the graph shows a rather low giveaway of 0.026# RVP on average. Hardly worth much attention. The problem with this metric was that it measured how well the blender could execute the plan with the blending components he was given. It did not quantify how well the refinery could perform.

Confused? The graph in Figure 12.3 is a bit more revealing. This graph plots giveaway compared with the release specification. The present condition is that we have almost 3# RVP giveaway. The difference between this graph and the previous graph submitted to management was that the long-range planner arranged for the monthly receipts of butane, whereas the chart from the blender only measured how well he could do with the butane he was supplied. It showed how well his "silo" was performing.

So Javier placed this question before the team: "What is preventing us from reaching the desired state?" It was immediately obvious that we did not have enough butane. But *why* did we not have the butane on hand? No one knew; the group was stumped by the first "why." Because of Javier's investigation, we found a flaw in the refinery's linear program for long-range planning. It had the wrong specs for gasoline RVP, and every single month they underordered butane. Now we

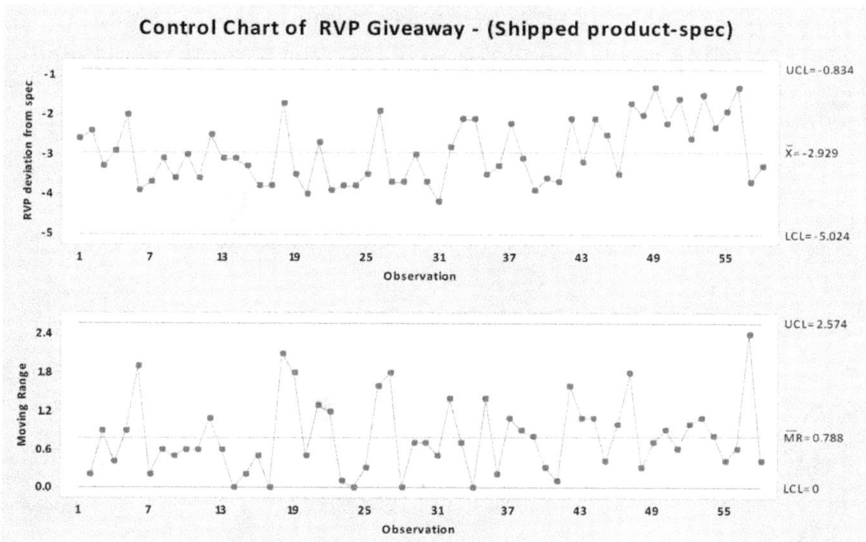

Figure 12.3 Total refinery gasoline RVP giveaway

knew at least part of the "why." The members of the blending group, tired of not being able to even come close to the RVP, simply reported based on what they could do. We traced this back for at least two years.

We established the target condition to be a maximum of 0.5 RVP giveaway in six months.

We ordered more butane, and then things started to happen. Our first problem was that we could not offload the butane cars fast enough. The operators on this team took that on, and by careful planning they could keep up with the higher demands. Next, we had the long-range planner change the RVP specs in his LP model and order the butane we needed to cover the period until the new plan kicked in. And this is what happened: As you can see in Figure 12.4, RVP giveaway dropped from 2.9 # to 0.4#. This process gain was worth $1.2 million per month to the refinery. Everyone was happy with this, and we congratulated the team.

This project intimately involved the operations, laboratory, and process engineering groups in the refinery as well as both planning functions. It took over four months to complete this task and achieve the earnings.

"Silos" in a refinery are a serious problem to be addressed. They come about by a small group of people focusing inwardly and making sure that the "silos" internal goals are met... even at the expense of the facilities goals.

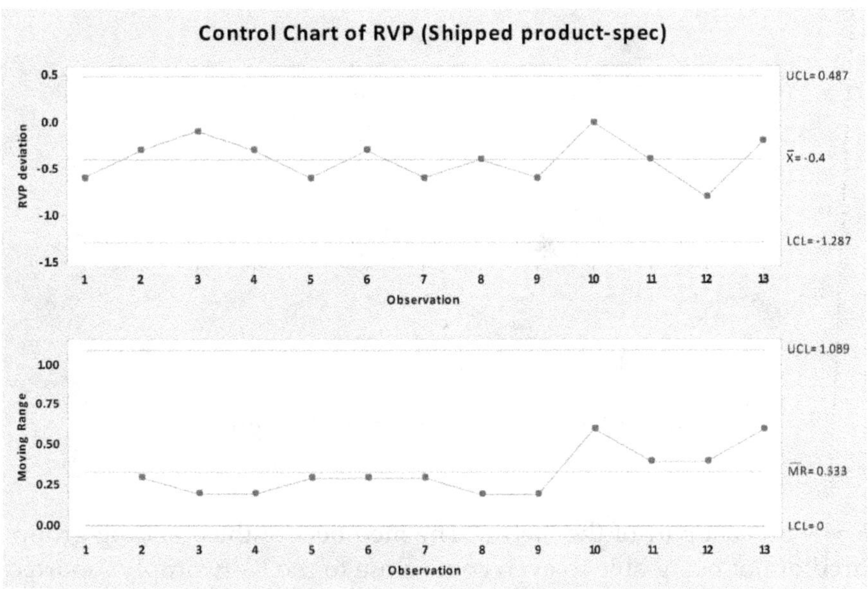

Figure 12.4 Total refinery gasoline RVP giveaway after lean

You must ask, "Why did this go on at all? And why did it go on for so long?" Both those questions can be answered if you think about the silos involved. At the meeting, it came up that the folks in each silo seldom talk with one another, as they are in different functional groups even though they worked in the same wing of the building. Second, you should ask, "Why is management not aware of this?" Well, this was in an FIOC (fully integrated oil company), and as I have told you, profits are near the bottom of the FIOC's priority list, so maybe the managers were just too busy with other priorities.

Nonetheless, this project was judged to be a tremendous success.

Equipment failing on start-up

My all-time favorite lean project in refining had to do with equipment, mostly pumps, failing on start-up. The problem was that when equipment failed on start-up after a repair, without exception, the operators would blame the mechanics for not doing a good repair job, and the mechanics would blame the operators for not starting the equipment properly. And there was seldom any information we could gather to unravel this issue.

We put together a team, headed by the refinery reliability supervisor, Kirk. I was his facilitator. A little background first. By now, at our refinery, we had been very successful using team-based problem solving. But when we needed operations or maintenance personnel on these teams, they were always handpicked, or "voluntold," by the manager, "so we would not have problems." The managers were sincere in this, but I wanted to get pure volunteers. And since we had several successes under our belts, a fact known refinery-wide, we were on pretty solid ground. The managers reluctantly agreed. To select volunteers from the list, we would pick the most senior person who was not already on a team. The request was sent out, and for the 3 positions from operations we solicited, we got 19 volunteers. We selected the 3 with the most seniority. For the maintenance support, we got 21 volunteers and selected 2 machinists and 1 pipefitter on the same basis. In addition to the 6 hourly employees, we had 1 reliability technician. Counting Kirk and me, there were 9 people on the team.

When the other two managers who worked with me saw the list of people we had chosen, they were flabbergasted. They said we had no chance at success, and I said, "We'll see." The truth is, I was worried as well. We had two employees who were at step 3 of our progressive discipline process; they had been given their final warning, and the next step was termination. On the team we also had the chairman of the worker's committee, probably one of the rowdiest employees in the refinery. He was always contentious and routinely disruptive. And finally we had one operator that no one liked. The supervisors did not like him because he would always turn down overtime. His peers in operations did not like him because he was not holding up his end of the bargain and everyone else had to work more overtime. And he more than once told off both groups. Anyway, this group looked like murderers' row, and Kirk and I had to make it work.

We planned our first meeting very well and had a great warm-up and discussed the data we had and the problem we needed to solve. At this meeting, one of the machinists thought we should, at the next meeting, spend time touring the shops area and the plants we were most concerned with. They wanted to "go to the gemba," and we did. For the second meeting, along with a warm-up and some training, we toured the machine shop. We spoke with the machinists and became acquainted with what they were doing. The machinists were quite proud to show the operators the complexity and skill it took to be a machinist. For the first time, some operators got to see what the people in the machine shop did and how they did it. On our third meeting, we went to one plant, and the operators took us through the start-up of some large pumps. The operators showed everyone the complexity of the piping and how the pump must be started up, including commissioning the auxiliary systems such as external lube oil systems and seal flushes. On both tours, the people showed with pride what they were doing as well as the breadth of skills that are needed to execute their job, as it was complex indeed. This process of going to the gemba, particularly to see each other's turf, turned out to be a great team-building exercise. The team started to bond at that time.

At our next meeting, we shared the data and started the problem solving. The control chart in Figure 12.5 displayed the present condition, the answer to question 1 of the Six Questions of Continuous Improvement ("What is the present condition?"). Kirk gave the team

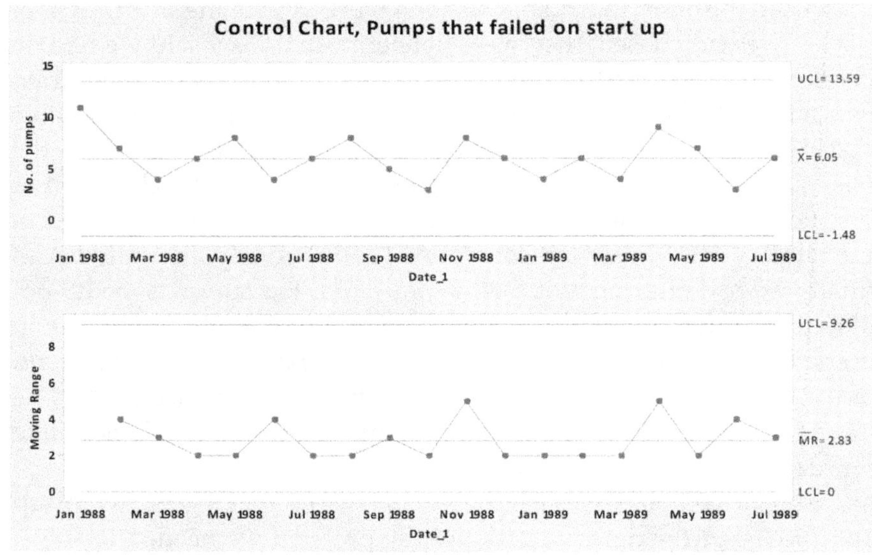

Figure 12.5 Pumps failing on start up—before lean

members a quick explanation of SPC. He explained to them that this process was in statistical control and that the average was 6 failures per month, but we could have anywhere from zero to 14 pumps fail on start-up (FOS) on any given month. They were very interested in the dynamics of SPC, so we returned to it on several subsequent meetings.

To answer question 2 of the Six Questions of Continuous Improvement ("What is the desired condition?") and decide upon a desired state, what we called the "target condition." We initially said it should be zero. Upon reflection, as a group we knew this would be unreasonable in the short term, so we elected to create a target condition of three, a 50% reduction in six months; hence we would plan to complete this in February 1990. That took us to the third of the six questions, "What is preventing us from reaching the desired condition?"

We had the detailed data, so as a group we made a Pareto diagram of the failure modes. We then reduced this to the top seven. Now everyone had a topic to research; the task was to gather data and take the information back to the group. We decided we'd analyze the data in a group format so we could "teach while we were doing." At our next meeting, we presented the data, and some wonderful things happened. Harold, one of the machinists, confessed that he knew two machinists who typically worked in the shop because they needed constant supervision. Not coincidentally, they were unskilled in installing certain mechanical seals. Imagine a machinist literally telling about some failings coming out of his own division. No sooner had this mechanic made that confession, but an operator said he knew of several new operators in his plant, the FCC, who were improperly starting up pumps. He did some investigation and found that the information in the training manual had not been updated.

Wow! That was openness and honesty you just never hear in a refinery. People may squeal on people in other divisions but never in their own, so this was quietly kept within the team. I now knew we had a team with high levels of trust.

We worked for a few weeks and came up with several countermeasures, including carefully and cautiously advising the machine shop supervisor about his people who needed training. Then one Monday morning I got a call from Manny, an operator who showed real leadership on the team. He said he and the other team members needed time at the Tuesday refinery management meeting. I asked, "What for?" and he said they needed the approvals from those people and wanted to make a presentation. Without knowing what is was, I arranged it so that at the end of the meeting they would get the audience they asked for. They showed up as a group and said they needed $1,200 to expedite the printing of the new operator training manuals. They explained what they had done, which was to rewrite the relevant portions in the manual, and since a new class was starting next Monday, they needed to expedite the printing of the training materials. The leadership team approved their request. However, more than approving the money, the leadership team was extremely impressed by the presentation made by this group of "rogues" and the poise they showed.

I was, however, a little baffled and after the meeting asked Manny how they got this done. He told me he became aware on Friday that a new operator training class was starting in 10 days. So they worked over the weekend at Manny's house—he and Leo, the operator who works no overtime and no one liked. They rewrote the relevant portions of the manual. The manual was over 50 pages long, and I asked him who typed it up (note the dates and you'll notice this is pre-PC days, so no Word, no Excel, just typewriters), and Manny told me that Roger, the third operator, got his wife to type it, while Roger proofed it.

Wow, I was amazed! We had two guys on their final warning and one operator no one liked who cooperated to do this work on their own time, with no pay . . . just to get it done. It would have been perfectly normal for them to bring this to the team and then get this ready for the next group of operators in six months. I learned a huge lesson in intrinsic motivation from these fine men.

In addition to this operator training, we found other areas needing training. We also set up awareness and technical training for the shop and field machinists. And the tide turned, slowly at first, but after eight weeks we were sure we had changed the system. That was in early 1990. Figure 12.6 shows the results. Our goal was to get to less than three failures per month by February 1990, and we killed it. It was amazing and enlightening.

We had been successful, very successful. We got the rowdiest group ever seen in El Paso to come together, work as a team, and solve a knotty, emotion-filled interdivisional problem. Just another example of problem solving to reach the ideal state through the total elimination of waste using a totally engaged workforce.

At Chevron, I worked with some of the best problem solvers I have encountered anywhere. However, the problems were always more re-

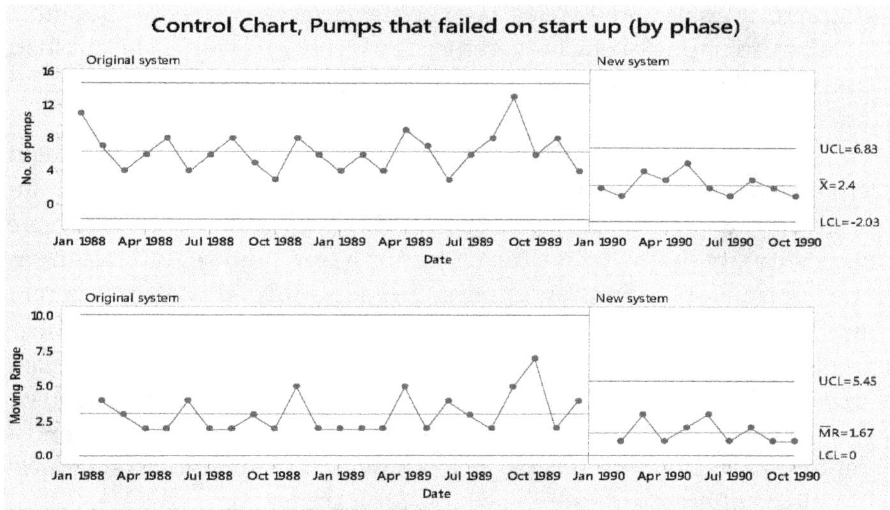

Figure 12.6 Pumps failing on start up—after lean

lated to process than people. The countermeasures to this problem were all people related. I know of two previous occasions that this problem had been assigned to someone to solve; both ended without success.

You see, this is not a problem to be solved by an engineer or a supervisor. Engineers, while they may be very bright and logical thinkers, in this case could not get the correct data. The data to solve this problem were largely in the heads of the operators and the mechanics. Consequently, if those people are not involved directly in the problem solving, and in the correct way, no one else can acquire the data; hence the problem becomes unsolvable.

The team declared victory and had a little luncheon party to celebrate the success. Each person got to say a few words, and nobody would have ever guessed that this rogues' gallery of employees could do this fine job of problem solving. Nobody outside our little team of nine gave them a chance, but they did very well.

On a personal note, this project, and how it matured, was a major factor in my retirement from Chevron. I could see very clearly how much could be accomplished by a group that is properly led and supported. I saw the power in proper leadership and in teaching, delegating, setting goals together, learning about each other together, but most of all, overcoming obstacles together. It was energizing. The group had become totally engaged, and the passion and energy was present all the time.

The sad and debilitating part was that while we did this in a team environment, we were not able to come close to that same level of engagement on a daily basis in operations and maintenance. I believe very strongly that what we developed as a team and set us apart from the "normal operations" was trust. The lack of trust was an inherent weakness in our refinery, and that was the weakness we had in utilizing the people. It is the weakness in every refinery I have ever worked in, as a Chevron employee or as a consultant to the refining industry.

You will note that the project disbanded in February 1990 and I retired from Chevron in April 1990 knowing I wanted to do statistical problem solving using small groups, for a starter.

The wax cloud issue

Let me give you another refinery example: improving winter diesel yields exacerbated by a strict wax cloud specification.

At our refinery, we made six grades of diesel, and in the winter, diesel was a high-margin product. However, in the wintertime we had two problems with the diesel yields. First, we had a very restrictive wax cloud specification that limited production on four grades of diesel. Second, the bottleneck to refinery throughput was the FCC (fluid catalytic cracker).

To make the restrictive wax cloud diesel, we needed to lower the diesel end point on all run-down streams. At the crude units, this, in turn, dropped otherwise salable diesel to FCC feed at a substantial downgrade as well as exacerbating the FCC bottleneck. However, we knew of no other way around our wax cloud limitations. We were dropping nearly 2,500 b/d of otherwise salable diesel to FCC feed.

We formed the team

The lead process engineer was the team leader, and I was the team facilitator. We had a mission and a set of team rules to guide our activities. The other team members consisted of one lab technician, one process engineer from blending support, and the operations shift supervisor in charge of blending. We were a team of five.

We did team building

We met to introduce the problem and get everyone up to speed by reviewing the materials flows and plant constraints. Each time we met, we did some team building and training, mostly on problem solving and SPC tools.

We attacked our first problem

We wanted to first analyze the limiting specification, wax cloud. And we ran smack dab into a huge problem. Our laboratory was very, very talented—however, a bit insecure. On any form of comparative evaluations, whether it was an audit or round-robin testing, our lab was always the best or one of the best performers. People from other refineries came to observe our lab management and techniques. In addition, we sent our octane engine technicians all around the corporation to teach others how to not only run the tests but maintain the equipment as well. On one occasion, an interesting story in itself, we sent one of our technicians to the Chevron Research facilities to show a group of PhD chemists how to evaluate some new sulfuric acid testers. (You can read the story on our website at www.wilsonleanrefining.com.)

We asked the lab people to do an MSA (measurement system analysis) on wax cloud testing, and they viewed that as questioning their skills. However, we had earlier taught several techs the skills of MSA, and they had already performed several MSA evaluations. After some discussion, the folks at the lab reluctantly agreed to do these tests under the guidance of one their own technicians. They did not want any engineers around, and certainly no supervisors. This plant had a strong union, but that was not the point; this was a matter of pride, and we knew it.

Anyway, slowly we got the MSA done. The lab ended up creating an entirely new method, and when all was said and done, we had reduced the variation due to measurement system error to less than one degree. In addition, due to the nature of the test, the laboratory release specification had a four-degree "guard band," effectively giving away four degrees on this specification. Once our accuracy, repeatability, and reproducibility were understood, we reduced the guard band to the one degree we now knew it could be. We also challenged the specification and found there were additional opportunities to increase the yield. This work alone increased the diesel pool by almost 800 b/d.

We evaluated the blending algorithms

Control charts were made of the blended diesels, and assignable causes were found and eliminated. There were problems with data input, problems when the blender was on manual, problems with lab results, and problems with the control loops on the blender. We literally found and solved dozens of issues with the blending and release of these products. We significantly reduced the variation of the finished product, with resultant yield gains.

Next, we created control charts for the component run-down streams and found and removed dozens of assignable causes. Our meeting room was wallpapered with control charts and capability histograms. The most common solution was to find a way to keep the run-down stream on automatic control. Streams set on manual control created numerous problems. Once we got the streams to be on automatic control, life was much better.

We also realized that although the blending algorithms were largely correct, some of the component run-down streams made a larger contribution to the wax cloud specification than others. We found that one crude unit stream had a huge impact, and first we reduced the 90% cut point temperature (our control parameter). This allowed us to increase the allowable 90% cut points on the other straight-run diesel streams as well the diesel from the hydrotreater and light cycle oil (LCO) from the cracker. The net effect was to increase the diesel pool another 700 b/d.

Once we realized the impact and importance of controlling each stream separately and to different cut points, we decided to add some more team members. We added a head operator from the crude units, a head operator from the FCC, and one from blending as well as the process engineer from the crude units. We were now a team of nine.

We learned more about the blending

The process engineers and blending operator got together and performed a designed experiment with the help of the lab tech. They found something very helpful. It seems that the FCC light cycle oil had a nice effect on wax cloud and seemed to reduce the wax formation; we called it a "solvency effect." In DOE terms, it showed up as a "positive interaction." Simply put, if you did a blend and calculated the linear average of the resultant wax cloud, the more LCO you added, the more it diverged from linearity, in a desirable way. LCO was a "wax formation depressant," that finding was serendipitous indeed.

The operators got engaged

At this point, it appeared that the engineers and managers had done all they could, and the gains were a very nice 1,500 b/d. However, the head operator from the crude unit came up with an idea. He wondered if we changed the way we controlled the side-cut streams, could we change the effect on the wax cloud? He did some tests and worked all night on this study. He found that if he changed the cut point between the #1 diesel draw, dropped more product to the #2 diesel draw, and adjusted the main fractionator heat balance, that would improve the #2 diesel significantly. This was the stream with the high contribution of wax cloud impact, even at low 90% cut points. It seemed that the D-86 distillation showed a long and ragged tail, and by changing the internal reflux and heat balance, he could largely clean up that long tail. And voilà, there was another gain of substantial proportions. The distillation operator at the FCC could get some of the same gains. In addition, the operator at the blender contributed several small kaizens, mostly in component tailoring, which again achieved gains. These three operators spent some time with each of the other three crews to train them in these activities. Some of the newly trained operators again submitted kaizens as well. And although the list of ideas slowed down, it never stopped. The gains achieved by these operators, largely on their own, yielded nearly 800 b/d of diesel recovery. Finally, the operators modified the operations procedures manuals, the training manuals, and the run sheets to help sustain these gains.

The overall project allowed us to increase the diesel pool by 2,300 b/d, and since we could relieve the bottleneck on the FCC, we could also increase the crude run at the same time. It was a huge win and demonstrated the benefits of a fully engaged team as we problem-solved our way to better operations.

The stationery project

Seriously, you got it, stationery: paper, forms, pencils, and pens. The office manager had come to me madder than a wet hen. It seems the stationery budget was overrun and he got in trouble. The stationery budget? Are you kidding me? I had no idea we even had one, especially in a multibillion-dollar business.

Anyway, I listened. He had a dotted-line relationship to someone in the home office, and I guess he got in hot water for spending too much on stationery supplies, including paper and forms. He also mentioned that his office had run out of space and needed to spend several thousand dollars on new shelving, for which no money was budgeted.

He was furious at the engineers because they were ordering several types of pens and mechanical pencils as well as several different colors and types of notepads, to name just a few of the problems. He had heard of our cross-functional problem-solving teams and their successes. He wanted to make this a project. I agreed; if he would lead the team, I would facilitate.

We put together a team that included the duplication clerk, one process lead engineer, the office manager who led the team, his clerk, and me.

We did some team building on the first visit, wrote our charter, and took a trip to the gemba. That was eye-opening, and everyone left with several ideas. The stationery storage room was an old conference room the size of two normal offices. It was packed with freestanding metal shelves, mostly spaced 15 inches apart, and most shelves had only 2 to 3 inches of forms stacked on them. There was lots of unused space.

Ideas were flying around when we regrouped. We reflected on what we had done and decided to meet in a few days.

During our next team meeting, we introduced the team to 5S and did a 5S learning activity. That was all we needed to do. The lead engineer and the clerk would perform the first "sort" and "set to order" activities; we agreed to meet in one week.

At our next meeting, the findings of the team were amazing. The "sort" found that over 50% of the form numbers were obsolete or duplicates. We simply put those forms in recycling. The forms on the shelves had been placed in numerical order, but because of size differences, this made for inefficient shelf usage. The team members decided to change the layout to one that matched forms of the same size on the same shelves; they created an index and color-coded the shelves so any form could be easily found. They also added shelves so the same floor space could accommodate more forms. At the same time, we required the engineers, and all others who used stationery supplies, to standardize on one of whatever it was they needed, such as writing pads and mechanical pencils. All told, this shrunk the required floor space by over 80% and the total inventory by 70%.

Finally, we addressed the ordering procedure. For example, if an engineer needed something, she would approach the clerk and request it. The clerk would then start an order, and when it got to over $100 (for free delivery), he would give the requisition to the office manager for approval. Sometimes this took several days, and then the order was placed. It would normally arrive the day it was ordered.

The lead engineer thought this was archaic and called the office supply shop and negotiated same-day delivery of any order at no cost. Next, we changed the procedure. We created a standard operating procedure just as we had in the plant. We set it up so that the clerk could order directly via phone. Min and max inventory values were set. The clerk would take inventory every Friday and order materials directly. We arranged to get the clerk the monthly costs, which he put on a graph; he notified the office manager of the budget status on a monthly basis.

The upshot: costs were cut by more than 30%; we saved enough space to create another two-person office; the engineers and everyone else got all their supplies, normally the same day, no wait and no approvals; and the office manager was pleased since he no longer had to worry about his office supplies budget.

However, the biggest lesson we all learned from this was what the office manager said at the final wrap-up as we closed the project. He said, "I learned that to gain real control I must give it away to the people really doing the work." He was right, and that was a huge takeaway for all of us. It is an interesting paradox . . . "to gain more control, you must give it away."

And although the savings were not earthshaking, the learning was.

And there were many more

While I was at El Paso, we also completed other projects; these were cross-functional integrated teams with engineering, management, mechanic, and operator involvement. A short list of topics includes:

- Improve electrical utility controls.
- Improve Av Gas product measurement.

- Upgrade jet A50 transfer equipment.
- Improve railroad diesel quality.
- Improve process of steno work.
- Improve acquisition of contract chemicals.
- Revise employee training.
- Revise plant monitoring.
- Reduce machinery failures in the cracker.

This group of problems using team-based, cross-functional teams taught me a great deal. First, well-led, engaged groups are powerful. Second, all these groups at some level needed management support other than me. Of the 21 projects we undertook, there were 3 complete failures. The commonality of the failures was a lack of management support; that fact stuck with me. Third, the problem in making faster progress in a refinery is not in the hands of the operators and mechanics; they are ready, willing, and able. The problem is in inadequate leadership and management exacerbated by serious trust issues.

When the elements of lean leadership come exploding on the scene and the management is willing to let go of some controls, wonderful things happen. It was mostly from these experiences when I was with Chevron that I confirmed these beliefs:

- The clear majority of skills needed to vault your enterprise into becoming a lean enterprise are already inherent within your organization.
- The key to success is largely an issue of unleashing these skills.
- The major and very critical tool needed to unleash these skills, and hence to become a lean enterprise, is to learn and employ the skills of lean leadership at all levels of the organization.

These are at the heart of the 15 Values and Beliefs of my firm, Quality Consultants.

CHAPTER SUMMARY

I wish I could give you a rundown of a true refinery transformation to a lean manufacturing system, but I can't. Like I said earlier, I can find no one doing lean in refining. The story of Aera is the best I could find, and it's an example in the oil industry and a very good one at that. However, on a smaller scale, we could get an engaged group of people to problem-solve their way to make the plant "better, faster, cheaper." So on a lesser scale I have convinced myself that making a lean transformation in a refinery is not only a possibility; it is a practicality, with nothing but upside potential.

Chapter 13

Distinguishing Lean from Other Improvement Methodologies

Aim of this chapter—*This chapter provides two specific examples to explain the distinguishing characteristics of lean manufacturing and to demonstrate in quantitative terms the advantage lean manufacturing has over other process improvement methodologies.*

IMPROVEMENT METHODOLOGIES

What does NOT *distinguish lean from other improvement strategies*

I frequently get asked to visit plants to "make a quick review of our lean manufacturing system and tell us what you think." We always start in the office of the plant manager (PM), and after the necessary pleasantries, I will ask the PM, "What makes yours a lean system?" Almost without exception I get drowned in Japanese terms. Words like "kamishibai," "nemawashi," "kanban," "hansei, "jishuken," "hoshin kanri," "sensei," and "jidoka" come flowing in the reply.

When the PM is done, I ask, "Where can we go to see these in action? I particularly would like to see an example of hoshin kanri planning and jidoka." So quickly we are at the gemba so we can see for ourselves exactly what is happening. This is the test if they are doing lean or just talking lean.

It is all too common to go to the floor and see that there is no connection between the floor activities and the hoshin kanri planning done at the executive level. The managers often just put together a couple of matrices, which they do not use, and then relabel their old planning system. It is not hoshin kanri planning at all. Likewise, when we review what they call jidoka, it is often just the same old problem solving done by engineers at a time so far after the fact that real-time problem solving is not occurring.

There is a catchy saying: "No matter how much lipstick you put on a pig, it is still a pig." Well, no matter how many Japanese terms and forms you use in your improvement methodology, that does not distinguish it from other improvement methodologies. The changed behavior of the people is the true litmus test.

There are many process improvement methodologies in use today in Western business. Almost all are focused on improving the company's financial position, although some are also designed to take on a wide range of business topics including safety, delivery, quality, and morale. In addition to lean manufacturing, there are Six Sigma, Lean Sigma, Lean Six Sigma, and reengineering programs. There are also smaller improvement efforts focused on one part or another of the business such as quality circles, value engineering, total productive maintenance, theory of constraints and many, many more.

All these methodologies are good. All, when executed properly, will make improvements in your business, and in particular, all have the strength to show early gains. However, when compared with lean manufacturing as a long-term manufacturing strategy, all fall short on three very critical aspects.

EXACTLY HOW DOES LEAN DO IT BETTER?

To explore these three critical areas, let's look at the examples of two projects we have discussed earlier: the Theta cell in Chapter 1 and the wax cloud issue in Chapter 12.

The Theta cell, summarized

First there was a kaizen project

This kaizen project was done using engineers and managers, and the gains were substantial. In less than one week, the process metric improved from 80 h/p/s (headrests per person per shift) to 188.

Then we got the operators engaged

The operators were trained in problem solving, given the cell's goals, and empowered to make changes. Over the next eight months, they implemented more than 100 kaizens, and the process not only did not deteriorate; instead the productivity improved to 304 h/p/s. In this cell, the process improvements would have continued except that the process was halted at the end of the three-year automobile model changeovers, and a new product was made in this cell. (See Chapter 1 to read the entire story of the Theta cell.)

And then there was the "wax-cloud" project team

At El Paso, we had a serious wintertime problem with wax cloud on our diesel. We first started looking at this problem because as we tried to meet the tough wax cloud spec in the winter, we would have to under draw diesel and let it fall into gasoil that was fed to our FCC (fluid catalytic cracker). The amount of diesel we dropped to the FCC feed was 2,500 b/d.

We undertook a megaproject to look at how we could optimize diesel production. The answer was contained in the question, "How can we produce more diesel, yet meet the wax cloud spec?"

I led a team that had five people—the lab supervisor, the process lead engineer, the designs lead engineer, the lead planner, and me as the facilitator. Our task was to answer the question above.

We evaluated the product

The first thing we looked at was the wax cloud testing. The lab technicians performed an MSA to statistically study the system and found a significant variation both between testers and within a single tester. The lab inspectors reviewed and revised the standard methods until the testing variation met MSA standards. In addition, we statistically evaluated the product shipments and improved the blending until we had a stable process with an acceptable Cpk for all diesel grades. This work, very engineering intensive, improved diesel yields by 800 b/d.

We evaluated the quality of the run-down streams and modeled the blending process

For each of the seven run-down streams, we evaluated their control properties and evaluated and improved them using SPC until they were statistically stable and met the required Cp and Cpk. This was a substantial engineering and laboratory effort. While we were doing this, we tested and carefully evaluated the composition of the diesel run-down streams. We would split off the tail of the distillation and check to see what part of the diesel created the wax to form. While doing this, we could isolate which streams contributed most to the wax cloud problems.

Now we had a good picture of how we could tailor the components. We established individual run-down specifications for each stream. The combined benefit of these engineering- and management-intensive efforts so far was to improve the diesel production by another 700 b/d. The combined effort of this largely engineering-intensive and management-driven project was now 1,500 b/d.

We engaged the operators

Unfortunately, even though these processes run-down streams were statistically stable, it was apparent there still were large variations in their quality. A little investigation confirmed that this was a problem with plant operations and controls. Further gains were possible. Consequently, we got the operators directly involved on the team. We added a head operator from the cracker, one from the crude units, and one from blending to the team. We spent some time to get them up to speed. Next, these three operators reviewed the operation and control of the run-down streams in their respective plants. The operators and engineers were working hand in hand to find the best set of operating conditions. They found many problems, both mechanical and process related, and after solving them, they created the best set of operating conditions to meet the final run-down specs for each stream. Once an optimum condition was found, the operators then updated their operating control sheets, the training manuals, the run sheets, and the audit check sheets. During this time, the operators made dozens of additional improvements that contributed to better control and improved yields. Because this last activity was largely operator driven and operator executed, we now had an engaged group of operators watching the streams 24/7.

Shortly thereafter, we were able to confirm we had achieved our goals. The combined effect of properly and consistently controlling each stream had increased the diesel pool by over 2,300 mb/d—and the contribution gained by engaging the operators in the problem solving totaled more than 800 b/d. (You can read the entire story in Chapter 12.)

COMPARING THE GAINS

The original question was, "Exactly how does lean do it better?" So let's compare the gains under two improvement methodologies.

What would the gains have been if these two improvement efforts had been done using a Lean Sigma or Six Sigma methodology?

Had the Theta cell project been a Six Sigma or Lean Sigma project, very likely the engineers and managers working on the project could have improved the productivity from 80 h/p/s to 188 h/p/s just as we did using the lean methodology. That is impressive! An improvement of 235% and the productivity would have substantially exceeded the value needed to be profitable. Recall that the cell would be "in the black" if it got to 100 h/p/s; it would have exceeded that nicely.

Likewise, if the wax cloud problem had been a Six Sigma project, the gains would have been an improvement of diesel yield of 1,500 b/d. Any refinery would drool at making another 1,500 b/d of light products with no additional crude. That is just like stealing.

Consequently, both would have been very successful projects in making money for the plant (see Table 13.1). Very likely the team involved would have gotten congratulations and proudly moved on to the next project.

The story of the Theta cell and the story of the wax cloud project are exactly the stories we read on the blogs and e-zines. And they are both impressive and true.

Fortunately, there is more to this story . . . well there is more if you use the lean methodology.

Table 13.1 Gains from the kaizen event

	Theta Cell Productivity	Wax Cloud Diesel gains
Initial situation	80 h/p/s	(2300 b/d of losses)
After "kaizen event"	188 h/p/s	+1500 b/d

What was the difference in using the lean methodology?

As I stated earlier in the book, there were three:

- The gains in the Theta cell did not stop at 188 h/p/s. As you read, they increased to 304 h/p/s. And the gains in the wax cloud issue did not stop at 1,500 b/d, but rather they improved to 2,300 b/d. Using the lean methodology, each one improved yield by 50% over the Six Sigma methodology (see Table 13.2).
- The gains would continue to increase.
- The gains not only increased; they were sustained.

Table 13.2 Incremental gains using a lean methodology

	Theta Cell Productivity	Wax Cloud Diesel gains
Initial situation	80 h/p/s	
After "kaizen event"	188 h/p/s	+1500 b/d
With engagement of total workforce	304 h/p/s	+2300 b/d
Incremental improvement using lean methodology	+116 h/p/s	+800 b/d
	+61%	+53%

Why does lean manufacturing have such an advantage?

Lean manufacturing is the creation of a culture of continuous improvement and respect for people. Other methodologies, like the Six Sigma methodology, are based on doing projects and projects only. They are events and as such are not continuous in nature. In addition, they have no, or very weak methodologies of sustaining the gains. In the lean methodology the initial gains are very often achieved in a kaizen project that is normally run like a Six Sigma project—and there the similarities of the two methodologies end.

After these initial gains, not only is the lean methodology far more successful in creating even greater gains, but it also has a strong mechanism to sustain them. You saw these dynamics in both the Theta cell and the wax cloud examples.

Finally, there is an often-overlooked aspect of lean—the learning involved. In a lean project the people who are making the changes stay around with the process; they do not leave to attend to the next project. Consequently, not only do you have a growing base of more talented workers, but the knowledge gained from the improvement efforts stay resident at the site and your knowledge base also grows. Some call the TPS the "thinking production system"—it is a truth.

The lean methodology is able to get these larger gains, continually improve on these gains, sustain these gains, and also grow the company. This is all because of two major differences that distinguish it from other methodologies.

On the surface the first and most obvious difference is the engagement of the operators. It is not just the operators who are engaged, there is also a marked difference in the engagement of the management. Lean uses a fully engaged workforce; this concept is grossly underestimated by those not familiar with lean. For example, in the case of the diesel recovery, not only did we use the intimate process knowledge of the engineers and the managers, we utilized the intimate process knowledge of the head operators and the operators.

And behind it all was a supportive, teaching management team. This was also a critical factor in the Theta cell improvements. We fully utilized the intellect of the entire workforce, not just the managers and engineers as is typical of a Six Sigma event. And the financial gains from this different approach are not only quantified but substantial. In addition, since the operators were deeply involved in this project, we could utilize their unique knowledge; that meant that our data input was substantially better. And because they were involved in the creation of the solutions, these solutions were theirs . . . not the engineers'. This leads to greater buy-in, commitment, and activation of the intrinsic motivators. You can say it many ways, but in the end the work gets done better, faster, and with less stomach acid. And this, in turn, leads to far greater sustainability of the improvements. Couple this greater "want to" with the lean techniques to sustain the gains, and you will have not only the short-term gains of the initial event but enlarged and sustained gains as well.

The second factor is not as obvious when viewed from the outside. All this engagement does not come naturally. It takes a concerted effort by management to make it happen. Go back to the list of the lean techniques used in the Theta cell and you will find that the initial kaizen event utilized only 8 of 54 techniques listed. The other skills, tactics, and techniques were taught to the workforce over the next year, and they were taught by the managers. The managers first needed to learn these techniques from their sensei, then they needed to use them until they were competent, and finally they then taught their subordinates. In a nutshell, it takes a large management commitment to get these gains.

As Paul Harvey used to say, "Now you know the rest of the story," and in the end this is what will separate you from the pack. This is what has separated, and continues to separate, Honda and Toyota from General Motors, Ford, and Chrysler.

Back to "lean basics"

Recall that:

- Lean is the creation of a culture of continuous improvement and respect for people.
- The means to lean is to:
    - ✦ Problem solve your way to the ideal state
    - ✦ Through the total elimination of waste
    - ✦ Using a fully engaged workforce

Parsing this, we find that:

- Lean like other methodologies is seeking improvement.
- Unlike other methodologies, lean is continuous, rather than event based.
- Lean makes great use of not only the mind but the body of all workers, as it values them as a critical asset and engages them in a joint problem-solving effort with engineering and management. Lean shows respect for all people. Other improve-

ment methodologies generally view labor as a variable cost to be minimized. These methodologies usually treat the forklift with more respect than they treat the forklift driver.

- Like all improvement methodologies, lean works to solve problems.
- However, lean does not stop at some predetermined goal; lean seeks the ideal state.
- Lean, like other methodologies, even if they don't call it that, is seeking to remove waste.
- However, lean is looking for the total elimination of waste, not stopping at some predetermined goal.
- Most process improvement methodologies use a select group of talented and well-trained problem solvers who move from one problem to the next, taking with them the knowledge and experience they learned in their last engagement.
- The largest difference is that lean uses a totally engaged workforce. Not only does it use a group of talented problem solvers, but everyone is taught problem solving, and then the experience is captured at the process level. The rank-and-file workers are included, and the type and degree of management engagement is also dramatically increased.

How lean distinguishes itself—from a different perspective

You can see quantitatively how the results from the Theta cell effort and the wax cloud issue were clearly augmented by using the lean methodology. The size of the gains, the continuity of the gains, and the sustainability of the gains were all measurable results.

For the example of the Theta cell as well as the wax cloud issue, behind all these results was a backdrop of "things done differently." Compared with "business as usual," the planning was done differently; the decision making was done differently; the problem solving was done differently; the organization was done differently; delegation was done differently; listening and talking were done differently;

work instructions were prepared differently; training was done differently; motivation was created differently; follow-up was executed differently. Literally everything was done differently, and literally everything changed—even the "way we change" was changed.

But most of all, we dramatically changed both how we managed and how we led. The critical importance of this aspect of any lean transformation cannot be understated. Dr. Deming was rather direct and very correct when he said,

"Institute leadership. The aim of leadership should be to help people, machines and gadgets to do a better job. Supervision of management is in need of overhaul, as well as supervision of production workers (Point 7) ("Deming Route to Quality and Productivity," CEEP Press, 1986)

And in these two examples, you can clearly see that the gains were increased by 50% when the Theta cell, using an engaged workforce improved productivity from 188 h/p/s to 304 and the wax cloud team improved diesel yield from 1500 b/d to 2300. This increment was solely due to the work of the engaged operators who were given superior support from their management. And every one of these changes was a change in the behavior of the people involved. Whether it was at the worker level, the supervisor level, or the manager level, it all was behavioral change.

We changed "just how we do things around here," which is the operational definition of culture. Yes, clearly and profoundly, we were changing to a culture of continuous improvement and respect for people as we were problem solving our way to the ideal state through the total elimination of waste using a fully engaged workforce.

Most improvement methodologies will help you improve your problem-solving skills and maybe even your organization for problem solving. But not one of them can create a culture of continuous improvement and respect for people. Only the lean methodology has the strategies, tactics, and skills, especially the management skills, to effect a true cultural change. When compared with other improvement methodologies, that is its strongest distinguishing characteristic.

CHAPTER SUMMARY

Lean, Six Sigma, Lean Sigma, reengineering, and others are all improvement methodologies designed to improve your business. And to some degree in some environments, if executed well, they will all create gains for your business. If they had no merit, we would not be talking about them today. However, we have shown the inherent advantages in the lean methodology. It is built on a foundation and supported by strategies, tactics, and skills that give you three huge advantages over other methodologies. First, you will find that the gains are larger; second, you will find that the gains continue to grow; and finally, you will find that the gains are sustainable. The incremental gains you achieve using the lean methodology are largely a matter of having a fully engaged workforce. The worker is far more engaged and the manager is likewise far more engaged as you create a culture of continuous improvement and respect for people—in a lean transformation done right.

Chapter 14

Implementing the Lean Transformation

Aim of this chapter—*We have discussed the history of lean and how it has been implemented, both well and poorly. We have described the tools of lean manufacturing and how they may apply in an oil refinery. And we have examined the relative merits of implementing lean in an FIOC versus a refinery-focused firm. We will now tie all this together and integrate it into an implementation plan, breaking it down into the phases of preparation, rollout, execution, and maturing.*

THE IMPLEMENTATION—AN OVERVIEW

The implementation will consist of four major phases: the preparation phase; the roll out phase; the execution phase; and the maturing phase. The preparation phase is very management intensive, the roll out phase gets everyone up to speed and demonstrates solidarity, the execution phase is where the real fun begins and the maturing phase is when you will get to fully appreciate the fruits of your labors.

The preparation phase

The preparation phase has five major steps. These all must be completed prior to the facility-wide rollout and are all top management intensive. Specifically:

1. The facility manager will select a sensei to use throughout the transformation.
2. The facility manager will select the lean leadership team (LLT).
3. The LLT will make five evaluations and create two sets of key documents to guide the facility.
4. The LLT will select the champions for the Six Initial Skill Focus Areas.
5. The LLT will design the facility rollout plan (FROP), ready for presentation to all employees at the facility.

The rollout phase

Members of the LLT will present the FROP to the entire facility in a short training presentation complete with question-and-answer time. The format will be small groups (~20) to facilitate dialogue and understanding.

The execution phase

The execution phase consists of four steps:

1. A small lean support group, the lean support team (LST), will be formed to assist in JIT training and support.
2. Goals and initial areas of interest will be deployed via hoshin kanri (HK) methods including catchball.
3. Local problem solving will begin.
4. Necessary facility-wide training will be deployed.

The maturing phase

The members of the LLT will meet monthly to assess systemwide and project progress, and they will act on the data obtained from the system assessments. They should make a formal evaluation of progress each quarter and adjust as necessary. In addition, through practice and direct involvement, the members of the LLT will grow and develop in their lean knowledge base.

In the remainder of the chapter we will focus on the first three phases—the preparation phase, the roll out phase, and the execution phase.

DETAILS OF THE PREPARATION PHASE

Step 1. The facility manager will select a sensei

Find someone with experience; there is no substitute for this. If your choice is to select a mid-level manager with some lean experience, my advice is simple and straightforward: quit this effort and move on to something else where you might succeed. Without a good sensei, this effort will fail.

How much time will you need from your sensei? That depends on how fast you wish to change—and how fast your culture is capable of changing. Your sensei will be able to give you advice on both topics. Almost surely your sensei will be an outside consultant. He or she will not be needed on a full-time basis initially, especially if you are only transforming one refinery. As the transformation progresses, this need may change.

In addition, you may need outside assistance for specific subject-matter training. Do not bring people in without discussing this with your sensei; from what I have observed, this practice of bringing in outside consultants is grossly overused. Almost without exception, it is much better to work hard to develop the expertise internally. The more internal expertise you create, the faster you will transform.

Step 2. The facility manager will select the lean leadership team

The facility manager will select the team with help from the sensei. The members of the LLT will provide the leadership to make the transformation a success. They will provide the guidance, direction, support, and oversight. Equally important, they will provide the proper modeling so the transformation can be successful. The refinery manager will lead the LLT, and the sensei will be the facilitator for the team and the mentor for the refinery manager as well as for the entire LLT.

Unlike many steering committees, this committee will have at least three major differences.

First, the makeup will be different. Normally this would be the refinery manager and his or her direct reports. The LLT, in addition to this group, will have significant representation from the unions, at least 30%. The representatives from the union must come from the union's own local leadership.

Second, most "steering" committees do just that; they steer. They set strategic direction and monitor progress. In so doing they can have very low levels of direct involvement, and yet they can still be relatively effective in that environment, because most steering committees are only interested in results. This "results-only" approach is not adequate for a lean transformation. The process is as important as the results, possibly more so. The LLT will not only charter the necessary changes; it will be the first to implement virtually all changes.

Third, most steering committee members can be effective by having a high-level understanding of the business, normally only financial. Not so with this LLT. Each member will need to read, study, and practice the lean strategies, the lean tactics, and most of the lean skills as they apply to their management levels. Not only will they lead, but they will lead by example. Lack of lean understanding by top management is Lean Killer No. 1. (See *How to Implement Lean Manufacturing*, 2nd ed., Chapter 2.)

Step 3. The LLT will make five evaluations and create two sets of key documents to guide the facility

These evaluations and guidance documents constitute most of the effort in the preparation phase. To guide the facility in its transformation and its future operations, team formation is as important as performing the five evaluations and creating the two necessary guidance documents.

Team formation

The purpose is to form a strong, functioning leadership group that can act in the best interests of the customers, the stockholders, and the employees as the leaders create a culture of continuous improvement and respect for people. The first necessary document that the LLT needs to create is a "Model of Effective, Efficient Meeting Management." This will be done at the initial meeting and become the refinery's meeting standard.

It is not an easy step to come together as a team, develop rapport and trust, and gain an understanding of the depth and breadth of the task, which is, as noted above, to create a culture of continuous improvement and respect for people. Trustful working relationships are never easily nor quickly accomplished. It will require some time and likely several facilitated brainstorming sessions as well as numerous team activities. Luckily, the key documents will be prepared using team activities, and this preparation step will provide a great opportunity to work together and not only cultivate but also practice good teamwork.

The rate at which this team progresses and the path it must take to become effective is dependent upon several factors.

First is the current culture. The progress will be very fast if the current culture in the company:

- Offers clear mission, vision, and values statements that are posted, understood, and lived, especially by management.
- Encourages a practice of teaching and rewarding lean leadership.
- Supports companywide values that state that the only true way to assure long-term success is to make all activities both learning and teaching opportunities for improvement.
- Allows for open, honest, and yet introspective communications that reach across organizational silos.
- Has a motivational system that uses both the intrinsic and the extrinsic motivators properly.
- Fosters an attitude to fully engage all employees in a joint effort to continuously improve the business.
- Provides "appropriate" transparency at all levels on all topics.
- Has reasonable levels of trust.
- Champions a leadership that is willing to go to the gemba and just listen.
- Champions a line leadership that really leads and a staff that really supports.

- Promotes a focus on strong processes to reach common goals.
- Uses goal-driven, data-based real-time problem solving.
- Is committed to assuring that the necessary data are correct and readily available.
- Not only supports all workers but challenges them to do better.
- Has a bias for action with an appropriate tolerance for "smart" errors.
- Develops mature, data-based, five-year and one-year plans that include continuous improvements.
- Integrates these plans with daily activities.
- Advocates "fighting fair."
- Provides a financial rewards system that is supportive of the facility vision, mission, values and facility goals.

If, on the other hand, your organization is highly bureaucratic, politicized, and inward looking; is filled with silos with incongruent goals and leadership going in divergent directions, each caring for itself with no commonality; is marked by poor communications where secrets and gossip, along with "he who has the gold makes the rules," are the hallmarks of power retention; has a cloudy vision typified by changing priorities for seemingly trivial reasons; has managers who lead from their office, are very top down directive, and only listen carefully to the success stories; where it takes "100 attaboys" to compensate for even one "ah shoot;" treats the rank and file as a pair of hands only, not letting the workers use their intellect; solves most problems relying on marginal data, using mostly experience and tribal knowledge as key problem-solving tools; and rewards those who are good firefighters and largely ignores those who able to "avoid the fires"—well . . . then things will take longer.

I have seen no organization that totally embodies the cultural traits from the bulleted list above, nor have I seen any as bad as those listed in the paragraph above; all are somewhere in between. All need to improve by some margin.

The second factor that helps determine the team's progress and direction is dependent upon how many duties the LLT has already completed or started. If the facility already has vision and values statements, has identified a "True North" using hoshin kanri planning, has worked on and defined the appropriate culture for its business addressing the Five Leading Indicators of Cultural Change (presented later in the chapter), and has a good start on the Five Precursors to a Lean Transformation (also presented later), the work will go very quickly. If some of these are not in place, then it will take longer.

My experience has shown that most firms have some of this preparation done quite well, especially the vision and values statements and the execution of some form of strategic planning, although most firms really do not have a well-defined and working "True North." Beyond this I find that almost no firms consciously work to define, measure, and change their culture, creating a clear opportunity to improve. Furthermore, the only firms I have found that rate well on the Five Precursors to a Lean Transformation are those that already are lean. And I have not seen any refinery that has scored well on the Five Precursors; this is a foundational area for improvement with large, often huge, financial gains to be made.

Performing the five necessary evaluations

There are five evaluations that need to be completed by the LLT. Individual members of the LLT will be assigned to initiate these evaluations. These will all be documented using an A3 format. The entire A3 will not be completed at this step. Rather, the sections on background, current condition, target condition, and impediments to the target condition will be completed. The completion of the A3 will be done in conjunction with the rest of the LLT, and many of the actions will become part of the HK plan. (At our website, www.wilsonleanrefining.com, you will find several examples of A3 forms and A3 problem solving.) These evaluations include:

1. *A thorough evaluation of the state of the current culture.* Which actions, thoughts, values, beliefs, artifacts, and language of any group of people need to change? The members of the team will need to determine the key cultural elements that should change, and not change, and decide how they want to forge those elements into the culture. A good "thought list" might

begin with the bulleted list in "Team Formation." This list will take more than one brainstorming session facilitated by your sensei before it is completed.

2. *An evaluation of management commitment.* This will be a very sensitive topic, and the sensei and the facility manager will need to meet and discuss exactly how this is done. The bottom line is that almost every manager truly believes he or she is committed and is almost always disappointed with the evaluation. We always find that an objective evaluation of the management of both the union and the facility yields scores that are very low initially. The second thing we find is that people who score themselves will give themselves a much higher score than they receive from their peers and their subordinates. It is not an easy evaluation to accomplish, but it is critically necessary. As common as it is for the scores to be very low initially, it is equally common that a second objective evaluation, done even six months later, shows significant improvement.

3. *The Five Precursors to a Lean Transformation:*

    1. High levels of stability and quality in both the process and the product
    2. Excellent equipment and line availability
    3. Talented problem solvers with a deep understanding of variation
    4. Mature continuous improvement philosophy
    5. Strong proven techniques to standardize

More information, including a scoring matrix, is available at our website, www.wilsonleanrefining.com.

4. *The Five Leading Indicators of Cultural Change:*

    1. Leadership
    2. Motivation

3. Problem solving

4. Whole-facility engagement

5. Learning-teaching-experimenting environment

5. *The Six Rollout Errors and the Ten Lean Killers*: These are discussed in *How to Implement Lean Manufacturing* (2nd ed.), and a scoring matrix can be found on our website, www.wilsonleanrefining.com.

Creating the two guidance documents

There are two sets of necessary documents to be completed by the LLT: the key guidance documents of the mission, vision, and values statements and the True North guidance documents.

Create mission, vision, and values statements

I find that most companies have mission and vision statements, but far fewer have a values statement. Although all three documents are intended to be statements to guide long-term behavior and decision making in particular, each has its purpose. The three, when taken together, should paint a very good picture of your culture. When we talk about a strong yet flexible culture, our values will be the least flexible of the three. They represent the true bedrock of what we consider to be really important, now and forever. Next on the least-flexible list would be the vision statement; it represents the future. And finally there is the mission statement. Because the mission is a discussion of the present, it might even change from year to year.

1. *Mission.* Its purpose is to explain why we are here today and what we are doing. The mission statement should answer the following questions:

    - Why are we here?

    - What are we to do on a daily basis?

    - What do we provide?

2. *Vision.* Its purpose is to provide a strategic future direction and activities. It is a statement of where we want to go and whom we want to be in the fullness of time. It must be:
    - Easily remembered.
    - Very challenging; it is almost better if it is "almost" impossible to accomplish.
    - Abstract and have what I call the "promised land" concept; that is, it is so good, it almost must be taken on faith that it is achievable.
    - Clear with an easily understood goal.
    - Inspiring.
3. *Values.* These are the ideas you hold in high regard, the ideas you think are very important, your core beliefs. There is a nice thought-provoking list in the story of Aero Energy in Chapter 4. The values statement is more important than either the mission or the vision statement, as it describes:
    - Who we are
    - What things we are prepared to do
    - What things we are not prepared to do
    - How we will go about doing those things

Better than your mission or your vision, your values help you to decide, in those seemingly tough situations, just how to behave.

Create a "True North" concept

Part of the "means to lean" is to problem solve your way to the ideal state. The ideal state is not often achieved in year one or two or even in year ten. The facility needs a set of goals to give it focus and direction. This is called "True North." There normally is a five-year plan, but as a minimum, there is a clear one-year plan to use as True North. The plan includes five elements: a descriptive phrase, a strong business and competitive evaluation, key metrics, master strategies, and substrategies. We'll examine each element:

FINDING TRUE NORTH SHOULD BE EFFORTFUL

Equally important to stating True North are the dialogue and discussions that precede the creation of the descriptive phrase. The descriptive phrase, like the other four elements, must be thoroughly discussed, questioned, and analyzed. If "True North" is completed in one session, very likely it is not a very good True North. Not only does a good True North have good "intellectual appeal," it must also have great "heart appeal." It must be both an intellectual and an emotional achievement to be an effective alignment and focusing mechanism. Don't be disappointed if it takes two, three, or even five sessions to arrive at True North; what is important is that it represents who you are, what you believe in, what you value, and where you are going.

1. *Descriptive phrase.* This is a pithy phrase that embodies your vision for the year or near term. For a refinery the pithy phrase might be something like: "Improvement; everybody, every day in everything we do," "Sustained Excellence Through Engagement," or "Flexibility-Safety-Excellence." The phrase is very important, as it should embody the entire vision and values statements.

2. *Strong business and competitive evaluation.* A thorough evaluation of the competitive nature of the refinery needs to be made, including a critical and comparative analysis of the other refineries in your distributive area. The case, of course, needs to address financials, but quality, safety, environmental, and morale issues need be addressed as well. A strong case needs to be made to show the strengths and weaknesses in your refineries profile and the improvements that need to be made in the future to survive and prosper. These weaknesses should point out the need to employ a more competitive business strategy with a more competitive operating system, a lean system that, among other things, places an emphasis on problem solving and total engagement by all. Most refineries have done some of this, and maybe just a renewal of your SWOT analysis will do. However detailed this evaluation is, it should highlight the areas of improvement

needed in the refinery. The rest of the HK planning process will focus on these areas.

3. *Key metrics.* These are end-of-the-pipe metrics that are strategic goals; they will be key measures of facility success—survival goals. Most businesses use the standard four of safety, quality, delivery, and cost (SQDC). For a refinery it would more likely be something like crude run, profitability, and compliance. The important thing is that they represent the key measures of success.

4. *Master strategies.* There are three ways to create master strategies. First is the standard method of SQDC, typical of Toyota firms. Then each functional group, such as production, maintenance, and engineering, would create master substrategies for each of the four areas of SQDC. But this creates several problems, the largest of which is the disparity in impact of the various refinery functions. The second method is to create master strategies by value streams. In a refinery, since there is so much common equipment and component fungibility in the products, it would be very hard to define the various value steams. The third and recommended approach is to create functional master strategies. A suggested list could be profitability, delivery, employee satisfaction, mechanical reliability, and environmental compliance. Profitability could then have metrics of operating costs, product specification giveaway, product upgrade, and opportunity crude, for example. Each of the master strategies will be managed by someone on the LLT using the A3 form for documentation. The basis for the A3 format is to document the process in the six-question format. Go to our website, www.wilsonleanrefining.com for samples of a blank and completed form.

5. *Substrategies.* These are normally the tactics (the small-group activities) to execute the master strategies. For example, for the profitability master strategy, there could be four tactics, one each for operating costs, product specification giveaway, product upgrade, and opportunity crudes. Each of these substrategies would have an owner. For very large and broad-reaching tactics, the owner would likely be a member of the LLT. How-

ever, for other tactics, such as product specification giveaway, this could be handled by someone not on the LLT. And again, each of these substrategies will utilize the A3 format for documentation.

Step 4. The LLT will select the champions for the Six Initial Skill Focus Areas

The Six Initial Skill Focus Areas are techniques that will have very broad application in the lean transformation. Certainly, this is not an exhaustive list of needed skills. Rather, these are skills that are often missing from most refineries; these skills will be used extensively and very early in the transformation, and they will be used by a very broad range of people. In the short term as well as the long term, these skills will have a deep and abiding impact on the success of the transformation. Recall from Chapter 8, the Six Initial Skill Focus Areas are:

1. Leadership
2. HK planning
3. Leader standard work
4. Problem solving
5. Statistical techniques
6. Meeting management and meeting facilitation

As the transformation becomes more mature, there may be other skill focus areas to include. A source of information will come from the "check" process done as a part of HK planning. If there are unmet targets, the check process will sometimes point to improvements needed in the skill focus areas and often new skills as well. Your sensei will be able to provide the initial training to the management team for all Six Initial Skill Focus Areas.

Remember from Chapter 8 that each of the Six Initial Skill Focus Areas will have a champion. The champion for the leadership and the HK planning skills will be the facility manager for a single refinery or the VP of refining for a multi-refinery business. The other four skill focus areas can be championed by any of the members of the lean

leadership team. It is best if the skill focus areas are assigned to what would normally be called an "illogical" choice. For example, it would be logical to assign statistical techniques to the technical manager, but the operations or environmental manager would be a better choice. I cannot explain it fully; it just works better.
The role of the champion is to:

- Become competent with the skill; he or she will do this by using the skill.
- Assess the facility-wide need for these skills and the levels at which they should be taught.
- Facilitate the acquisition and delivery of these skills for the needed persons. This includes:
    - ✦ Arranging for the needed training for the managers
    - ✦ Assuring the "appropriate competency" of the managers
    - ✦ Assuring that the managers deploy these skills, including training their subordinates
    - ✦ Assuring that the skills are being effectively utilized
- Evaluate the need to expand upon these competencies.

Step 5. The LLT will design the facility rollout plan, ready for presentation to all employees at the facility

By now, there has been a lot of preparation work done, and the entire workforce knows that a lot is in the wind. Many people have met the new sensei, the LLT has formed, and it will not go unnoticed that this is a joint union-management team. The LLT will have met several times to complete the five evaluations and create the two key guidance documents. While preparing the HK plan, the LLT will have included many people in their respective divisions, so people will be extremely aware that things are changing. At this time, it is also normal for the champions of the Six Initial Skill Focus Areas to have had several classes, certainly a class in meeting management for the entire facility and one in SPC for the technical people. There will be a very positive buzz in the air. So the rollout is more to formalize the transformation, make sure everyone understands the scope and direction, and to provide a forum for questions and answers by everyone in the facility.

DETAILS OF THE ROLLOUT PHASE

What is the rollout?

The rollout will consist of a series of small-group meetings using a common presentation. The presentation should be short, approximately 20 minutes, with plenty of time for Q&A. Most sessions will have several questions and could last up to one hour. It is important that the same presentation be given to all groups. The rollout must address three basic questions. They are the three questions of change management:

1. *Why is the present condition no longer acceptable?* This usually prompts a discussion of changes in the industry, our competitive position, and improvements that need to be made, emphasizing the point that we can no longer do business as usual. The world of refining is changing, and we must change with it.

2. *Where do we want to go?* True North must become a target for the entire facility. The direction may also state qualitative objectives such as the need to become more competitive and the need to use the entire workforce more fully to do that. By now there are some clear objectives we have enumerated in our HK planning, such as reducing operating costs from $15/bbl to $12/bbl, increasing jet production, etc. "Where we want to go" must include a balance of end-of-the-pipe *results,* along with a discussion of the *means* to achieve those results.

3. *How are we going to get there?* This is a discussion focused on working together to maximize the utilization of all people—problem solving by all.

Very important note: The rollout must also make it clear that people will not lose their jobs due to productivity improvements. This is a must! It needs to be well thought out and discussed in depth by the LLT prior to the rollout.

When should the rollout begin?

The first two evaluations—the cultural evaluation and the evaluation of management commitment—need to be completed before the rollout begins. The other three evaluations must be well on their way but need not be completed. It is best if the HK planning is completed, but at a minimum, catchball must be completed down to the supervisory level and across all functions. The six champions for the Six Initial Skill Focus Areas have already been chosen and likely are very active by now. With a dedicated effort, all this can be achieved in a three- to six-month range.

How is the rollout done?

Two, three, or four members of the LLT will make the presentations to each individual group. This cannot be delegated!

In addition, each meeting should have the functional manager for that group make the presentation; that means the operation and maintenance managers likely will make several presentations. In addition, every meeting needs to have one person from company management and one person from union management present. You need to send a unified and consistent message.

I suggest that this presentation be practiced several times, with the managers presenting it to the LLT as a role-playing experience. It is absolutely critical that there is a unified and a consistent message sent; lack of either will create untold problems.

Once the first rollout meeting is done, every effort should be made to get the word to everyone as quickly as possible.

After the regular rollout meetings have been completed, a special meet-

> **THE ROLE OF THE LEAN SUPPORT TEAM**
>
> The LST members will provide just that—support. Like all staff functions, they should not initiate, nor should they manage. They will be a small group, probably three to six people even for a large refinery, who will become the in-house experts on the lean transformation and all its strategies, tactics, and skills. They will supply three basic support functions:
>
> 1. Basic program design support and advice for the entire facility
> 2. Initial training to leadership for most topics
> 3. Specialized subject-matter support

ing should be held for all first-line supervisors to address their specific concerns. I have found that the first-line supervisors face the greatest day-to-day pressures at the start of any culture-changing transformation. This will be particularly true for those who supervise the union workers, for their jobs will have the most changes. We need to make sure they understand there is a short line of communications to address their concerns.

DETAILS OF THE EXECUTION PHASE

Step 1. A small lean support group, the lean support team (LST), will be formed to assist in JIT training and support

After the rollout, the leader of the LST should be selected. It may be another six months before any other members are added. As needs dictate, the group will expand, but very slowly. The LST leader will report directly to a member of the LLT. A small LST of three to six people, even for a very large refinery, is sufficient. They will provide four basic functions.

First, they will supply any technical support needed by the LLT.

Second, for any technical matter, they will be the in-house experts. They will not execute these techniques; rather, they will teach them, primarily to the LLT. The members of the LST will have direct support from the sensei.

Third, for some very deep and narrow problem-solving methods that are not widely used, they may be called in to assist. This would not include such topics as SPC (statistical problem solving), MSA (measurement system analysis), Kepner-Tregoe problem solving, or even DOE (designs of experiments), which should be widely used and understood. It is reserved for such narrow topics as response surface methodology, autonomous maintenance or multivariate analysis.

Fourth, they will do periodic reviews and the annual audits on overall system performance.

It is equally important to define what they do *not* do. They do not initiate activities, they do not lead kaizen events, they do not do routine problems-solving events, and they do not do mass teaching except to the LLT. All teaching is designed to be directed to the top management, and it is the managers' responsibility to teach their respective teams.

Step 2. Goals and initial areas of interest will be deployed via hoshin kanri methods including catchball

These include master strategies and substrategies. The normal monthly oils plan, as it has been in the past, will guide what the facility does. There will be some changes because of the HK planning, but for the most part, what was done in the past will continue to be done in the future. The key difference will be how those things are done. And that key area will depend on the "abnormal" things that occur in the refinery. Recall HK planning that was explained in step 7 of Table 8.2 in Chapter 8: "All control items need to have both normal results and abnormal results defined in quantitative terms." And in step 9, in the "check" phase, it is "To check annually, monthly, weekly, daily, and hourly."

Step 3. Local problem solving will begin

Then any variable that is abnormal will have action taken and, it is hoped, in real time. If it cannot be handled at the operator or mechanic level, it may need to be escalated to a support group. This escalation procedure needs to be quantified, explained, and monitored, and it needs to be visual (transparent).

Step 4. Necessary facility-wide training will be deployed

The next step of HK planning is to *act*. Acting on the results does not necessarily mean to change anything, although that may be the right answer. The first step of "action" is to understand. After checking and finding an abnormal situtation, this then needs to be resolved, which precipitates problem solving. At this point, new skills may need to be pulled in to assist the problem solving, and very often this comes in the form of training.

Using HK planning, we have now completed the plan, do, check, and act steps, and it is now a matter of sorting through priorities to assess what we need to do first. In addition, we may find a mismatch of the problems and the resources, i.e., people to complete the policy deployment of our HK planning.

There will also be checks on a daily and weekly basis and probably formal checks on a monthly, possibly quarterly, and certainly annual basis as we mark progress toward True North (see step 9 of the HK model in Chapter 8). Anytime these checks find a gap between the actual and the desired state, the responsible manager will begin corrective actions, first seeking the answer to question 4 of the Six Questions of Continuous Improvement, "What is something we can do, right now, to get closer to the desired condition?" The corrective actions, the countermeasures defined in the answer, are then designed to be responsive to the three parts of question 5:

1. What do we expect to happen?
2. How much of it will happen?
3. When will it happen?

And we start the cycle once again. And so it goes on forever in our quest for the ideal state. . .

CHAPTER SUMMARY

A lean transformation in any facility is a very effortful, management-intensive effort, and nowhere is it more intensive than in the preparation phase. There will be a great deal of time expended to do the five evaluations and probably more effort to complete the two sets of guidance documents. There will be a great deal of discussion and disagreement before the documents are finally agreed upon, and there is nothing wrong with this. Quite frankly, if these guidance documents are prepared quickly and easily, they will likely not be very useful. The rollout step is an opportunity to show solidarity and get everyone up to speed. Finally, we need to execute the True North and begin the process of changing the culture—our ultimate goal.

APPENDIX

CHAPTER 2
THE STORY OF TERRY'S DELI

I had stopped for lunch at Terry's when he asked me, "What do you do?" So I told him. Then he asked me, "How could that apply to my business?" I asked Terry if he had a few minutes. The lunch crowd had cleared and the deli was almost empty, and he said, "Sure." I then asked him what he would like to do to make his business more profitable. Immediately he replied, "I lose a lot of customers at lunch. They see a line at the register and just go somewhere else." Terry's Deli was in a small mall near a lot of other small businesses.

Most of the day, Terry had one person assigned to making sandwiches; at lunch there were two. This was Terry's lunchtime bottleneck: making sandwiches. I asked Terry to gather some data and tell me how long a line he could develop before people would pass up his business. I gave him a little data sheet that he diligently filled out, and it was pretty clear that with three or less in the line, business would continue to flow into his deli. And at six, for sure people were taking a peek in and bypassing his business.

Terry was very quality conscious, and all the meats, cheeses, vegetables, and breads were all freshly cut. He had one full-time prep person working most of the day, until the lunch rush was over; then he did other maintenance in the deli, including washing all the tables and chairs, cleaning the bathrooms, straightening up inventory, and placing orders for the next day.

With this in mind we devised a countermeasure for Terry. Whenever the line got to three people, he was to turn the prep guy into a sandwich maker, and when the line got to zero, he could go back into prep. Terry was a bit worried, because this meant the prep guy had to prep more and do it earlier, building up more WIP, anticipating the lunch rush. Terry was worried that the food might get stale. On this topic of freshness, Terry was adamant, and he said in a rather down-home fashion, "If it ain't fresh as daisies, we don't serve it!" Anyway, Terry proceeded with our plan.

And the results were amazing. Over the next month, he averaged $700 per week more revenue, almost all coming at lunch. And Terry confided in me that less than 30% of his costs were raw materials, so his gains were nearly $500 a week [$700 * (1.00 – 0.30)]. There was no more labor, no more building costs, no more utilities, and no more overhead. Since he had just broken the bottleneck to his lunchtime business, he was turning raw materials into finished goods at no incremental cost. It just doesn't get any sweeter. Later Terry told me that he unfortunately had to scrap around $50 per week of "stuff that wasn't as fresh as daisies." He said it bothered him deeply to scrap the food . . . but "values are values," as he told me.

And I never had to pay for another sub at Terry's deli.

Because either he was so successful or he just wanted to move, he relocated to the other side of town, where there is a much larger population, and he moved into a much larger location. As near as I could tell, Terry was prospering nicely; lean had helped him. And I had a cool vignette for my book.

CHAPTER 4
AN EXCERPT FROM
"LEAN MANUFACTURING IN THE OIL REFINERY"*

By Lonnie Wilson with Jason Farley

This paper offers a basic description of Lean and its origins along with a brief history explaining why Lean has yet to be fully embraced by the oil and gas industry.

*Originally published in June 2010 by B ridgeGap Consulting.

Figure A1

For years now Lean Manufacturing has provided discrete parts industries with a systems approach, tools and methods for delivering high quality products at lower costs on time, every time. Because the results have been fantastic on a large scale, many have adapted the methodology for service industries and government. However, migration into continuous process manufacturing has been slow, if not absent all together. In particular, energy sector businesses have been reluctant to investigate the potential of Lean Manufacturing. No doubt there have been legitimate reasons for continuous process operations, such as oil refining, to take such a position of reluctance even with the promise of tremendous benefits.

This paper suggests there are now, more than ever, compelling reasons for refineries and other continuous process operations to reconsider implementing Lean Manufacturing.

In the early 1980's US companies began to recognize that the TPS and similar methodologies offered a superior manufacturing model. What had become obvious to those paying attention is that these systems produced higher quality products with fewer resources. Soon, numerous interpretations and adaptations of what was being done in Japan on the whole began to spring up in US operations, primarily in discrete parts manufacturing. Much of the focus was on quality, the flagship of this movement being Total Quality Management based on The Deming Philosophy. However, with the publication of *The Machine That Changed the World,* significant attention was drawn to productivity and efficiency gains offered by the TPS. Pioneers began paving the way to what we know today as Lean Manufacturing.

Why Lean has not been more fully exploited in oil refining

Numerous case studies revealed significant improvements for those willing to adopt or adapt to the TPS. These case studies showed significant bottom line gains as well as the expansion of their business with greater job security for the rank and file and much happier customers. Many of these examples were almost too good to be true . . . but they *were* true, and the number of success stories multiplied in the 90's. However, some industries weren't buying it.

Seeking competitive advantage

So, if Lean was proving to be so effective, why does it appear that whole industry sectors seemed unimpressed and, generally disinterested? The many influences and factors that shape industries, determine business priorities and focus strategies are beyond the scope of this paper. However, one particular factor is poignant and worth mentioning as it moved several industries distinctly down a path toward Lean and left others seemingly apathetic. Consider the automotive and the electronics markets of the 80's and their highly competitive environments, especially with respect to international threats. Foreign competition forced them into an "up or out" situation, improve or else! It stands to reason that participants would be keen on any method, tool or strategy promising and then delivering competitive advantage. By contrast, the oil industry did not face the same level of foreign competition during the same period. Has the landscape changed? Truthfully, while global markets have changed, the competitive situation for American oil has not been affected much. However, while refineries are not facing the kind of survival issue that compelled American auto makers, other industry dynamics are evolving to create similar pressure. Changes in the economy and in management performance standards have created pressure to get more from less. And perhaps more influential, the threat of alternate energy sources has become substantial as society demands greener, more environment friendly processes for creating energy.

Defectives (PPM)

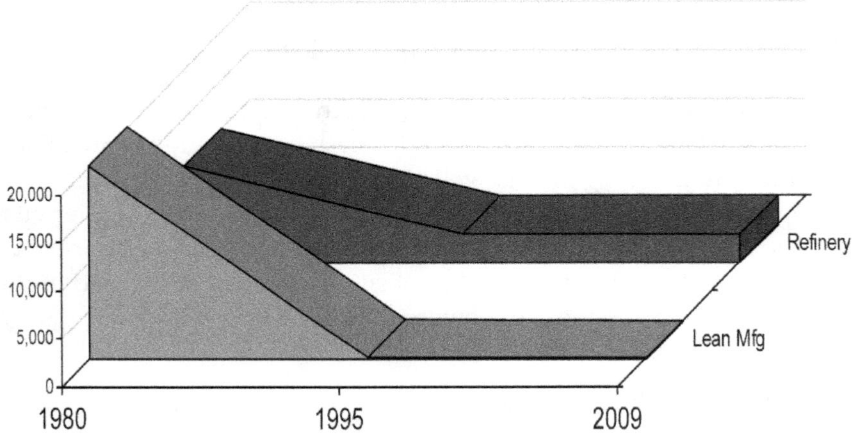

Inventory Turns

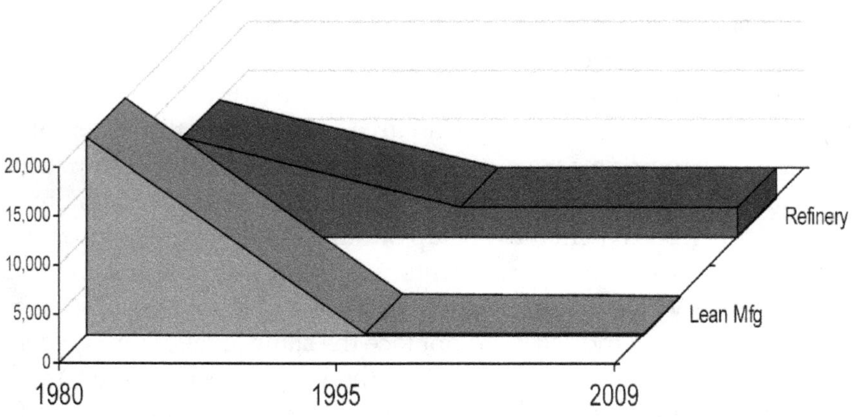

Figure A2 Continous process vs. Lean

Discrete vs. continuous

By 1995, it was clear that the TPS was a model that had broad application in not only manufacturing but the service sector as well. However, even given its popularity and promise, few continuous process manufacturing operations, such as the petroleum industry, had fully embraced the movement. For some reason, TPS was almost isolated to the discrete parts industry and the continuous process industries, including oil refining, remained basically unchanged. As it turns out, some of the initial rationale formed a pretty solid argument for avoiding the risk.

Consider two key business characteristics that are indicative of manufacturing performance: delivery speed as measured by inventory turns and quality as measured by defective product. In the 1980s,

Table *A1 Production Performance: Lean Discrete Parts vs. Continuous Process Manufacturing*

Lean Mfg (Discrete Parts)	1980	1995	2009
Inventory Turns	2–5	10–50	75–400
Defective Product (%)	2–10	0.01–1	25–500 ppm
Refining (Continuous Process)			
Inventory Turns	10–50	10–50	10–50
Defective Product (%)	1–3	0.3–2.0	3,000–20,000 ppm

typical oil industry processes actually were as good as, if not better than processes featured in a Lean model for both of these performance metrics. However, by the mid 90s the Lean operations had made profound improvements and delivered processes that had caught up to and even surpassed existing refinery process performance standards. Today, discrete manufacturing operations that employ Lean consistently outperform those found in typical oil refining models. *See Table A1—Production Performance.*

CHAPTER 8
THE FIVE TESTS OF MANAGEMENT COMMITMENT TO LEAN MANUFACTURING

1. Are you actively studying about and working at making your facility leaner and hence more flexible, more responsive, and more competitive? (All must continue to learn and must be actively engaged; no spectators allowed!)

2. Are you willing to listen to critiques of your facility and then understand and change the areas that are not lean? (We must be intellectually open.)

3. Do you honestly and accurately assess your responsiveness and competitiveness on a global basis? (We must be intellectually honest.)

4. Are you totally engaged in the lean transition with your:

   - Time
   - Presence
   - Management attention
   - Support (including manpower, capital, and emotional support)

   (We must be doing it; we must be on the floor, observing, talking to people, and imagining how to do it better. Lean implementation is not a spectator sport.)

5. Are you willing to ask, answer, and act on the question "How can I make this facility more flexible, more responsive, and more competitive?" (We must be inquisitive; willing to listen to all including peers, superiors, and subordinates alike, no matter how painful it may be; and then be willing and able to make the needed changes.)

If you have answered yes to all five questions, you have passed the commitment tests. If you have answered no to any question, there is an opportunity for management improvement.

You will find several forms to use in this evaluation, including evaluation of the refinery leadership team as well as evaluation of the union leadership. Go to www.wilsonleanrefining.com.

CHAPTER 10
THE MOTHERLOAD

There is a motherload of financial opportunities to be found using statistical process control and measurement system analysis.

Product specification is one such area that is very ripe for picking. Gasoline, diesel, and jet fuel, since they are high-volume products, are clear opportunities. After these have been addressed, you can use this pattern for any product.

The general pattern to capture these large financial opportunities is:

1. Evaluate the product.
2. Evaluate the run-down streams.
3. Model the system to optimize the specific streams and capture system interactions.

Specifically, this would look like:

Evaluate the product

1. For the specifications in question, carefully evaluate the specification and the test method. Do a rigorous evaluation of the measurement method including sampling integrity and all three MSA evaluations.
2. Place the data from each specification on a control chart and evaluate for stability. If not stable, problem solve until it is.
3. Perform a process capability study comparing the process target and variation with the specification.
4. Look for random causes of variation and reduce this variation using statistical problem solving.
5. Look for opportunities to "recenter" the process; problem solve to do so.

Evaluate the run-down streams

1. Repeat the above steps 1 to 5 for each component run-down stream. They will give you information to use if you need to change specifications for the component streams. Evaluating the run-down streams will give you clear opportunities to improve. My experience is that by tailoring these component streams, there are attractive economic opportunities that are frequently as much as were gained by the overall blend improvements when you "evaluated the product."

2. Review the major contributors to the specification in question; problem solve to reduce their impact while increasing volumes. Alter the specifications on each stream as needed.

Model the system to optimize the specific streams and capture system interactions

1. At the blender, perform an MSA on each analyzer.

2. Perform hand blends to confirm the blending algorithms.

3. Make a thorough check of all control loops and make sure they are on automatic. Check the feedback loops to assure they are operating properly and rapidly. My experience has been that they have significant errors. While they protect the blends from going off test, they are unnecessarily conservative with excessive giveaway and can be improved significantly.

4. Perform DOE (designs of experiments) to find new run-down optima and new blending optima.

This motherload is often called low-hanging fruit. Using SPC, MSA, and The Six Questions Of Continuous Improvement in this methodology, you can identify a number of very attractive projects in a refinery. The benefits are usually a huge financial gain that can almost always be achieved with zero or very little capital. To gain the full benefit of these projects, refer to Chapter 13, as it describes how the typical gains are augmented by using the entire toolkit of lean manufacturing.

BIBLIOGRAPHY

Akao, Yoji, *Hoshin Kanri: Policy Deployment for Successful TQM* (Productivity Press, Portland, OR, 1991).

Carroll, Joe, and Jessica Resnick-Ault, "Record Glut of Refineries Selling at 80% Discount: Real M&A," Bloomberg, February 21, 2011, Global Exchange, http://www.globalexchange.org/news/record-glut-oil-refineries-selling-80-discount-real-ma.

Deal, Terrence, and Allan Kennedy, *Corporate Cultures: The Rites and Rituals of Corporate Life* (Addison Wesley, Reading, MA, 1982).

Deci, Edward L., "Self-Determination Theory and the Facilitation of Intrinsic Motivation, Social Development, and Well-Being," *American Psychologist,* January 2000, vol. 55. PP 68-78

Deci, Edward L., and Richard M. Ryan, "The 'What' and 'Why' of Goal Pursuits: Human Needs and the Self-Determination of Behavior," *Psychological Inquiry*, 2000, vol. 11, no. 4, pp. 227–268.

Deming, W. Edwards, *Out of the Crisis* (MIT CAES, Cambridge, MA, 1982).

"Exclusive: After Motiva Split, Saudi Aramco Aims to Buy More U.S. Refineries—Sources," http://www.reuters.com/article/us-saudi-aramco-exclusive-idUUSKCN0WK2HX.

Hall, Robert, *Attaining Manufacturing Excellence: Just in Time, Total Quality, Total People Involvement* (McGraw-Hill, New York, 1987).

Hall, Robert, *Zero Inventories* (Business One Irwin, Homewood, Il, 1983).

Hersey, Paul, *The Situational Leader* (Center for Leadership Studies, Escondido, CA,1984).

Hopp, Wallace, and Mark Spearman, *Factory Physics*, 3rd ed. (McGraw-Hill, New York, 2008).

Ishakawa, Kaoru, *Guide to Quality Control* (Asian Productivity Organization, White Plains, NY, 1986).

Justin, Jules, *How to Manage with a Union* (Industrial Relations Workshop Seminars, New York, 1969), vols. 1 and 2.

Kepner, Charles, and Benjamin Tregoe, *The Rational Manager* (McGraw-Hill, New York, 1965).

King, Bob, *Hoshin Planning: The Developmental Approach* (GOAL/QPC, Methuen, MA,1989).

Kotter John, *Leading Change* (Kotter, HBS Press, Boston, MA, 1996).

Ohno, Taiichi, *Toyota Production System: Beyond Large-Scale Production* (Productivity Press, Portland, OR, 1988).

Oncken, William, Jr., *Managing Management Time: Who's Got the Monkey?* (Prentice Hall, Edgewood Cliff, NJ, 1984)

Rother, Mike, and John Shook, *Learning to See: Value Stream Mapping to Add Value and Eliminate MUDA* (Lean Enterprise Institute, Cambridge, MA, 1998).

Scherkenbach, William W., *The Deming Route to Quality and Productivity* (CEEP Press, Washington, D.C..1986)

Shingo, Shigeo, *A Revolution in Manufacturing: The SMED System* (Productivity Press, Portland, OR, 1985).

Shingo, Shigeo, *A Study of the Toyota Production System* (Productivity Press, Portland, OR, 1989).

Scholtes, Peter, and Brian Joiner, *Team Handbook*, 1st ed. (Joiner, Madison, WI, 1988).

Schonberger, Richard, *Japanese Manufacturing Methods: Nine Lessons in Simplicity* (Free Press, New York, 1982).

Schonberger, Richard, *World Class Manufacturing: The Lessons of Simplicity Applied* (Free Press, New York, 1986).

The Toyota Way, 2001 (Toyota Motor Corporation, 2001).

Wilson, Lonnie, *How to Implement Lean Manufacturing* (McGraw-Hill, New York, 2009).

Wilson, Lonnie, *How to Implement Lean Manufacturing,* 2nd ed. (McGraw Hill, New York, 2015).

Wilson, Lonnie, with Jason Farley, "Lean Manufacturing in the Oil Refinery," June 2010, by BridgeGap Consulting, http://www.bridgegapconsulting.com/sites/bridgegapconsulting.com/files/lean-oil.pdf.

Womack, James, *From Lean Tools to Lean Management* (Lean Enterprise Institute, Cambridge, MA, 2006).

Womack, James, *Lean Thinking* (Simon & Schuster, New York, 1996).

Womack, James, Daniel Jones, and Daniel Roos, *The Machine That Changed the World* (Free Press, New York, 1990).

INDEX

MEDICINAL PLANTS

WITH MACRO AND MICROSCOPIC STUDY

THE AUTHORS

Dr. Rajesh Sharma has 21 years of experience in teaching, research and extension in the discipline of Herbal field. He graduated in Ayurveda from Panjab University, Chandigarh. He did his M.D in the field of Dravyaguna (Herbal medicine) from university of Mumbai, did his Ph.D in Dravyaguna from Pune University, Maharashtra.

He has been teaching Dravyaguna (Herbal medicine) for the past 18 years. Presently he is working as Associate professor in "Ayurvedic and Unani Tibbia College" University of Delhi. He has published a number of articles/papers in reputed national and international journals.

Dr. Indrajit Singh Kushwaha has 11 years of experience in teaching and research in the discipline of Herbal medicine. He graduated in Ayurveda from Sampoornand Sanskrit University, Varanasi. He did his M.D in the field of Dravyaguna (Herbal medicine) from Dr. S. R. Rajsthan Ayurved University, Jodhpur, did his Ph.D in Dravyaguna from Dr. S. R. Rajsthan Ayurved University, Jodhpur, Rajasthan.

He has been teaching Dravyaguna (Herbal medicine) for the past 8 years. Presently he is working as Assistant professor in "Ayurvedic and Unani Tibbia College" University of Delhi. He has published a number of articles/papers in reputed national journals.

MEDICINAL PLANTS

WITH MACRO AND MICROSCOPIC STUDY

DR. RAJESH SHARMA
Associate Professor

DR. INDRAJIT SINGH KUSHWAHA
Lecturer
Ayurvedic & Unani,
Tibbia College,
Karol Bagh, New Delhi

2017

Daya Publishing House®

A Division of

Astral International Pvt. Ltd.

New Delhi-110 002

ISBN: 978-93-5130-666-5 (International Edition)

Published by : **Daya Publishing House®**
A Division of
Astral International Pvt. Ltd.
– ISO 9001:2015 Certified Company –
4736/23, Ansari Road, Darya Ganj
New Delhi-110 002
Ph. 011-43549197, 23278134
E-mail: info@astralint.com
Website: www.astralint.com

FOREWORD

It is gratifying to see the book has finally come up, beautifully well meeting the requirements of book lovers. Dravyaguna is a comprehensive discipline, which embraces various aspect of knowledge about the identification, collection and classification of drugs, their physicochemical properties, bio-transformation and metabolic activities, uses, dosage schedule, therapeutic actions, interaction, synergism, toxicity, etc. In present scenario majority of world population in developing countries relies on herbal medicines to meet the health need. The WHO is fully aware of the importance of herbal medicines for the health of many people, throughout the world and continuing extensive researches in this field for better results.

Dravyaguna Shastra has undergone a wide range of developments from *Vedic* period to the present Modern period. *Rig-Veda*, the oldest drug formulary of *Ayurveda* dealt with about 67 plants. Numbers of drugs have increased from Rig-Veda to *Atharvaveda.* At the end of post Vedic period scientific study of drugs, i.e. morphological and pharmacological studies were started and were recorded in *Samhita* and *Samgrah* respectively. About 526 and 576 drugs have been described in *Charak* and *Sushruta Samhita* respectively. 902 drugs have been mentioned in *Ashtanga Hridaya*. Pharmacodynamic principles of Ayurveda i.e. the concepts of Rasa, Guna, Veerya, Vipaka and Prabhava were established in the *Samhita* period. These are the basic concepts of Ayurvedic pharmacology which were relevant even today.

The pharmacological action of any drug is based on the pharmacodynamic properties of the drug such as Rasa, Guna, Virya, Vipaka and Prabhava, which are inherently present in the dravya. To know about the actions of drugs the knowledge of properties are very necessary.

This book contains sketch detail description of 100 plants like botanical name, family, synonyms, botanical description along with its macroscopic and microscopic study, habitat, chemical constituents, action and uses in various diseases, parts used, dose and formulation etc with monographs. References are provided from Ayurvedic texts and Ayurvedic pharmacopoeia.

This book is written by Dr. Rajesh Sharma, Associate Professor, A&U Tibbia College, Karolbagh, New Delhi and Dr. Indrajit Singh Kushwaha, Lecturer A&U Tibbia College, Karolbagh, New Delhi. They have critically examined the entire script and thoroughly spruced it up in the light of suggestions from seasoned fraternity. They have presented vivid description of each plant with its macroscopic and microscopic study to further elevate students' learning capacity. Today macroscopic and microscopic study is the need of student as per new guidelines of CCIM syllabus. After reading the book, student or reader would be fully geared to tackle any issue related to the syllabus and beyond.

Though this book is very much useful for B.A.M.S students, Research scholars and casual readers, they will definitely gain comprehensive knowledge about herbal drugs.

So I sincerely hope that this book will prove to be a boon to the readers.

Dr. Tanuja Nesari
Additional Director &Professor/H.O.D
Department of Dravyaguna
C.B.P. Charaka Ayurveda Sanshthan
Najabgarh, Delhi, India.

ACKNOWLEDGEMENT

It is a matter of great pleasure to express my heartfelt gratitude to my competent, committed honorable and adorable guide Dr. Mita Kotecha, Professor/HOD, Post graduate Department of Dravyaguna, N.I.A. Jaipur, for this scientific direction, scholastic critical suggestions, encouragement and parental affection to execute the literary work with excellent expedient supervision during the whole period. She enlightens me in the present shape with his majestic touch and imparts confidence in me to combat with any situation in the field of Dravyaguna.

We are greatly indebted to esteemed and honorable Prof. Tanuja Nesari, Head of Department, Dravyaguna, C.B.Charak Ayurvedic Sansthan for give me the necessary direction for writing the present work. We are soliciting our due regards to Prof. M. C. Sharma, Ex-Director, N. I. A. for his generous support and guidance for time to time.

Our sincere thanks to Dr. Sakshi Sharma for their unflinching support, invaluable help, suggestions & extremely humble attitude to write this book.

The friends are the essence of life and it is beyond the reach of my language to oblige the affection and support of my respected seniors Dr. H.C. Gupta.

The words can express only thoughts but cannot express feelings; We feel here for my loving teacher Dr. A. N. Singh for his continuous encouragement and mental support during entire period of book writing.

Pay my heartful adoration to my most beloved and respected parents Shri. R. S. Kushwaha and Smt. Bhagawani Devi for their love and care thoughts, their constant preaching sacrifices, attention and blessing, we have able to attain my present position.

Lastly we wish to acknowledge our obligation and thanks to Dr.......Shrivastava for his pain taking effort to complete this book in present shape. We are extremely grateful to all those who directly or indirectly rendered their service to complete this dissertation.

Dr. Indrajit Singh Kushwaha & Dr. Rajesh Sharma

PREFACE

Since ages herbs and herbal products are being used as medicines. The World Health Organization (WHO) estimates that 80% of the World's population still relies on herbal medicines as its major source of medicinal products. India is one of the richest Biodiversity and herbal wealth in the world.

We are delighted to present this book worth mentioning that the textual matter of this book contains only important characters/findings under each subhead, to give a general and broad idea of the whole plant and its properties. The species described here are arranged in a chronological order bases on Ayurvedic nomenclature. Care has been taken to give Order. Classical names include only those mentioned in ancient Ayruvedic literature. The controversial names, particularly those which are used for more than one botanical entity have been omitted. Vernacular names include only those which are most commonly used and limited to important regional languages and trade.

Botanical description includes habit of the plant and important morphological features, which would be helpful in tentative identification of the plant in the field. As It present a brief and concise account of parts used, classical and vernacular names, botanical characters, distribution in India, important action and uses, Ayurvedic properties, pharmacognostic characters, chemical constituents, pharmacological activities, therapeutic evaluation, along with photographs of the plants and parts used.

Pharmacognosy mostly includes important macro and microscopic characters and Physical constants. Major chemical constituents, important pharmacological activities, and findings related to therapeutic evaluation as and where available have been included under the respective subheads. Under formulation and preparations, only classical Ayurvedic preparations are include, patented and proprietary items are not included.

We are confident that this book will also be appreciated by the students, researchers of various disciplines like Botany, Chemistry, Pharmacology, Pharmacognosy, Agriculture, Forestry apart from Ayurveda, Siddha and other traditional systems of Medicine as well as Modern Medicine. It would be useful for the persons/agencies interested in Medicinal Plants and Ayurveda including Pharmaceutical Industries.

We express our sincere gratitude to Dr. Sakshi Sharma Research Officer (Ay.) for her keen interest and who acted as driving force in this publication.

Dr. Rajesh Sharma

Dr. Indrajit Singh Kushwaha

CONTENTS

TULSI

Botanical name - *Ocimum sanctum* Linn.

Family - Labiatae

Synonyms - Tulsi, Surasa, Gauri, Bhutaghni

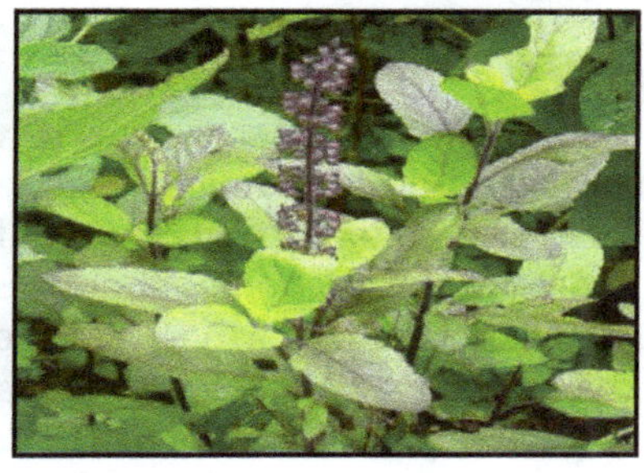

Botanical description- It is an erect, much branched, soft hairy, annual shrub, 30-75 cm hight. Leaves entire or serrate, minutely gland dotted. Flower purplish or crimson, in close whorled racemose. Seeds are small, pale black.

a) Macroscopic

Root - Thin, wiry, branched, hairy, soft, blackish-brown externally and pale or violet internally.

Stem - Erect, herbaceous, woody, branched; hairy, sub quadrangular, externally purplish-brown to black, internally cream, coloured; fracture, fibrous in bark and short in xylem; odour faintly aromatic.

Leaf - 2.5-5 cm long 1.6 - 3.2 cm wide, elliptic oblong, obtuse or acute, entire or serrate, pubescent on both sides; petiole thin, about 1.5-3 cm long hairy; odour, aromatic; taste, characteristic.

Flower - Purplish or crimson coloured, small in close whorls; bracts about 3 mm long and broad, pedicels longer than calyx, slender, pubescent; calyx ovoid or campanulate 3-4 mm bilipped, upper lip broadly obovate or suborbicular, shortly apiculate, lower lip longer than upper having four mucronate teeth, lateral two short and central two largest; corolla about 4 mm long, pubescent; odour, aromatic; taste, pungent.

Fruit - A group of 4 nutlets, each with one seed, enclosed in an enlarged, membranous, veined calyx, nutlets sub-globose or broadly elliptic, slightly compressed, nearly smooth; pale brown or reddish with small black marking at the place of attachment to the thalamus; odour, aromatic; taste, pungent.

Seed - Rounded to oval; brown, mucilaginous when soaked in water, 0.1 cm long, slightly notched at the base; no odour; taste, pungent, slightly mucilaginous.

b) Microscopic

Root - Shows a single layered epidermis followed by cortex, consisting of seven or more layers of rectangular, round to oval polygonal, thin-walled, parenchymatous cells, filled with brown content, inner layers of cortex devoid of contents; phloem consisting of sieve elements, thin-walled, rectangular parenchyma cells and scattered groups of fibres, found scattered in phloem; xylem consists of vessels, tracheids, fibres and parenchyma; vessels pitted; fibre tracheides, long, pitted with pointed ends; fibres thick walled and with pointed ends.

Stem - Shows a single layered epidermis with uniseriate, multicellular covering trichomes having 5-6 cells, occasionally a few cells collapsed; cortex consists of 10 or more layers of thin-walled, rectangular, parenchymatous cells; phloem consists of sieve elements, thin-walled, rectangular parenchyma cells and fibres; fibres found scattered mostly throughout phloem, in groups and rarely in singles; xylem occupies major portion of stem consisting of vessels, tracheids fibres and parenchyma; vessels pitted; fibres with pointed ends; centre occupied by nan-ow pith consisting of round to oval, thin-walled, parenchymatous cells.

Leaf

Petiole - Shows somewhat cordate outline, consisting of single layered epidermis composed of thin-walled, oval cells having a number of covering and glandular trichomes; covering trichomes multicellular 1-8 celled long,rarely slightly reflexed at tip; glandular trichomes short, sessile with 1-2 celled stalk and 2-8 celled balloon-shaped head, measuring 22-27 in dia; epidermis followed by 1 or 2 layers and 2 or 3 layers of thin-walled, elongated, parenchyma cells towards upper and lower surfaces respectively; three vascular bundles situated centrally, middle one larger than other two; xylem surrounded by phloem.

Midrib - Epidermis, trichomes and vascular bundles similar to those of petiole except cortical layers reduced towards apical region.

Lamina - Epidermis and trichomes similar to those of petiole; both anomocytic and diacytic type of stomata present on both surfaces, slightly raised above the level of epidermis; palisade single layered followed by 4-6 layers of closely packed spongy parenchyma with chloroplast and oleo-resin; stomatal index 10-12-15 on upper surface and 14 - 15 - 16 on lower surface; palisade ratio 3.8; vein islet number 31 - 35.

Powder - Greenish: shows thin-walled, parenchymatous cells, a few containing reddish brown contents, unicellular and rnulticellular-trichomes either entire or in pieces; thin walled fibres, xylem vessels with pitted thickenings, fragments of epidermal cells in surface view having irregular shape, oil globules, rounded to oval, simple as well as compound starch grains having 2-5 components, measuring 3-17 μ in diameter.

Chemical composition- It contains pale yellow-green essential oil which becomes crystslline over a period.

Habitat- It is found throughout India ascending up to 1800 mt in the Himalaya, commonly cultivated in gardens.

Properties- Rasa- katu, tikta, Guna- laghu, ruksha Veerya- ushna, Vipaka-katu

Doshakarma- Kaphavatashamaka

Action and Uses- The plant is bitter aromatic, digestive, diuretic, vermifuge, stomachic. It is useful in cardiopathy, blood disorder, leucoderma, asthma, bronchitis, vomiting, gastropathy, ring worm, genito-urinary disorders & skin diseases.

Useful parts - Whole plant, leaf, seed.

Dose: juice- 5 - 10 ml; powder - 1-3 g.

Main preparation - Tulsi swarasa.

YASHTIMADHU

Botanical name - *Glycyrrhiza glabra* Linn.

Family - Fabaceae

Synonyms - Yashtimadhu, Yashti, Kleetaka, Madhusrava.

Botanical description - It is a hardy herb or undershrub attaining a height up to 2mt. Root is long, reddish yellow and bark give multiple branches. Leaves are compound multifoliate, flower in axillary spikes, containing reniform seeds.

a) Macroscopic

Stolon consists of yellowish brown or dark brown outer layer, externally longitudinally wrinkled, with occasional small buds and encircling scale leaves, smoothed transversely, cut surface shows a cambium ring about one-third of radius from outer surface and a small central pith, root similar without a pith, fracture, coarsely fibrous in bark and splintery in wood, odour, faint and characteristic, taste, sweetish.

b) Microscopic

Stolon- Transverse section of stolon shows cork of 10-20 or more layers of tabular cells, outer layers with reddish-brown amorphous contents, inner 3 or 4 rows having thicker, colourless walls, secondary cortex usually of 1-3 layers of radially arranged parenchymatous cells containing isolated prisms of calcium oxalate, secondary phloem a broad band, cells of inner part cellulosic and outer lignified, radially arranged groups of about 10-50 fibres, surrounded by a sheath of parenchyma cells, each usually containing a prism of calcium oxalate about 10-35 μ long, cambium form tissue of 3 or more layers of cells, secondary xylem distinctly radiate with medullary rays, 3-5 cells wide, vessels about 80-200 μ in diameter with thick, yellow, pitted, reticulately thickend walls, groups of lignified fibres with crystal sheaths similar to those of phloem, xylem parenchyma of two kinds, those between the vessels having thick pitted walls without inter-cellular spaces, the remaining with thin walls, pith of parenchymatous cells in longitudinal rows, with inter-cellular spaces.

Root-Transverse section of root shows structure closely resembling that of stolon except that no medulla is present, xylem tetrarch , usually four principal medullary rays at right angles to each other, in peeled drug cork shows phelloderm and sometimes without secondary phloem, all parenchyrnatous tissues containing abundant, simple, oval or rounded starch grains, 2-20 µ in length.

Constituents - Glycyrrhizin, glycyrrhizic acid, glycyrrhetinic acid, asparagine, sugars, resin and starch.

Habitat- Afganistan, China, Europe, In India it is found in Deheradun & Jammu.

Properties: Rasa- madhura, Guna- guru, snigdha, Veerya- sheeta, Vipaka - -madhura.

Doshakarma - Vatapittashamaka

Action and Uses - The root are sweet, emetic in large doses, tonic, diuretic, mild laxative and intellect promoting. They are useful in hyperdypsia, cough and gastric ulcer. Decoction of root is good for falling and greying of hair.

Part used - Root.

Doses - Powder - 3- 5 g.

Main preparation - Yashtyadi churna, Shatavaaryadi ghrita.

JEERAKA

Botanical name - *Cuminum cyminum* Linn.

Family- Umbelliferae

Synonyms- Ajajika, Dipya, Dipyaka.

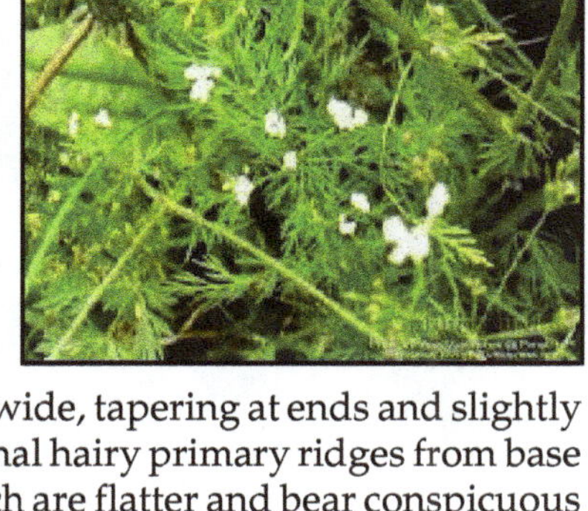

Botanical description- Smaller slender annual herb up to 35 cm hight with much branched angular weak stem. Leaves are 5-10 cm long alternate. Flowers are small white or pink. Fruits 5-7 mm long cylindrical, greyish, brownish, tapering towards both ends and compressed laterally.

a) Macroscopic

Fruit, a cremocarp, often separated into mericarps, brown with light coloured ridges ellipsoidal, elongated, about 4-6 mm long, 2 mm wide, tapering at ends and slightly compressed laterally, mericarps with 5 longitudinal hairy primary ridges from base to apex, alternating with 4 secondary ridges which are flatter and bear conspicuous emergences, seeds orthospermous, odour umbelliferous characteristic, taste, richly spicy.

b) Microscopic

Transverse section of fruit shows epidermis consisting of short polygonal, tabular cells densely covered with short, bristle hairs on ridges, mesocarp with few layers of parenchyma and five vascular bundles under five primary ridges, six vittae under secondary ridges, four on dorsal and two on commissural surface, endocarp consists of polygonal cells containing fixed oil and aleurone grains carpophore consists of slender fibres.

Chemical composition- It contains 3.5 to 5.2 % volatile oil called as thymine, because of which it has peculiar taste and fragrance.

Habitat- Cultivated as a cold season crop on the plains and as summer crop on the hills in Northern India, Himalayas, Himachal, Punjab and Kashmir.

Properties- Rasa- katu, Guna- laghu, ruksha, Vipaka- katu, Veerya- ushna.

Action and uses- The fruit is aromatic, acrid, sweet, astringent, carminative, anthelmintic, anti- diarrhoel, anti-dysenteric, anti-inflammatory, stomachic, uterine and nervine stimulant.

Part used- Fruits

Doses-: Powder- 1-3 g, decoction- 10-25ml.

Main preparation- Jeerakadi modak, Jeerakadi churna.

CHITRAK

Botanical name - *Plumbago zeylanica* Linn.

Family- Plumbaginaceae

Synonyms- Chitraka, Agni, Ushana, Pathi, Vyal.

Botanical description -It is a perennial shrub, 60-120 cm high. Leaves are alternate, ovate. Flowers are white, in bracteates, glandular and elongated spikes, 10- 30 cm long. Fruits are legume, oval shaped with cover. It is sticky to touch. Each fruit has one oval seed. Roots are finger like thick red from outside but white inside. Roots are pungent in test.

a) Macroscopic

Roots 30 cm or more in length, 6 mm or more in diameter as also as short stout pieces, including root stocks reddish to deep brown, scars of rootlets present, bark thin and brown, internal structure striated, odour-disagreeable-taste, acrid.

b) Microscopic

Transverse section of root shows outer most tissue of cork consisting of 5 -7 row, of cubical to rectangular dark brown cells, secondary cortex consists of 2-3 rows of thin-walled rectangular, light brown cells, most of the cortex cells contain starch grains, secondary cortex followed by a wide zone of cortex, composed of large polygonal to tangentially elongated parenchymatous cells varying in size and shape, containing starch grains and some cells with yellow contents, fibres scattered singly or in groups of 2-6, phloem a narrow zone of polygonal, thin-walled cells, consisting of usual elements and phloem fibres, similar to cortical zone, phloem fibres usually in groups of 2-5 or more but occasionally occurring singly, lignified with pointed ends and narrow lumen, similar in shape and size to those of secondary cortex, cambium indistinct, xylem light yellow to whitish, vessels radially arranged with pitted thickenings, medullary rays straight, 1-6 seriate, cells radially elongated starch filled with starch grains, stone cells absent.

Chemical Composition - Chitrak contains number of chemicals these are Plumbagin.3-chloroplumbagin, 3, 3–biplumbagin, Ellptinone, Chitranone, Zeylinone 3, Isozeylinone, Droswrone, Plumbagic acid, Plumbazeylanone, Naphthelenone, Isoshinanolone.

Habitat- Wildly grows throughout India, cultivated in gardens.

Properties: Rasa- katu, Guna- laghu, ruksha, Veerya- ushna, Vipaka- katu.

Action and uses - It acts as antipyretic, appetizer, uterotonic, antibacterial, antifungal, anticancer (plumbagin), anticoagulant, hepatoprotective and Central nervous system depressant. It is used in intestinal troubles, dysentery, leucoderma, inflammation, piles, bronchitis, diseases of liver and ascitis.

Part used- Root, root bark.

Dose- 1-2 g.

Main preparation- Chitraka Haritaki, Chitrakadi Vati.

KAPIKACHCHHU

Botanical name - *Mucuna pruriens* Linn.

Family- Leguminosae

Synonyms - Atmagupta, Kandura, Markati.

Botanical description- It is an herbaceous twining annual, leaves trifoliate. Flowers are purple in colour. Pods 5-10 cm long, curved. Seeds are 4-6 in a pod, ovoid, black or dull black.

a) Macroscopic

Seed ovoid, slightly laterally compressed, with a persistent oblong, funicular hilum, dark brown with spots; usually 1.2-1.8 cm long, 0.8-1.2 cm wide, hard, smooth to touch, not easily breakable; odour, not distinct; taste, sweetish-bitter.

b) Microscopic

Mature seed shows a thin seed-coat and two hard cotyledons; outer testa consists of single layered palisade-like cells; inner testa composed of 2 or 3 layers, outer layer of tangentially elongated, ovoid, thin-walled cells, inner 1 or 2 layers of dumb-bell or beaker-shaped, thick-walled cells; tegmen composed of a wide zone of oval to elliptical, somewhat compressed, thin-walled, parenchymatous cells; some cells contain starch grains; cotyledons composed of polygonal, angular, thin-walled, compactly arranged, parenchymatous cells, containing aleurone and starch grains; starch grains small, simple, rounded to oval measuring 6-41 µ in dia., but not over 45 μ in dia.; a few vascular bundles with vessels showing reticulate thickening or pitted present,

Powder - Pale cream coloured; shows fragments of testa with palisade-like cells thin-walled parenchyma, reticulate and pitted vessels, aleurone and starch grains small, simple, rounded to oval measuring 6-41 μ in dia., but not over 45 μ. in dia.

Chemical composition – The seeds contains L-dopa, proteins, oil and manganese in small amount. The trichimes contain 5-HT and an enzyme protease- mucunain. Fixed Oil, Alkaloid and 3,4-Dihydroxyphenylalanine.

Habitat- It is found allover India at the height up to 1000m in Himalaya.

Properties: Rasa-madhur, tikta Guna-guru, snigdha Veerya-ushna, Vipaka- madhura.

Doshakarma- Vatashamaka

Action and uses - It is stimulant, purgative, aphrodisiac, diuretic, anthelmintic and tonic. It is useful in constipation, nephropathy, dysmenorrhoea, amenorrhoea, elephantiasis, dropsy, neuropathy, ulcers, fever and delirium.

Part used- Root, leaf, seed.

Dose: Seed powder- 3-6 gm, Root decoction- 50- 100ml.

Main preparation- Kaunch paak, Vanari gutika.

APAMARGA

Botanical name - *Achyranthes aspera* Linn.

Family - Amaranthaceae

Synonyms - Adhashalya, Kharamanjari, Mayuraka.

Botanical description - It is annual herb 30-90 cm tall, often with a woody base. Leaves are variable ovate or obovate- rounded. Flowers are greenish white. Seeds are subcylindrical redish brown.

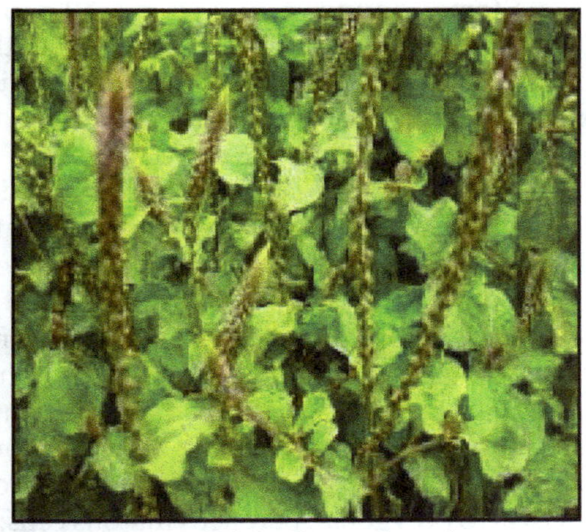

a) Macroscopic

Root - Cylindrical tap root, slightly ribbed, 0.1-1.0 cm in thickness, gradually tapering, rough due to presence of some root scars, secondary and tertiary roots present, yellowish-brown; odour, not distinct.

Stem - 0.3 - 0.5 cm in cut pieces, yellowish-brown, erect, branched, cylindrical, hairy, solid, hollow when dry.

Leaf - Simple, subsessile, exstipulate, opposite, decussate, wavy margin, obovate, slightly acuminate and pubescent due to the presence of thick coat of long simple hairs.

Flower - Arranged in inflorescence of long spikes, greenish-white, numerous, sessile, bracteate with two bracteoles, one spine lipped, bisexual, actinomorphic, hypogynous; perianth segments 5,free, membranous, contorted or quincuncial, stamens 5, opposite, the perianth lobes, connate forming a membranous tube-like structure, alternating with truncate and fimbriate staminodes, filament short; anther, two celled, dorsifixed; gynoecium bicarpellary, syncarpous; ovary superior, unilocular with single ovule; style, single; stigma, capitate.

Fruit - An indehiscent dry utricle enclosed within persistent, perianth and bracteoles.

Seed - Sub-cylindric, truncate at the apex, round at the base, endospermic, brown.

b) Microscopic

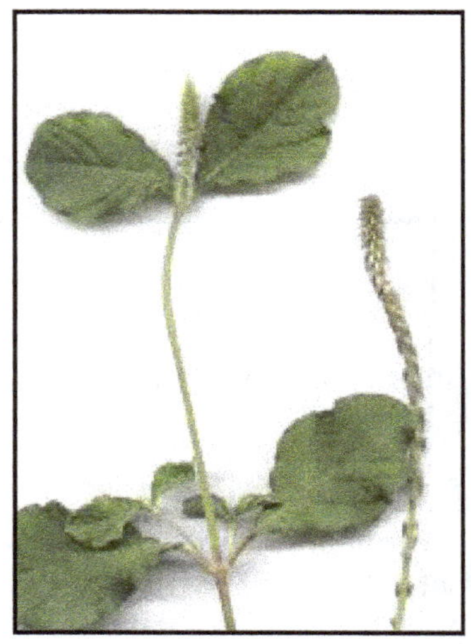

Root - Mature root shows 3-8 layered, rectangular, tangentially elongated, thin-walled cork cells; secondary cortex consisting of 6-9 layers, oval to rectangular, thin-walled, parenchymatous cells having a few scattered single or groups of stone cells; followed by 4-6 discontinuous rings of anomalous secondary thickening composed of vascular tissues; small patches of sieve tubes distinct in phloem parenchyma, demarcating the xylem rings; xylem composed of usual elements; vessels simple pitted; medullary rays 1-3 cells wide; small prismatic crystals of calcium oxalate present in cortical region and numerous in medullary rays.

Stem - Young stem shows 6-10 prominent ridges, which diminish downwards upto the base where it becomes almost cylindrical; epidermis single layered, covered by thick cuticle having uniseriate, 2-5 celled, covering trichomes and glandular with globular head, 3-4 celled stalk; cortex 6-10 layered, composed of parenchymatous cells, most of them containing rosette crystals of calcium oxalate; in the ridges cortex collenchymatous; vascular bundles lie facing each ridge capped by pericyclic fibres; transverse section of mature stem shows lignified, thin-walled cork cells; pericycle a discontinuous ring of lignified fibres; vascular tissues show anomalous secondary growth having 4-6 incomplete rings of xylem and phloem; secondary phloem consisting of usual elements form incomplete rings; cambial strip present between secondary xylem and phloem; secondary xylem consisting of usual elements, fibres being absent; vessels annular, spiral, scalariform and pitted, fibres pitted, elongated, lignified; pith wide consisting of oval to polygonal, parenchymatous cells; two medullary bundles, either separate throughout or found in some cases, present in pith; micro-sphenoidal silica crystals present in some epidermal, cortical and pith cells.

Leaf

Petiole - Shows crescent-shaped outline, having single-layered epidermis with thickcuticle; ground tissues consisting of thin-walled, parenchymatous cells containing rosette crystals of calcium oxalate; 4-5 vascular bundle situated in mid region.

Midrib - Shows a single layered epidermis, on both surfaces; epidermis followed by 4-5 layered collenchyma on upper side and 2-3 layered on lower side; ground tissue consisting of thin-walled, parenchymatous cells having a number of vascular bundles; each vascular bundle shows below the xylem vessels, thin layers of cambium, followed by phloem and a pericycle represented by 2-3 layers of thick-walled, non-lignified cells; rosette crystals of calcium oxalate found scattered in ground tissues.

Lamina - Shows single layered, tangentially elongated epidermis cells covered with thick cuticle having covering trichomes which are similar to those of stem found on both surfaces; mesophyll differentiated into palisade and spongy parenchyma; palisade 2-4 layered of thick parenchyma larger, slightly elongated in upper, while smaller and rectangular in lower surface; spongy parenchyma 3-5 layers thick, more or less isodiametic parenchymatous cells; idioblast containing large rosette crystals of calcium oxalate distributed in palisade and spongy parenchyma cells; stomata anisocytic and anomoacytic in both surface; stomatal index 4.5-9.0 on upper surface, 9.0-20.0 on lower surface; palisade ratio 7.0-11; vein islet number 7-13 per sq. mm.

Powder - Light yellow; shows fragments of elongated, rectangular, thin-walled epidermal cells, aseptate fibres, vessels with annular, spiral, scalariform and pitted thickening, uniseriate hair with bulbous base, rosette and prismatic crystals of calcium oxalate.

Habitat – It is found throughout India, upto an altitude of 2100 m and in the ground.

Constituents – The seeds and panchang contain potassium salts as well as Saponins.

Properties – Rasa- katu, tikta, Guna- laghu, ruksha, tikta, Veerya- ushna, Vipaka -katu.

Doshakarma – Kaphavatashamaka.

Action and uses – The plant is acrid, bitter, thermogenic, expectorant, carminative, digestive, laxative, anthelmintic, diuretic, haematinic and anti-inflammatory.

Part used - Whole plant.

Dose - Root powder - 3-6 g, seeds- 3g, kshar- 1/2-2g.

Main prepation - Apamarg kshar.

ARKA

Botanical name - *Calotropis procera* Ait.

Family- Asclepiadaceae

Synonyms - Arka, Raktark, Toolaphal, Ksheerparna

Botanical description – It is a small, erect and compact shrub, 1-2 m height. Leaves sub sessile, broadly ovate. Upper surface of the leaf is smooth and dorsally it is hairy. Flowers are white, purple spotted. Flower stalk grows in the acute angle of leaf and it has many branches. Fruits are long, curved and break open on drying.Soft cottom come out when it gets perforated automatically.

a) Macroscopic

Root- rough, fissured longitudinally, corky and soft, externally yellowish-grey while internally white, central core cream coloured, bark easily separated from xylem, odour, characteristic: taste, bitter and acrid.

b) Microscopic

Transverse section of root shows outer most cork tissue consisting of 4-8 rows of tangentially elongated and radially arranged cells followed by 3-6 rows of moderately thick-walled, irregular cells of secondary cortex devoid of calcium oxalate crystals and starch grains, cortex composed of large polyhedral parenchymatous cells containing abundant rounded starch grains, some cortical cells contain rosette crystals of calcium oxalate, scattered laticifer cells with brown contents, phloem consists of sieve elements and phloem parenchyama, sieve tubes thick-walled, cells more prominent towards inner region of phloem traversed by uni to tetraseriate medullary rays, phloem cells contain crystals of calcium oxalate, starch grains and laticifers similar to these found in cortex: cambium present just within the phloem consisting of 2-5 rows of thin-walled, tangentially elongated cells xylem forms the central part of root composed of vessels. tracheids, fibres and xylem parenchyma, vessels present throughout xylem region and arranged radially in groups of 2-7, sometime single vessels also occur, usually cylindrical having bordered pits on their walls, xylem fibres long, lignified with wide lumen, tapering on ends and have simple pits on walls, medullary rays 1-4 seriate and triseriate in outer region and uni or biseriate in inner region: cells of medullary rays radially elongated, filled with starch similar to those present in cortical cells.

Habitat - It is often found throughout India.

Chemical Constituents - Glycosides (calotropin).

Varieties- Shweta and Rakta.

Properties: Rasa- katu, tikta, Guna- laghu, ruksha, tikshna, Veerya- ushna, Veepaka-katu.

Doshakarma - Kaphavatashamaka.

Action and uses - The whole dried plant is bitter, thermogenic, laxative, anthelmintic, anticarcinogenic, expectorant and good tonic. It is useful in intestinal worms and ulcers. The root bark is bitter, anthelmintic, laxative and is useful in cutaneous diseases, intestinal worms, cough, ascitis and anasarca.The powdered root promotes gastric secretions and is useful in bronchitis, dyspepsia, gastroenteritis, dysentery, piles, boils, scrotal enlargement, filariasis and cancer.

Part used- Root, root bark, leaf, latex.

Dose: Root bark - 0.5 - 1 g, latex- 250 – 750 mg, flower -1-3g.

Main preparation - Arka kshar.

ASHOKA

Botanical name - *Saraca ashoka* Roxb De wild.

Family - *Leguminosae*

Synonymes - Madhupuspa, Rakta pallav, Hemapuspa, Gatashoka.

Botanical description - It is a small evergreen tree, 6-10 mtr in height. This tree is resembled to mango tree. Bark -with warty surface, dark brown to grey. Leaves are green peripinnate 15-20cm long like mango tree. Flowers - vermillion red and attractive, arranged in a cluster. Fruit- pods are flat and 8 to 25 cm long.

a) Macroscopic

Bark channelled, externally dark green to greenish grey, smooth with circular lenticels and transversely ridged, sometimes cracked, internally reddish-brown with fine longitudinal strands and fibers, fracture splintery exposing striated surface, a thin whitish continuous layer is seen beneath the cork layer, taste, astringent.

b) Microscopic

Transverse section of stem bark shows periderm consisting of a wide layer of cork, radially flattened, narrow cork cambium, secondary cortex wide with one or two continuous layers of stone cells with many patches of sclereids, parenchymatous tissue contains yellow masses and prismatic crystals: secondary phloem consists of phloem parenchyma, sieve tubes with companion cells and phloem fibres occuring in groups, crystal fibres present.

Chemical Composition - In ashoka so many phytochemicals are found in different parts these are sapogenetic glycoside, sterols and aliphatic alcohols (plant);Tannins and catechol, ester, free primary alcohol in (wax); Leucopelargonidin and leucocyanidin (bark); Oleic, linoleic, palmitic and stearic acids (seeds); Palmitic, stearic, linolenic, leucocyanidin and gallic acid (flowers) and Catechol, quercetin in wood .

Habitat - Grown in shady evergreen forests in Central and Eastern Himalayas, Abundant in South India.

Properties: Rasa - kashaya, tikta, Guna- laghu, ruksha, Veerya-sheeta, Veepak-katu.

Action and uses- This plant is used in thirst, burning sensation, worms, poison disorder, cardiac diseases, leucorrhoea, renal stones, urinary disorders and in female uterine disorders.

Part used - Stem, bark, seeds.

Doses - Decoction- 50 -100 ml, seed powder - 3-6 g.

Main preparation - Ashoka ghrit, Ashokaristha.

AMALAKI

Botanical name - *Emblica officinalis* Gaertn.

Family - Euphorbiaceae

Synonyms - Amalaki, Dhatri, Amla.

Botanical description - A tree of medium height ranging from 20 - 25 feet. The leaves are small and arranged on both side of main stem in a wing like order. The leaves are basically oval in shape. Flowers are pale yellow coloured in the form of inflorescence. Fruits are round, fleshy, green in colour and on ripping turned yellow reddish.

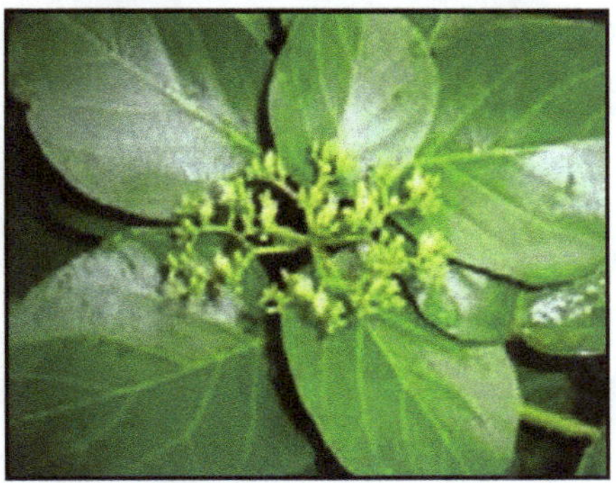

a) Macroscopic

Fruit, globose, 2.5-3.5 cm in diameter, fleshy, smooth with six prominant lines; greenish when tender, changing to light yellowish or pinkish colour when mature, with a few dark specks: taste, sour and astringent followed by delicately sweet taste.

b) Microscopic

Transverse section of mature fruit shows an epicarp consisting of single layer of epidermis and 2-4 layers of hypodermis; epidermal cell, tabular in shape, covered externally with a thick cuticle and appear in surface view as polygonal; hypodermal cells tangentially elongated, thick-walled, smaller in dimension than epidermal cells; mesocarp forms bulk of fruit, consisting of thin-walled parenchymatous cells with intercellular spaces, peripheral 6-9 layers smaller, ovoid or tangentially elongated while rest of cells larger in size, isodiametric and radially elongated; several collateral fibrovascular bundles scattered throughout mesocarp consisting of xylem and phloem; xylem composed of tracheal elements, fibre tracheids and xylem fibres; tracheal elements show reticulate scalariform and spiral thickenings; xylem fibres elongated with narrow lumen and pointed end; mesocarp contains large aggregates of numerous irregular silica crystals.

Chemical Composition- A good source of vitamin C, carotene, nicotinic acid, riboflavine, D-glucose, D-fructose, myoinositol and a pectin with D-galacturonic acid, D-arabinosyl, D-xylosyl, L-rhamnosyl, D-glucosyl, D-mannosyl, and D-galactosyl residues, embicol, mucic, Indole acetic acid, phyllembic acid and phyllembin (fruits) and fatty acids (seed oil), leucodelphinidin, procyanidin, 3-0-galleted prodelphinidin and tannin (bark), ellagic acid, lupeol, oleanolic aldehyde and 0-acetyl, oleanolic acid (root), tannins, polyphenolic compounds; 1,2,3,6- trigalloylglucose, terchebin, corialgin, ellagic acid, alkaloids, phyllantidine and phyllantine (leaves and fruits).

Habitat - It is found all over India.

Varieties - Wild & Domestic varieties are two basic varieties in the fruit.

Properties - Guna - guru, ruksha, sheeta, Rasa – Panchrasa (without lavana), Veerya - sheeta Vipaka - madhur.

Doshakarma - Tridoshashamaka.

Action and uses - Paste is applied locally in burning, headache due to Pitta and retention of urine. Juice is used in eye disorder. It acts in loss of appetite, anorexia, constipation, liver disorders, hyperacidity, peptic ulcer, eructations, ascities and piles. Leaf juice is useful in haemorrhagic dysentery. It is useful in spermatorrhoea, menorrhagia and uterine debility. It acts as Rasayana and rejuvenator.

Part used - Fruit, leaf juice.

Doses - Powder - 3 to 6 g, fruit juice – 10-20 ml.

Main preparation - Chyavanprash, Amalaki rasayan, Dhatriloha.

ARAGVADHA

Botanical name - *Cassia fistula* Linn.

Family - Leguminosae

Synonyms - Aragvadha, Rajavriksha, Shampak, Chaturangula.

Botanical description - It is a medium sized handsome tree, 8 - 15 mtrs. in height with greenish grey smooth bark when young and rough when old. Compound leaves with 4 - 8 pairs leaflets. Flowers are bright yellow in colour. Pods are cylindrical, pendulous, smooth, dark brown or black in colour.

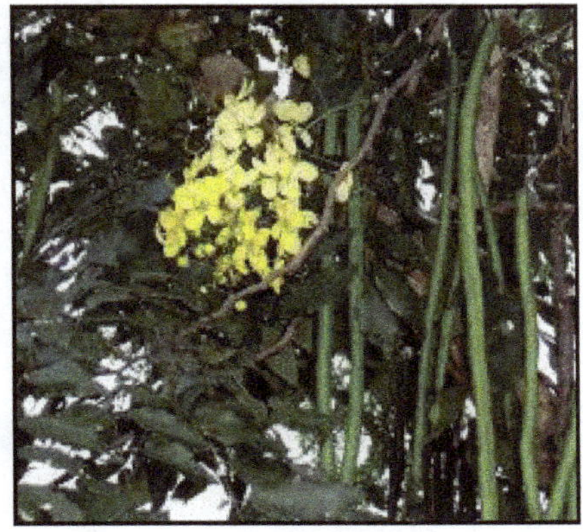

a) Macroscopic

Fruit, a many celled, indehiscent pod, 35-60 cm long and 18-25 mm diameter, nearly straight and subcylindrical, chocolate-brown to almost black in colour, pod surface smooth to naked eye, but under lens showing minute transverse fissures, both dorsal and ventral sutures evident, but not prominent, short stalk attached to base of fruit and rounded distal end mucronate, pericarp thin, hard and woody, fruit initially divided by transverse septa about 5 mm, apart, each containing a single seed attached to ventral suture by a long dark, thread-like funicle about 8-12 by 6-8 mm, circular to oval, flattened, reddish-brown, smooth, extremely hard and with a distinct dark brown line extending from micropyle to base, seed initially embedded in a black viscid pulp consisting of black, thin,

shining, circular disc like masses having central depression of seed on both surfaces or as broken pieces adhered with each other, when dipped in water makes yellow solution which darkness to brownish-yellow to dark brown, on keeping, pulp fills the cell but shrinks on drying and adheres to both sides of testa, seeds often lye loose in their segments, odour faint, sickly, taste, sweet.

Habitat - All over India.

Chemical constituents - Sugar, mucilage, pectin and anthraquinone.

Properties - Guna - guru, snigdha, mridu, Rasa - madhur, Veerya - sheeta, Vipaka - madhur.

Doshakarma - Vatapittashamaka.

Action and uses - The leaves are laxative antiperiodic, gouty arthritis, boils, carbuncles, ulcers and skin diseases. The fruit pulp is sweet, cooling with pleasant taste. It is smooth laxative, blood purifier and antipyretic. It is used in flatulence, anorexia, constipation, pruritus, leprosy, allied skin infections and blood coagulation disorder.

Part used - Leaf, fruit pulp, root bark.

Doses - Fruit pulp - 5-10 g, Root bark decoction - 50 -100 ml.

Main preparation – Aragwadhadi lepa.

ERANDA

Botanical name - *Ricinus communis* Linn.

Family - Euphorbiaceae

Synonyms - Eranda, Panchangula, Vardhamana, Deerghadanda.

Botanical description - It is an annual growing shrub 2-3 meter height. Leaves green, broad, interrupted, pointed like five fingers. Stalk 10-30 cm long. Flower unisexual male flower in circumferences, many stamens. Fruit a pickly capsule of three valve cocci.

a) Macroscopic

Root light in weight almost straight with few rootlets, outer surface dull yellowish brown, nearly smooth but marked with longitudinal wrinkles, some places whitish-yellow and soft, odourless, taste, acrid.

b) Microscopic

Transverse section of root shows thin layer of cork of squarish to tangentially elongated, thin-walled cells, beneath cork, secondary cortex of thin-walled, tangentially elongated cells, narrow cortex of rounded to tangentially elongated thin-walled parenchymatous cells, some containing large oil globules, rosettes of calcium oxalate crystals and round simple or compound starch grains, phloem a broad zone, consisting of sieve tubes, phloem parenchyma and phloem fibres, fibres long, mostly septate, highly thickened, having narrow lumen, some fibres surrounded by concentric rows of cells containing crystals of calcium oxalate, sieve tubes, thin-walled with companion cells and phloem parenchyma in the inner region of phloem more prominent, some phloem parenchyma cells contain crystals of calcium oxalate, cambium 3-5 layered, cells rectangular in shape, xylem occupies major part of root, pentarch, five groups of primary xylem distinct in the centre of the wood, xylem consists of vessels, parenchyma and fibres, vessels uniformly scattered throughout the xylem region, either solitary or in groups, larger in size towards phloem, with bordered pits, xylem parenchyma less in number around vessels containing starch grains, xylem fibres long and thick-walled, medullary rays uni-to-biseriate, more or less straight, 4-5 seriate rays, sometimes found near protoxylem groups, ray cells, thin-walled, slightly radially elongated in phloem region, thick-walled in xylem region, all ray cells contain starch grains.

Varieties - Two varieties are found- 1) white 2) red.

Habitat - All over India, Tamil Nadu, Bengal, Maharashtra, Bihar, Odisha.

Chemical Composition - The alkaloids ricinine, albumin, ricin and b-sitosterol, octacosanol, ricininie, gallic acid, rutin and b-sitosterol, b-D-glucopyranoside, the flavonoid glycosides-hyperoside and quercetin- 3- glucoside (leaves); lupeol, lipids, a glycoprotein, phosphatides and two hemicelluloses, first yielded glucose and mannose and second yielded xylose, arabinose glucose, galactose, glucuronic acid and 4-0- methyl glucuronoic acid (bean coat); germanicol ester derivative and an unidentified triterpene (root); inorganic material like potassium, sodium, magnesium, chloride, nitrate, iron, aluminium, manganese, calcium carbonate and phosphate including gallotannins (root, root bark); hyperoside, rutin, ricinine, apigenin and chlorogenic acid (flowers); flavonoid- lucenin (seeds); arachidic (12-hydroxyoleic), chlorogenic, oleic, palmitic, ricinoleic, stearic and dihydrostearic acids, hexadecanoic, hydrocyanic and uric acid (oil), squalene and tocopherols (plant).

Properties - Rasa - madhur, Guna - snigdha, tikshna, sukshma, Vipaka – madhur Veerya - ushna.

Doshakarma - Kaphavatashamaka.

Action and uses - The root is astringent, powerful purgative, carminative, diuretic, aphrodisiac and expectorant. They are uses in gastric disorders, constipation, inflammations, fever, ascites, cough, bronchitis and leprosy, pain in chest, rheumatic affections, lumbago and colic. Leaves are useful in burns, night blindness and used for fomentation in painful joints & rheumatoid arthritis.

Part used - Root, leaf, seed oil.

Doses - Seed oil -4 to 16 ml, root paste - 3 to 6 g.

Main preparation - Erand paak, Erandamuladi kwatha.

KANTAKARI

Botanical name - *Solanum virginianum* Linn.

Family - Solanaceae

Synonyms - Kantakari, Duhsparsha, Kshudra.

Botanical description - It is a thorny shrub of almost 0.75 to 1.25 meters height and green in colour. Leaves are ovate; they have thorns on the dorsal aspect. Flowers are blue coloured. Fruit- round, small and thorny. They are green with white streaks when raw and turn yellow on ripening.

a) Macroscopic

Root -10-45 cm long, few mm to two cm in diameter, almost cylindrical and tapering, bearing a number of fine longitudinal and few transverse wrinkles with occasional scars or a few lenticels and small rootlets, transversely smoothened surface shows a thin bark and wide compact cylinder of wood, fracture, short, taste, bitter.

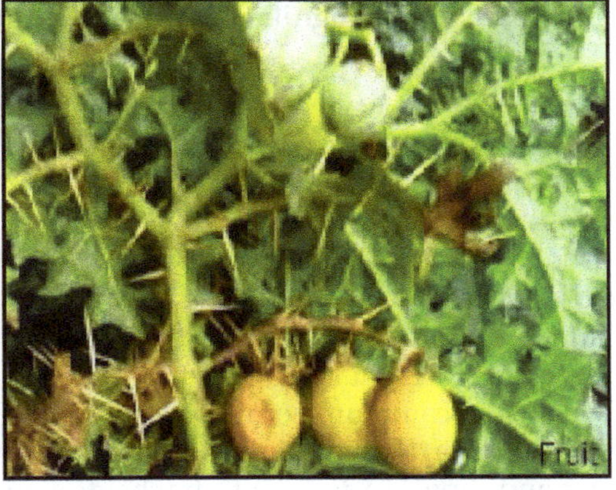

Fruit

Stem - Herbaceous, prickly with prominent nodes and internodes, green when fresh, young branches, covered with numerous hairs, mature ones glabrous, furrows more prominent in young stem appearing almost circular towards basal region, stem pieces

8-10 mm thick of variable length, external surface light green, when dry, surface yellowish green and smooth, transversely smoothened surface shows a very thin bark and prominent wood, centre shows a large and distinct, pith, mature and dry stem often with hollow pith, fracture short to slightly fibrous.

Leaves - Petiolate, exstipulate, ovate—oblong or elliptic, sinuate or sub-pinnatifid, sub-acute hairy, 4-12.5 cm long and 2-7.5 cm wide, green, veins and midrib full with sharp prickles, odour and taste not distinct.

Flower - Ebracteate, pedicellate, bisexual, pentamerous, regular, complete, bright blue or bluish purple, calyx-persistent, gamosepalous, tube short, globose, linear-lanceolate, acute, hairy, 0.5-1 .3 cm long and densely prickly, corollagamopetalous, lobes deltoid, acute, hairy, 1-2 cm long and purple in colour, stamens 5, epipetalous, basifixed, filament short 1-1.5 mm long, anther, oblong lanceolate, 0.7-0.8 cm long, ovary superior, ovoid, glabrous, bilocular with axile placentation having numerous ovules.

Fruit - Berry globular , measuring 0.8-1 cm in diameter, surrounded by persistent calyx at base unripe fruits variegated with green and white strips, ripe fruit shows different yellow and white shades.

Seeds - Circular, flat, numerous, embedded in a fleshy mesocarp about 0. 2 cm in diameter, glabrous taste, bitter and acrid.

b) Microscopic

Root - Transverse section of mature root shows cork composing of 3-6 layers of thin-walled, rectangular and tangentially elongated cells, cork cambium single layered followed by 6-15 layers of thin-walled, tangentially elongated to oval or circular parenchymatous cells, stone cells either single or in groups of 2-20 or even more present in this region, secondary phloem composed of sieve elements and phloem parenchyma traversed by medullary rays, stone cells present in singles or in groups of 2-20 or more in outer, and middle phloem regions, phloem rays 1-4 cells wide and 2-22 cells high, cambium 3-5 layered of thin-walled rectangular cells, xylem composed of vessels, tracheids, fibre trachieds, parenchyma and transversed by medullary rays, all elements being lignified, vessels and trach.

Stem - Transverse section of mature stem, 1.5-2 cm thick consists of 6-12 layers of cork of thin- walled somewhat rectangular cells, epidermis remains intact for a long time, secondary cortex consists of 7-11 layers of parenchymatous cells, some cells thickened and lignified forming stone cells primary cortex remains intact even in quite mature stage but later gets crushed, pericyclic fibre, occur singly or in small groups of 2-3, secondary phloem consists of sieve elements, parenchyama, a few fibres, stone cells and traversed by phloem rays, fibres found scattered in singles or in small groups in outer and middle phloem region, inner phloem devoid of fibres, stone cells present in singles or in small groups of 2-4, phloem rays, 1-2 or rarely 3 cells wide, cambium composed of 2-3 layers, xylem consists of vessels, tracheids, parenchyma, fibres and traversed by xylem rays, vessels vary greatly in shape and size and show bordered pits, tracheids elongated with irregular walls and bordered pits, fibres much elongated, thick-walled and lignified with tapering and pointed ends, some having truncated ends or bifurcated at one or both ends with a few simple pits, trancheids fibres smaller than fibres, with both ends tapering and have reticulate thickening, xylem parenchyma cubical to rectangular with simple or bordered pits or reticulate thickening, xylem rays conspicuous by their pitted thickenings, longer size and radial elongation of cells, 1-2 or rarely 3 cells wide and 2-25 cells high, internal phloem composed of sieve elements and parenchyma, forming more or less continuous band and embedded in perimedullary zone, a few phloem fibres similar to those of outer phloem region also present, central region occupied by a large pith, microsphenoidal crystals of calcium oxalate as sandy masses and simple starch grains present in cortex, secondary cortex, phloem, medullary rays and pith cells.

Leaves

(i) Petiole - Transverse section of petiole shows circular to wavy outlines, epidermis single layered, covered externally by a thick cuticle, hypodermis consists of 3-4 layers of collenchymatous cells, one large-crescent-shaped, bicollateral, central vascular bundle and two small lateral bundles present, rest of tissue of petiole composed of polygonal, angular, thin-walled, parenchymatous cells, epidermis shows mostly stellate.

(ii) Midrib - Transverse section of midrib shows a biconvex structure, epidermis on either side covered externally by a thick cuticle, below epidermis 3-4 layers of collenchyma present, stele composed of crescent-shaped, bicollateral, central vacscular bundle and two small lateral vascular bundles, rest of tissue composed of thin-walled, parenchyma, some stellate hair present on epidermis.

(iii) Lamina - Transverse section shows dorsiventral structure, epidermis on either side, wavy in outline, covered externally by a thick cuticle, on upper side mesophyll composed of a single layered palisade and 4-6 layers of loosely arranged spongy parenchyma, some stellate hairs (4-8 armed) present on both sides of epidermis, anisocytic stomata present on both surfaces, vein-islet number 46-80 on lower epidermis (mean 63), 61-80 on upper epidermis (mean 70), stomatal index 20-25 (mean 22.5) on lower epidermis, 14-24 (mean 19) on upper epidermis, palisade ratio 1.7-4 (mean 2 .85).

Fruit - Transverse section of mature fruit shows single layered epidermis, covered externally by a thin cuticle, 1-2 layers of collanchyma present below epidermis, mesocarp composed of thin-walled, oval to polygonal cells, some fibre., vascular bundles present scattered, seed consists of thick-walled radially elongated testa, narrow endosperm with embryo, some cells of endosperm contain oil globules.

Powder - Greenish, under microscope shows single or groups of stone cells, groups of aseptate fibre with tapering ends, pitted vessels, groups of spongy parenchyma, fragments of palisade tissue, anisocytic stomata, stellate hairs and simple, rounded to oval starch grains measuring 2.75-11 μ in dia.

Chemical composition - Glucoalkaloids and sterols.

Habitat -It is found throughout India.

Varieties - It is of two types - 1) - Brihat 2) - Laghu.

Properties - Rasa - katu, tikta, Guna - laghu, ruksha, Veerya - ushna, Vipaka -katu.

Doshakarma - Kaphavatashamaka.

Action and uses - The plant is bitter, anti - inflammatory, digestive, carminative, appetizer, stomachic, anthelmintic, expectorant, laxative, diuretic, rejuvenating, emmenagogue and aphrodisiac. It is useful in dental caries, inflammation and pain in joints, flatulence, colic, constipation, dyspepsia, hypertention, fever, chronic coryza, cough, asthma and bronchitis.

Part used - Whole plant.

Dose - Decoction - 50 - 100 ml.

Main preparation - Kantakari ghrita, Vyaghri haritaki.

BAKUCHI

Botanical name - *Psoralia corylifolia* Linn.

Family - Fabaceae

Synonyms - Bakuchi, Somraji, Kalameshi, Chadralekha.

Botanical description -It is an erect herbaceous annual seasonal plant 0.5 to 1.5 mt. in height with grooved and gland dotted stems and branches. Leaves simple, broadly, rounded. Flowers blue 1-30 flowers in branches appear on a long stalk. Fruits are blackish in bundles.

a) Macroscopic

Fruits, dark chocolate to almost black with pericarp adhering to the seed-coat, 3-4.5 mm long, 2-3 mm broad, ovoid-oblong or bean shaped, some what compressed, glabrous rounded or mucronate, closely pitted, seeds campylotropous, non-endospermous, oily and free from starch, odourless, but when chewed smell of a pungent essential oil felt, taste, bitter, unpleasant and acrid.

b) Microscopic

Transverse section of fruit shows periocarp with prominent ridges and depressions, consisting of collapsed parenchyma and large secretory glands containing oleo-resinous matter testa, an outer layer of palisade epidermis, layer of bearer cells which are much thickened in the inner tangential and basal radial walls and 2-3 layers of parenchyma, cotyledons of polyhedral parenchyma and three layers of palisade cells on the adaxial side.

Habitat - All over India, especially in Rajasthan & Punjab.

Chemical constituents - Essential oil, fixed oil, psoralen, psoralidin, isopsoralen and bakuchiol.

Properties - Rasa - katu, tikta Guna - laghu, ruksha Veerya - ushna Vipaka-katu.

Doshakarma - Vatakaphashamaka.

Action and uses - It acts as deepan, pachan, anuloman, anthelmintic and liver stimulant. It is useful in indigestion and constipation. It is useful in all types of worms especially round worm. Powdered bakuchi and its oil prepared from it are used locally in leucoderma, dermatoses, wounds and alopecia.

Part used - Seeds, oil.

Doses - Powder- 1 to 3 g.

Main preparation - Avalgujadi yoga, Triphala modak, Maheshwara ghrita.

BHRINGARAJA

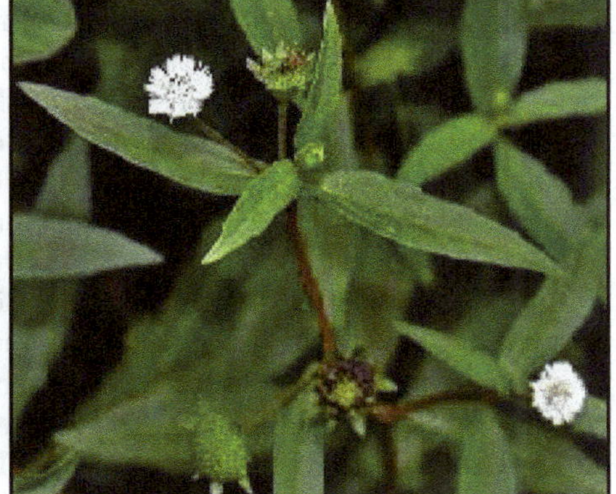

Botanical name- *Eclipta alba* Hassk.

Family - Compositae

Synonyms - Bhringaraja, Markava, Angaraka, Keshranjana.

Botanical description - A small herb having height 30-60 cm. Trunk is black. Leaves are opposite, very variable, sessile having serrated margins. Flower stalk-elongated having white flower on its tip. A single fruit has many seeds. Black coloured elongated seeds like black cumun seed.

a) Macroscopic

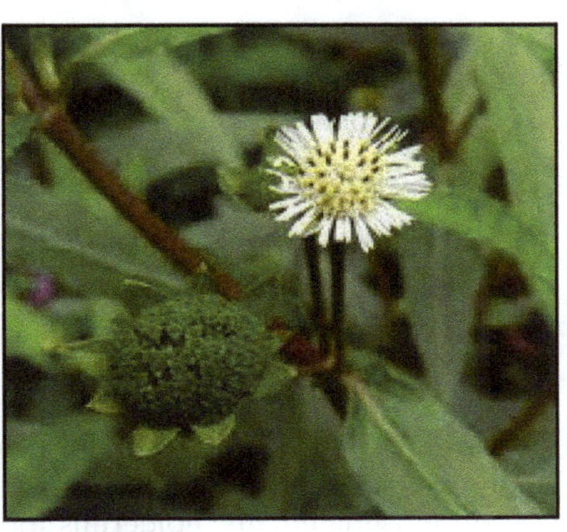

Root - Well developed, a number of secondary branches arise from main root, upto about 7 mm in dia., cylindrical, greyish.

Stem - Herbaceous, branched, occasionally rooting at nodes, cylindrical or flat, rough due to oppressed white hairs, node distinct, greenish, occasionally brownish.

Leaf - Opposite, sessile to subsessile, 2.2 - 8.5 cm long, 1.2 - 2.3 cm wide, usually oblong, lanceolate, sub-entire, sub-acute or acute, strigose with appressed hairs on both surfaces.

Flower - Solitary or 2, together on unequal axillary peduncles; involucral bracts about 8, ovate, obtuse or acute, herbaceous, strigose with oppressed hairs; ray flowers ligulate, ligule small, spreading, scarcely as long as bracts, not toothed, white; disc flowers tubular, corolla often 4 toothed; pappus absent, except occasionally very minute teeth on the top of achene; stamen 5, filaments epipetalous, free, anthers united into a tube with base obtuse; pistil bicarpellary; ovary inferior, unilocular with one basal ovule.

Fruit - Achenial cypsella, one seeded, cuneate, with a narrow wing, covered with warty excrescences, brown.

Seed - 0.2 - 0.25 cm long, 0.1 cm wide, dark brown, hairy and non endospermic.

b) Microscopic

Root - Mature root shows poorly developed cork, consisting of 3-5 rows of thin-walled, tangentially elongated cells; secondary cortex consists of outer one or two rows of tangentially elongated or rounded cells with air cavities, inner secondary cortex of tangentially elongated to irregular shaped, parenchymatous cells with conspicuous air cavities; stone cells found scattered in secondary cortex and cork, in singles or in groups of various shape and size; pericyclic fibres in tangentially arranged bands of many cells or in singles; secondary phloem consists of sieve elements including phloem fibres traversed by multiseriate phloem rays; phloem rays broader towards periphery, consisting of rounded cells; xylem composed of vessels, fibre tracheids, fibres and xylem parenchyma, traversed by xylem rays; vessels numerous, found scattered throughout wood, in macerated preparation vessels small, drum-shaped, cylindrical elongated with pitted walls and perforations, simple, rarely slightly oblique; fibre tracheids, pitted, with very pointed tips, xylem fibres long with pointed tapering ends and short lumen, a few fibres show peg-like outgrowths towards the tapering ends; xylem parenchyma sparse usually squarish to rectangular having simple pits on their walls, xylem ray distinct, run straight in tangential section, generally 5-32 cells in height and 3-5 cells in width although very rarely uniseriate and biseriate rays also found, ray cells pitted.

Leaf

Petiole - Shows single layered upper and lower epidermis consisting of tubular cells, covered with striated cuticle; trichomes of two types, non-glandular, uniseriate, 1-5 celled, warty, and with pointed apical cell; epidermis followed by wide cortex, consisting of 2-5 layered collenchyma on both, upper arid lower side with distinct angular thickening; parenchyma 4-6 layered on upper side and 5-8 layered on lower side consisting of isodiametric, thin-walled cells with intercellular spaces; five vascular bundles central one largest while four others small flanking to either side of central bundle, consists of xylem on dorsal side and phloem on ventral side; xylem vessels arranged in radial rows traversed by xylem rays.

Midrib - Cut at basal region shows both upper and lower single layered epidermis, externally covered with cuticle, a few epidermal cells elongate outwards to form uniseriate hairs; epidermis followed by cortex, consisting of 3-5 layered collenchymatous cells on both sides; section cut at middle region shows 3-4 layered collenchymatous cells on dorsal and 1-3 layered on ventral side, while the section cut at apical region, shows 2 layered collenchymatous cells on both sides, similarly transverse section cut at a basal, middle and apical regions shows 4-6 layered parenchymatous cells on dorsal side and 6-9 layered parenchyma on ventral side, in section cut at basal region 4-6 layered parenchyma on both the sides in the middle region with thin-walled cells and intercellular spaces, 2-3 layered parenchymatous cells on both side in the apical region; in the basal region section shows vascular bundle similar to that of petiole while in the section cut at middle and apical region section shows 4 smaller bundles shifting towards lamina.

Lamina - Shows a dorsi ventral structure, epidermis single layered, externally covered with cuticle, followed by single layered palisade parenchyma containing chlorophyll contents; spongy parenchyma irregularly arranged with distinct intercellular spaces and filled with chlorophyll contents; mesophyll traversed by number of veins; anisocytic and anomocytic stomata present on both surface, more abundant on lower surfaces; stomatal index 20.0-22.5 on upper and 23.5 -26.0 on lower surface; palisade ratio 3.8 -4.5; hairs stiff, pointed, wide at the base, about 3 celled, uniseriate, middle cells longest, uppermost generally not exceeding the basal cell in length, septa thick-walled.

Stem - Mature stem shows single layered epidermis, externally covered with cuticle, a few epidermal cells elongate to form characteristic non-glandular trichomes, the cork where formed, poorly developed consistsing of rectangular cells; secondary cortex composed of large, rounded or irregular shaped parenchymatous cells having wide air spaces; endodermis single layered consists of tangentially elongated cells; pericyclic fibres distinct, arranged in tangential strands; vascular bundles in a ring, collateral, endarch, of varying sizes traversed by medullary rays; phloem a narrow strip composed of sieve elements and phloem parenchyma; xylem consists of large number of vessels, xylem fibres and xylem parenchyma; xylem vessels appear evenly distributed throughout the xylem; in macerated preparation vessels barrel-shaped, some elongated with simple perforations, pitted with spiral thickening; xylem fibres with wide lumen, pointed tips and pitted walls, a few often bifurcate and a few other large, peg-like outgrowth; xylem parenchyma rec-tangular with pitted thickening; xylem rays triseriate to pentaseriate, normally biseriate and uniseriate, 8-15 cells in height and 3-5 cells in width; centre occupied by a wide pith consisting of isodiametric cells of parenchyma.

Powder - Dark green; shows vessels in large groups or single broken pieces with pitted walls, numerous fibres entire or in pieces, trichomes entire or in pieces, warty, a few attached with epidermal and subsidiary cells, anomocytic and anisocytic stomata.

Chemical constituents - Alkaloids, Ecliptine and Nicotine.

Varieties - There are three varieties, mentioned in texts according to the colour of the flowers.

1) White 2) Yellow 3) Blue

Habitat- Throughout all over India, in water lodging areas.

Properties - Rasa - katu, tikta Guna - ruksha, laghu Veerya - ushna Veepaka - katu.

Action and uses - The plant is an excellent appetizer, digestant and liver stimulant. Its main action is on liver and spleen. It cures loss of appetite, indigestion, liver and spleen enlargement. It is also an anti-inflammatory, ophthalmic, digestive, diuretic, hair tonic and absorbent. It is useful in hepato- splenomegaly, skin diseases, wounds, ulcers, jaundice and in blood pressure.

Part used - Whole plant, seeds.

Dosage - Juice - 5-10 ml, seeds 3-5 g.

Main preparation - Bhringaraja tail, Bhringaraj ghrit.

PASHANABHEDA

Botanical name - *Bergenia ligulata* Wall.

Family – Saxifragaceae

English name- *Iris* sp.

Synonyms- Pashanabheda, Ashmabheda, Ashmaghna, Shilagarbhak, Pashan.

Botanical description - This small herb survives for many years and is found in hilly and rocky regions. Its stalk emerges from the cracks in the rocks. Leaves are flashy, round, dentate, greenish on upper surface and reddish on lower surface. Flowers are whitish or bluish.

a) Macroscopic

Rhizome, solid, barrel shaped, cylindrical, 1.5-3 cm long and 1-2 cm in diameter with small roots, ridges, furrows and root scars distinct, tranversely cut surface shows outer ring of brown coloured cork, short middle cortex, vascular bundles and large central pith, odour, aromatic, taste, astringent.

b) Microscopic

Transverse section of rhizome shows cork divided into two zones, outer a few layers of slightly compressed and brown coloured cells, inner zone multilayered consisting of thin-walled tangentially elongatd and colourless cells, followed by a single layered cork cambium and 2-3 layers of secondary cortex composed of thick-walled, tangentially elongated, rectangular cells with intercellular spaces, some cells contain rosette crystals of calcium oxalate and simple starch grains cortex a narrow-zone of parenchymatous cells containing a number of simple starch grains, most of cortical cells also contain large rosette crystals of calcium oxalate. Vascular bundles, arranged in a ring, collateral, conjoint and open, phloem tissues composed of sieve elements and parenchyma, in outer region found as compressed masses while in inner region intact. a number of rosette crystals of calcium oxalate also found as crystal fibres, cambium present as continuous ring composed of 2-3 layers of thin-walled, tangentially elongated cells, xylem consist of fibres, tracheids, vessels and parenchyma, with centre occupied by large pith composed of circular to oval, parenchymatous cells, varying in size and containing starch grains with crystals of calcium oxalate similar to those found in cortical region.

Habitat - Himalayas at the height of 2 thousand mtrs.

Chemical constituents - Tannic acid, gallic acid and glucose.

Properties - Rasa- kashaya, tikta Guna - laghu, snigdha Veerya - sheeta, veepak-katu.

Doshakarma - Tridoshashamaka.

Action and uses - It is anti-inflammatory, wound healing, hemostatic, cardiotonic, antipyretic and antitoxic. Its local application is useful in conjunctivitis. Internally it is used in diarrhoea, dysentery, tumours, piles, splenomegally, heart diseases, bleeding disorders, cough, leucorrhoea and dysmenorrhoea. But it is especially useful in urinary calculi and dysuria.

Part used - Rhizome, leaf.

Dosage - Powder - 3 to 6 g, decoction - 50-100 ml.

Main preparation - Pashanabhedadi kwath, Pashanabhedadi ghrita.

PIPPALI

Botanical name - *Piper longum* Linn.

Family - Piperaceae

Synonyms - Kana, Kola, Chapala, Tandula.

Botanical description - It is a creeper which spreads on the ground or climbs up nearby trees for support. Leaves are resembled to betel leaves and have 5 veins. Fruits long, reddish on ripening and turn black when dried.

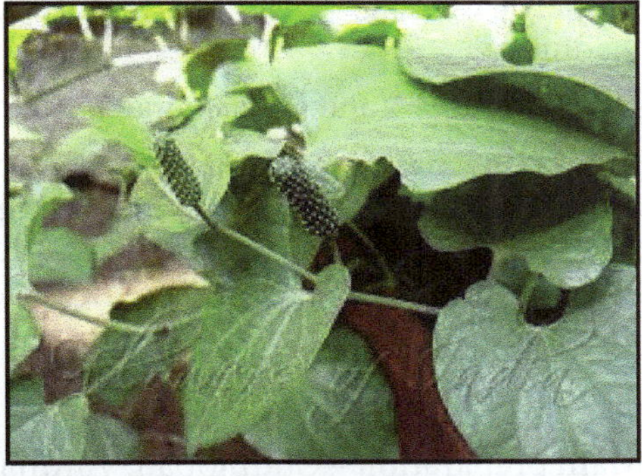

a) Macroscopic

Fruit greenish-black to black, cylindrical, 2.5 to 5 cm long and 0.4 to 1 cm thick, consisting of minute sessile fruits, arranged around an axis; surface rough and composite; broken surface shows a central axis and 6 to 12 fruitlets arranged around an axis; taste, pungent producing numbness on the tongue; odour, aromatic.

b) Microscopic

Catkin shows 6 to 12 fruits, arranged in circle on a central axis, each having an outer epidermal layer of irregular cells filled with deep brown content and covered externally with a thick cuticle; mesocarp consists of larger cells, usually collapsed, irregular in shape and thin-walled; a number of stone cells in singles or in groups present; endocarp and seed coat fused to form a deep zone, outer layer of this zone composed of thin-walled cells and colourless, inner layer composed of tangentially elongated cells, having reddish-brown content; most of endocarp filled with starch grains, round to oval measuring 3 to 8 μ in dia.

Powder - Deep moss green, shows fragments of parenchyma, oval to elongated stone cells, oil globules and round to oval, starch grains, measuring 3 to 8 μ in dia.

T.L.C.

T. L. C. of alcoholic extract of the drug on Silica gel 'G' plate using Toluene: Ethylacetate (90: 10) as mobile phase. Under U.V. (366 nm) six fluorescent zones are visible at Rf. 0.15, 0.26, 0.34, 0.39, 0.50 and 0.80. On exposure to Iodine vapour seven spots appear at Rf. 0.04, 0.15, 0.26, 0.34, 0.39, 0.50 and 0.93 (all yellow). On spraying with Vanillin-Sulphuric acid reagent and heating the plate at 105°C for ten minutes five spots appear at Rf. 0.04, 0.22, 0.35, 0.43 and 0.82. On spraying with Dragendorff reagent three spots appear at Rf. 0.15, 0.26 and 0.34 (all orange).

Chemical constituents - Volatile oil, starch, gum, fatty oil, inorganic matter and resin piperine as well as alkaloids.

Varieties – Its four varieties-- 1. Pippali 2. Gajpippali 3. Saimhalee 4.Vanpippali

Habitat -It is grown chiefly in the following places- Bengal, Bihar, Assam and Travancore, Nepal.

Properties - Rasa - katu Guna - laghu, snigdha, tikshna Veerya - ushna Vipaka - madhur.

Doshakarma - Vatakaphashamaka.

Action and uses - It is an appetizer, carminative, analgesic and mild laxative. It is used in anorexia, loss of appetite, indigestion, colic pain, piles, liver disorder and ascites. It is good rejuvenator for raktadhatu and regulates the function of liver and spleen. Pippali is an excellent medicine for cough, asthma and hiccough.

Part used - Fruit, root.

Dose - 0.5-1.0 g.

Main preparation – Gudpippali, Vardhamanapippali.

PUGA

Botanical name - *Areca catechu* Linn.

Family - Arecaceae

Synonyms - Puga, Ghoranta, Pugi, Kramuka.

Botanical description - A tree 10-15 mtrs high has no branches. Leaves- made up of group of leaflets. Flower stalk- hard and branched. Fruits- appear in bunches, 3-6 cms, round, smooth and hard. Green when unripe and yellow red when ripes. Catechu is obtained after removing the hard rind.

a) Macroscopic

Ovoid, externally pale, reddish-brown to light yellowish-brown, marked with a net work of paler lines, frequently with adhering portions of silvery brittle endocarp and adhering fibres of mesocorp at base of seed, seed hard with ruminate endosperm of brownish tissue alternating with whitish tissue, odour, characteristic, taste, astringent.

b) Microscopic

Transverse section of seed shows a seed coat consisting of several rows of cells, tangentially elongated, with inner walls more or less thickened, whitish cell of endosperm tissue with thick porous walls containing oil globules and aleuronic grains, brown peri sperm tissue with thick walled cells and delicate tracheae.

Powder-Reddish brown to light brown, under microscope shows fragments of endosperm tissue with porous walls, irregularly thickened and small stone cells of seed coat, a few aleurone grains and oil globules and a few delicate tracheae, starch absent.

Habitat - Tropical region, Karnataka, Malabar, Kerala, Tamil Nadu.

Constituents - Alkaloid (arecoline), tannins and fats.

Properties - Rasa - kashaya, madhur Guna - ruksha, guru Virya - sheeta Veepaka - katu.

Doshakarma - Kaphapittashamaka.

Action and uses - It is haemostatic and wound healing. Used for gargling in oral disorders. It reduces blood pressure and is useful in bleeding disorders, vata disorders, stomatitis and bad breath.

Useful part- Fruit, leaf

Doses - Seed powder - 1 to - 3 g.

Main preparation - Pugakhand.

PUNARNAVA

Botanical name - *Boerhavia diffusa* Linn.

Family - Nyctaginaceae.

Synonyms - Punarnava, Shothaghni, Kathillaka.

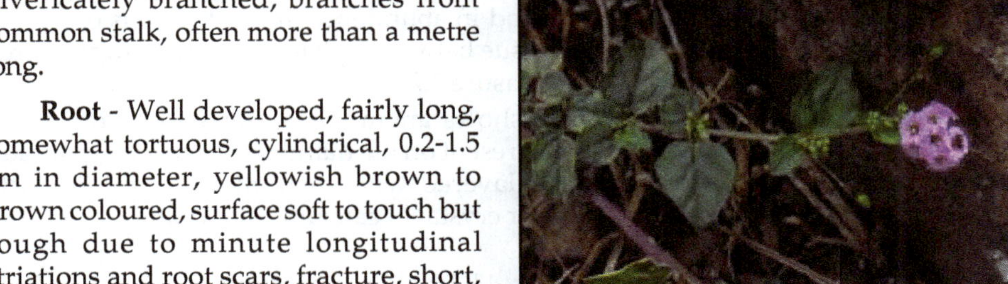

Botanical description - Perennial creeper 0.75 to 1 mt in length, sometimes grows up to 4mtrs in rainy season and dried in summer. Leaves are fleshy, with soft hairs, at distance and whitish ventrally. Flowers- small, white coloured. Root - thick, twisted on drying.

a) Macroscopic

Stem - Greenish purple, stiff, slender, cylindrical, swollen at nodes, minutely pubescent or n early glabrous, prostrate divericately branched, branches from common stalk, often more than a metre long.

Root - Well developed, fairly long, somewhat tortuous, cylindrical, 0.2-1.5 cm in diameter, yellowish brown to brown coloured, surface soft to touch but rough due to minute longitudinal striations and root scars, fracture, short, no distinct odour, taste, slightly bitter.

Leaves - Opposite in unequal pairs, larger ones 25-37 mm long and smaller ones 12-18 mm long ovate-oblong or suborbicular, apex rounded or slightly pointed, base subcordate or rounded, green and glabrous above, whitish below, margin entire or sub-undulate, dorsal side pinkish in certain cases, thick in texture, petioles nearly as long as the blade, slender.

Flowers - Very small, pink coloured, nearly sessile or shortly stalked, 10-25 cm, in small umbells, arranged on slender long stalks, 4-10 corymb, axillary and in terminal panicles, bracteoles, small, acute, perianth tube constricted above the ovary, lower part greenish, ovoid, ribbed, upper part pink, funnel-shaped, 3 mm long, tube 5 lobed, stamen 2-3.

Fruit - One seeded nut, 6 mm long clavate, rounded, broadly and bluntly 5 ribbed, viscidly glandular.

b) Microscopic

Stem - Transverse section of stem shows epidermal layer containing multi cellular, uniserite glandular trichome consisting of 9-12 stalked cells and an ellipsoidal head, 150-220 µ long, cortex consists of 1-2 layers of parenchyma, endodermis indistinct, pericycle 1-2 layered, thick-walled often containing scattered isolated fibres, stele consisting of many small vascular bundles often joined together in a ring and many big vascular bundles scattered in the ground tissue, intra fascicular cambium present.

Root - Transverse section of mature root shows a cork composed of thin-walled tangentially elongated cells with brown walls in the outer few layers, cork cambium of 1-2 layers of thin walled cells secondary cortex consists of 2-3 layers of parenchymatous cells followed by cortex composed of 5-12 layers of thin-walled, oval to polygonal cells, several concentric bands of xylem tissue alternating with wide zone of parenchymatous tissue present below cortical regions, number of bands vary according to thickness of root and composed of vessels, tracheids and fibres, vessels mostly found in groups of 2-8 in radial rows, having simple pits and reticulate thickening, tracheids, small, thick-walled with simple pits, fibres aseptate, elongated, thick-walled, spindle shaped with pointed ends, phloem occurs as hemispherical or crescentic patches outside each group of xylem vessels and composed of sieve elements and parenchyma, broad zone of parenchymatous tissue, in between two successive rings of xylem elements composed of thin-walled more or less rectangular cells arranged in radial rows, central regions of root occupied by primary vascular bundles, numerous raphides of calcium oxalate, in single or in group present in cortical region and parenchymatous tissue in between xylem tissue, starch grains simple and compound having 2-4 components found in abundence in most of cells of cortex, xylem elements in paren-chymatous tissue between xylem elements, simple starch grains mostly rounded in shape and measure 2.75-11 µ in diameter.

Leaves-Transverse section of leaf shows anomocytic stomata on both sides, numerous, a few short hairs, 3-4 celled, present on the margin and on veins, palisade one layered, spongy parenchyma 2-4 layered with small air spaces, idioblasts containing raphides, occasionally cluster crystal of calcium oxalate and orange-red resinous matter present in mesophyll.

Palisade ratio 3.5-6.5, stomatal index 11-16 , vein islet number 9-15.

Assay-Contains not less than 0.1 per cent of total alkaloids, when assayed by the following methods.

Take accurately about 100 g of the drug (60 mesh powder) and moisten with dilute solution of Ammonia. Extract continuously in a soxhlet apparatus for 18 hours with 95 per cent Alcohol. Remove the alcohol by distillation. Extract the residue with five 25 ml portions of 1 N Hydrochloric acid till complete extraction of the alkaloid is effected. Transfer the mixed acid solutions into a separating funnel and wash with 5 ml of Chloroform, runoff the Chloroform layer. Make the acid solution distinctly alkaline with Ammonia and shake with five 25 ml portions of Chloroform or till complete extraction of alkaloids is effected. Wash the combined chloroform extracts with two portions each of 5 ml of water. Filter the chloroform layer in tared flask and evaporate to dryness. Add to the residue 5 ml of Alcohol, evaporate to dryness, repeat the process once again and weigh the residue to constant weight in a vacuum desiccator.

Varieties - Two varieties - 1) brown 2) white.

Habitat - All over India.

Constituents - Alkaloid (Punarnavine).

Properties - Rasa - Madhur, tikta, kashaya Guna - katu Vipaka - katu Veerya - -sheeta

Doshakarma - Tridoshashamaka.

Action and uses - The plant is bitter, astringent, cooling, anthelmintic, diuretic, aphrodisiac, cardiac stimulant, diaphoretic, emetic, expectorant, laxative and tonic. It is useful in all types of inflammations, leucorrhoea, opthalmia, dyspepsia, myalgia, scabies, general debility and bronchitis. Punarnava is a very useful drug for the treatment of inflammatory renal diseases and nephritic syndrome.

Useful part - Whole plant, root.

Dose - Juice of whole plant - 5-10 ml, decoction of roots - 5-10 ml.

Main preparation - Punarnavastak kwath, Punarnavashava.

KARANJA

Botanical name - *Pongamia pinnata* Pieree.

Family - Fabaceae

Synonyms - Karanja, Ghritapura, Naktamala

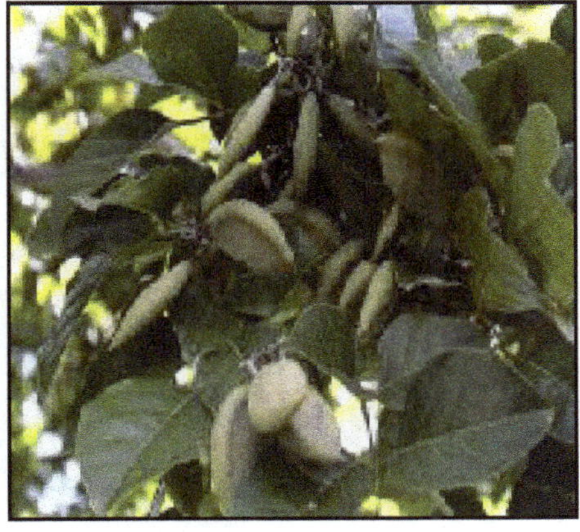

Botanical description - A medium sized tree, up to 18m high. Leaves are slimy, smooth and green in colour. Flowers - pink or violet in auxiliary racemes. Fruit - 4 to 5 cm long, flat, oval, very hard and concave on dorsal surface.

a) Macroscopic

Seed usually one and rarely two, elliptic or reniform in shape, 1.7-2.0 cm long and 1.2-1.8 cm broad, wrinkled with reddish leathery testa, micropylar end of cotyledons slightly depressed while other side semi-circular in shape.

b) Microscopic

Transverse section of seed shows, testa composed of a layer of palisade like outer epidermis, filled with brown pigment, covered externally with a thick cuticle, a layer of large, thin walled, somewhat rectangular cells, 2-4 layers of thick-walled parenchyma cells, a few rows of cells with small inter-cellular spaces, 2-3 layers of thick-walled elongated cells, a few layers of spongy parenchyma having large inter-cellular spaces, a number of parenchyma cells containing brown pigment, cotyledons composed of outer layer of epidermis with cylindrical cells, externally covered with thin cuticle, epidermis followed by rectangular to polygonal cells of mesophyll, filled with globules, also present scattered in this region.

Varieties -Its four varieties are described. Namely 1) Karanj 2) Putikaranja 3) Chirbilva 4) Karanji.

Habitat - It is found almost throughout India.

Chemical Composition – It contains a Karanjin, pongapin, pongamol, four furanoflavones viz., Karanjin, pongapin, kanjone and pongaglabrone and a diketone pongamol. Leaves are contains carotene, triterpenoids, furanoflavones, chromenoflavone, simple flavones. Amino acids viz.,a-alanine, arginine, aspartic acid, glumatic acid, histidine, isoleucine, leucine, lysine, methionine, phenylalahine, tryptophan, tyrosine, glycine and valine.

Properties - Rasa- tikta, katu Guna- laghu, tikshna Veerya - ushna Veepaka -katu.

Doshakarma - Kaphavatashamaka

Action and uses - Its bark, leaves, and seeds are antiseptic, antipruritic and analgesic. Powder of the seeds is useful in various skin diseases and wounds. Leaves and bark are used in loss of appetite, digestive disorders, constipation, helminthiasis and haemorrhoids.

Part used - Bark, leaves, seeds.

Dose - Stem bark- 3-6 gm, leaf juice- 10-20 ml, seed powder- 1-3 g.

Preparation - Karanjadi churna, Karanjadi ghrita.

KUMARI

Botanical name - *Aloe vera Tourn ex* Linn.

Family - Liliaceae

Synonyms - Kumari, Ghritakumari, Deerghapatrica.

Botanical description - Its shrub are 30-60 cm hight. Leaves - thick, filled with juice. The margins of the leaves have spikes. Leaves contain a thick and mucilaginous juice. Once the shrub become old, a straight stalk emerges from its center which bears red flower in cluster.

a) Macroscopic

Dark chocolate brown, to black, compact, irregular masses: surface dull, opaque with slightly vitreous appearance, odour, characteristic, taste, nauseous and bitter.

b) Microscopic

Powder when mounted in glycerin or lactophenol and examined under the microscope shows innumerable crystalline, yellowish-brown to chocolate coloured particles of varying size and shape.

Identification

Mix 0.5 g with 50 ml of water, boil until nearly dissolved, cool, add 0.5 g of *Kieselguhr* and filter, to the filtrate apply the following tests-

(i) Heat 5 ml of filtrate with 0.2 g of *Borax* until dissolved, add a few drops of this solution to a test-tube nearly filled with water, a green fluorescence is produced.

(ii) Mix 2 ml of filtrate with 2 ml of a freshly prepared solution of Bromine, a pale yellow precipitate is produced.

Varieties - Kumari is available in four varieties- 1) Sacrotine 2) Arabian 3) Jaffarabadi 4) Maisuri.

Habitat - Kumari is predominantly found in Africa, Arabia and India. In India it is abundantly seen in the south.

Chemical constituents - Anthraquinone, glycoside.

Properties - Rasa- tikta, katu Guna- guru, snigdha Veerya - sheeta Veepaka- -katu.

Doshakarma - Kaphapittashamaka.

Action and uses - In small doses, kumari has deepan, pachan, bhedan and uttejak properties. In large doses, it is virachak and krimighna. Juice of kumari is useful in loss of appetite, ascites, tumour, liver and spleen enlargement and abdominal colic. It is a known diuretic. It is used in splenomegaly in combination with haridra.

Part used - Leaves, juice from leaves.

Doses - Juice of leaves - 10-20 ml, dried powder 1 to 3 g.

Main preparation - Kumaryasava, Kumaripaka, Rajahpravartani vati.

KHADIRA

Botanical name - *Acacia catechu* Willed.

Family - Leguminosae

Synonyms - Khadira, Balpatra, Raktasara, Gayatri.

Botanical description - This is a medium size tree. Trunk is rough, spiny and yellowish from outside and reddish from inside. Leaves- compound leaf, leaflets are seen in pairs. Flowers are small, yellow, having three petals. Fruit- legume.

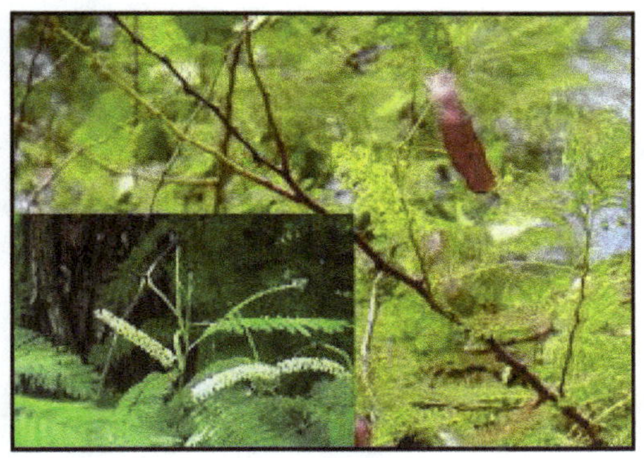

a) Macroscopic

Heart-wood, light red, turning brownish-red to nearly black with age, attached with whitish sapwood, fracture hard.

b) Microscopic

Transverse section of heart-wood shows, numerous, uni-to bi-seriate medullary rays, vessels occurring isolated or in small groups of two to four, xylem fibres with narrow lumen occupying major portion of wood, xylem parenchyma usually predominantly paratracheal, forming a sheath around vessels, wood consists of crystal fibres with 14-28 segments, each having one prismatic crystal of calcium oxalate, a few tracheids with scalariform thickening, some of cells, including vessels, filled with brown content, prismatic crystals of calcium oxalate present in a number of cells throughout the wood.

Powder - Brown coloured, under microscope shows a number of xylem fibres, vessels, crystal fibres, and prismatic crystals of calcium exalate.

Habitat - In Himalaya at the height of 5000 feet and dry climate.

Constituents - Catechin, catechu-tannic acid and tannin.

Properties - Rasa - tikta, kashaya Guna - lagnu, ruksha Veerya - sheeta Vipaka - katu.

Doshakarma - Kaphapittashamaka.

Action and uses - The bark is used in conjunctivitis and haemoptysis. Its heartwood is bitter astringent, cooling, antiseptic and antipyretic, haemoptatic and antileprotic. Being astringent and haemostatic, powder is used in wound healing and in dental condition. Gargling is useful in cough and hoarseness of voice.

Part used - Bark and resin.

Doses - Bark powder - 1-3 g, Decoction – 50 - 100ml.

Main preparation – Khadiradi vati.

GUGGULU

Botanical name - *Commiphora mukul* Hook ex stocks.

Family - Burseraceae

Synonyms - Guggulu, Deodhoop, Mahishaksha.

Botanical description - It is a small and thorny tree, with spine scent branches, 1.5- 2.0 mtrs heigh. Leaves - composite, leaflets sessile to sub sessile. Flowers are brownish, five petals. Fruits - pulpy, round and red coloured. Gum-thick, scented, multicoloured, burnt on fire, liquefies in sun heat.

a) Macroscopic

Drug occurs in vermicular or stalactitic pieces of pale yellow or brown coloured mass, makes milky emulsion in hot water and readily burns, when fresh viscid and golden coloured, odour, aromtic, taste., bitter and astringent.

Varieties - Five varieties based on colour -1) Mahishaksha 2) Mahaneel 3) Kumud 4) Padma 5) Kanak.

Habitat - Marwar, Sindha, Kutch, Mysore.

Chemical Constituents - Essential oil, gum, resin, steroids

Properties - Rasa - tikta, katu Guna –laghu, ruksha Veerya - ushna Vipaka - katu.

Doshakarma - Tridoshashamaka.

Action and uses - It is anti-inflammatory, analgesic, cleaning of wound and healing due to antibacterial action. Paste of guggul is locally applied in rheumatoid arthritis, cervical lymphadenitis, skin diseases and piles. It reduces foul smell and swelling of wounds. Its analgesic and nervine tonic action is useful in neuralgia, rheumatoid arthritis, sciatica, facial paralysis, hemiplagia and gout.

Parts used - Gum (resin gum).

Doses - 2-4 g.

Main preparation - Yogaraja guggulu, Kaishore guggulu.

GUDUCHI

Botanical name - *Tinospora cardifolia* willd.

Family - Menispermeaceae.

Synonyms - Guduchi, Madhuparni, Amrita, Chhinna.

Botanical description - It is a long lasting creeper climbing over the trees like mango, neem etc. The stem is covered by transparent layer. It has many tentacles hanging down. Leaves are heart shaped, individualized, pointed at the tip and slimy. Flowers are small yellow appearing in clusters. Fruits are bean shaped.

a) Macroscopic

Drug occurs in pieces of varying thickness ranging from 0.6-5 cm in diameter, young stems green with smooth surfaces and swelling at nodes, older ones show a light brown surface marked with warty protuberances due to circular lenticels; transversely smoothened surface shows a radial structure with conspicuous medullary rays traversing porous tissues, taste bitter.

b) Microscopic

Transverse section of stem shows outer-most layer of cork, differentiating into outer zone of thick-walled brownish and compressed cells, inner zone of thin walled colourless, tangentially arranged 3-4 rows of cells, cork broken at some places due to opening of lenticels, followed by 5 or more rows of secondary cortex of which the cells of outer rows smaller than the inner one, just within the opening of lenticels, groups of sclereids consisting of 2-10 cells found in secondary cortex region, outer zone of cortex consists of 3—5 rows of irregularly arranged, tangentially elongated chlorenchymatous cells, cortical cells situated towards inner side, polygonal in shape and filled with plenty of starch grains, simple, ovoid, or irregularly ovoid-elliptical, occasionally compound of 2-4 components, several secretory cells, found scattered in the cortex, pericyclic fibres lignified with wide lumen and pointed ends, associated with a large number of crystal fibres containing a single prism in each chamber, vascular zone composed of 10-12 or more wedge-shaped strips of xylem, externally surrounded by semi-circular strips of phloem, alternating, with wide medullary rays, phloem consists of sieve tube, companion cells and phloem parenchyma of polygonal or tangentially elongated cells, some of them contain crystels of calcium oxalate, cambium composed of one to two layers of tangentially elongated cells in each vascular bundle, xylem consists of vessels, tracheids, parenchyma and fibres, in primary xylem, vessels comparatively narrow devoid of tyloses, secondary xylem elements thick-walled, lignified, vessels cylindrical in shape bearing bordered pits on their walls some large vessels possess several tyloses and often contain transverse septa, meduallry rays 15-20 or more cells wide containing rounded, hemispherical, oblong, ovoid, with faintly marked concentric striations and central hilum appearing like a point, starch grains of 5.5-11.20 μ in diameter and 6-11.28 μ in length, pith composed of large, thin-walled cells mostly containing starch grains.

Habitat - Throughout all over India.

Chemical composion - Tinosporine, tinosporon, tinosporic acid, tinosporol, tinosporide, tinosporidine, columbin, chasmanthin, palmarin, berberine, giloin, giloinisin and b-sitosterol, cordifolide, unosporin, heptacosanol, cordifol, cordifolon, magnoflorine, tembetarine, cardiofoliosides A & B.

Properties - Guna- guru, snigdh Rasa - tikta, katu Veerya - ushna Vipaka - -madhur.

Action and uses - It act as a appetizer, digestive, anthelmintic, antacid, cardiotonic, haemopoetic, anti-diabetic, aphrodisiac and acts specifically on vata- rakta. It is useful in thirst, vomiting, loss of appetite, abdominal pain, liver disorder, jaundice, acid-peptic disorder, dysentery, sprue and worms. Guduchi satva is effective in cardiac debility, rheumatoid arthritis, spleenomegaly, gout and anaemia.

Part used - Root, stem, leaf.
Doses - Decoction - 50-100 ml, Powder - 3-6 g, Satva - 1-2 g.
Main preparation - Guduchyadi kwath, Amritarishta.

TRIVRITTA

Botanical name - *Operculina turpethum* Linn.

Family - Convolvulaceae.

Synonyms - Rechani, Sarala, Triputa, Suvaha.

Botanical description - It is a large, perrineal twining herb with milky juice. Stem is erect, triangular and hairy. Leaves are of various shapes, some are oval and some are triangular. Flowers are white, bell shaped. Fruits – oval shaped.

a) Macroscopic

Roots occur in pieces, 1.5-15 cm long, 1-5 cm dia., usually unbranched, cylindrical, elongated, bearing thin rootlets; thicker pieces, occasionally split and show central wood portion; surface dull grey, reddish-grey to light brown, showing deep furrows or longitudinal wrinkles giving a rope-like or columnar appearance; transversely cut surface shows thick, whitish bark and light yellow centre; fracture in bark, short; in wood, fibrous; odour, indistinct; taste, slightly acrid and nauseating when kept in mouth for some time.

b) Microscopic

Mature root shows thin cork, consisting of 3-5 rows of brown cells; secondary cortex 4-6 layered, composed of tangential elongated, thin-walled cells; some of the cortical cells become thick-walled appearing as isolated, oval to subrectangular sclerenchymatous cells having wide lumen; secretory cavities surrounded by subsidiary cells and resin canals found scattered in secondary cortex; secondary phloem, a wide zone, consisting of sieve elements and phloem parenchyma; vascular bundles arranged in a continuous and a discontinuous ring, traversed by uni and biseriate medullary rays; numerous resin cells also seen in phloem in longitudinal rows; xylem shows 3-5 radiating arms; small patches of intraxylary phloem often formed; xylem vessels in singles or 2-3 in groups, having simple pits on their walls; calcium oxalate crystals as prisms and rosettes found scattered in cortex, phloem parenchyma, xylem parenchyma and medullary ray cells; starch grains, both simple .and compound, simple ones elliptical to spherical with central cleft hilum, compound grains consisting of 2-4 components, size vary from 5-44 μ in dia., found scattered in cortex, phloem parenchyma, xylem parenchyma and medullary ray cells.

Powder - Greyish to light brown; shows parenchymatous cells, cellulosic fibres with pointed tips, vessels with simple pits, simple and compound starch grains elliptical to spherical with central cleft, measuring 5-44 μ in dia., having 2-4 components, rosette and prismatic crystals of calcium oxalate.

T.L.C.

T.L.C. of the alcoholic extract on Silica gel 'G' plate using Toluene : Ethylacetate (9:1) shows under UV (366 nm) three fluorescent zones at Rf. 0.08, 0.21 (both light blue) and 0.58 (blue). On exposure to Iodine vapour seven spots appear at Rf. 0.21, 0.41, 0.49, 0.58, 0.71, 0.90 and 0.97 (all yellow). On spraying with Vanillin-Sulphuric acid reagent and heating the plate for ten minutes at 110°C seven spots appear at Rf. 0.21, 0.41, 0.49 (all light violet), 0.58, 0.70, 0.90 and 0.97 (all violet).

Chemical constituents - Resinous Glycosides.

Varieties - Charak has mention two varieties- 1) White 2) Black.

Habitat - All over India.

Properties - Rasa - katu, tikta, madhur Guna -laghu, ruksha Veerya – ushna Vipaka - madhur

Action and uses - The roots are bitter, acrid, sweet, thermogenic, purgative, carminative, anthelmintic, expectorant, antipyretic and hepatic stimulant. They are useful in colic, constipation, dropsy, paralysis, myalgia, obesity, fever, leucoderma, ulcers, haemorrhoids, tumours, jaundice and opthalmia.

Part used - Root.

Doses - Root powder - 1 - 3 g.

Main preparation - Trivritadi churna, Trivritadi kwath.

BILVA

Botanical name - *Aegle marmelos* Corr.

Family - Rutaceae

Synonyms - Bilva, Shriphala, Sadaphala, Maloora.

Botanical description - Bilva is a medium size thorny tree 8 - 10 meters hight. Leaves are compound, trilaminal, having odour. Flowers are white, having 4-5 petals. Fruits are large, round and rigid.

a) Macroscopic

Root cream yellow or pale yellowish-brown, thin, irregularly and shallowly ridged due to formation of longitudinal and transverse lenticels, surface ruptured, peeling off in layers, internal surface cream to light yellow; fracture, short; taste, sweet.

b) Microscopic

Root shows lignified and stratified cork consisting of 3 or 4 alternating bands of 4-14 layers of smaller cells and a few layers of larger cells having golden yellow contents; secondary cortex, a wide zone, consisting of large, polyhedral, parenchymatous cells and stone cells of varying shapes and sizes, thick-walled, lignified, scattered throughout region; secondary phloem consists of sieve elements, fibres, parenchyma and crystals fibres traversed by phloem rays; some sieve elements compressed, forming tangential bands of ceratenchyma alternating with bands of lignified phloem fibres in outer phloem region, but intact in inner phloem region; phloem parenchyma radially and transversely elongated; phloem fibre groups arranged in concentric rings, fibre groups in inner phloem region extend tangentially from one meduallary ray to another, each group consisting of 2-35 or more cells; fibres long, generally with tapering ends but occasionally forked, lignified, some others have wavy walls; crystal fibres numerous, long, about 9-30 chambered, each containing a prismatic crystal of calcium oxalate; medullary rays uni to triseriate in inner region while bi to pentaseriate in outer region of phloem; cambium consists of 3-7 rows of tangentially elongated to squarish cells; secondary xylem consists of vessels tracheids, fibres and xylem parenchyma; vessels scattered throughout xylem region, in groups of 2-5, single vessels also found, varying in shape and size, mostly drum-shaped, with bordered pits some having a pointed, tail-like process at one end; fibres thick-walled with blunt or pointed tips; xylem parenchyma rectangular in shape; medullary rays uni to triseriate, bi and triseriate rays more common, triseriate rays 12-40 cells high, uniseriate rays 4-10 cells high; prismatic crystals of calcium oxalate present; starch grains simple, 5-19 ¼ in dia., mostly round to oval with centric hilum; compound starch grains having 2-3 components present in inner few layers of cork cells, secondary cortex, phloem and xylem rays.

Powder - Grey to greyish-brown; shows thick-walled, angular cells of cork, numerous prismatic crystal of calcium oxalate, crystal fibres, starch grains simple, 5-19 ¼ in dia., mostly round to oval with centric hilum; compound starch grains having 2-3 components, fragments of xylem vessels with bordered pits and thick-walled xylem fibres.

Chemical constituents - Aegelin, Aegelinin, Marmelosin.

Varieties - It has of 2 variities wild and cultivated.

Habitat - All over India. This tree is cultivated in many gardens.

Properties - Rasa - kashaya, tikta Guna - ruksha, laghu Veerya- ushna Vipaka -katu.

Doshakarma - Kaphavatashamaka.

Action and uses - Unripe fruit is act as an appetiezer, digestant and astringent. Ripe fruit is astringent and a sweet, mild laxative. Root bark, unripe fruit are useful in loss of appetite, diarrhoea, dysentery, sprue and pain in abdomen.

Parts used - Root, root bark, leaf and unripe fruit.

Doses - Powder - 0.75 - 2 g, juice -10 - 20 ml.

Formulation - Bilvadi churna, Bilvadi ghrita.

VIBHITAK

Botanical name - *Terminalia belerica* Roxb.

Family - Combretaceae.

Synonyms - Bibheetaka, Aksha, Karshaphala, Kalidrum.

Botanical description - A large tree grows up to a height of 16 to 32 mtrs. The bark is brownish in colour. The wood of trunk is hard. The leaves resemble those of banyan tree and are 8 to 16 cms in length. At the base of the leaves, where the lamina ends, there are two small nodules. Flowers are very small and yellow. Fruits are round in shape, brownish and hairy.

a) Macroscopic

Fruit nearly spherical to ovoid, 2.5-4.0 cm in diameter, fresh ripe fruits slightly silvery or with whitish shiny pubescent surface, mature fruits grey or grayish brown with slightly wrinkled appearance, rind of fruit shows variation in thickness from 3-5 mm, taste, astringent.

b) Microscopic

Transverse section of fruit shows an outer epicarp consisting of a layer of epidermis, most of epidermal cells elongate to form hair like protuberance with swollen base, composed of a zone of parenchymatous cells, slightly tangentially elongated and irregularly arranged, intermingled with stone cells of varying shape and size, elongated stone cells found towards periphery and spherical in the inner zone of mesocarp in groups of 3-10, mesocarp traversed in various directions by numerous vascular strands, bundles collateral, endarch, simple starch grains and some stone cells found in most of mesocarp cells, few peripheral layers devoid of starch grains, rosettes of calcium oxalate and stone cells present in parenchymatous cells, endosperm composed of stone cells running longitudinally as well as transversely.

Varieties - Depending on the size of the fruits, it is classified into two, having small fruits and having large fruits.

Habitat - Throughout all over India.

Chemical Constituents - Gallic acid, tannic acid and glycosides.

Properties - Rasa - kashaya Guna - ruksha, laghu Veerya - ushna Vipaka -madhur.

Doshakarma - Kaphavatashamaka.

Action and uses - The bark is mildly diuretic and is useful in anaemia and leucoderma. Fruits are astringent, acrid, sweet, ophthalmic, antipyretic, anti-emetic and rejuvenating. They are useful in cough, asthma, bronchitis, cardiac disorder. The mature and dry fruit is constipating and is useful in diarrhoea.

Part used - Fruit.

Dose - 3-6 g.

Main Preparation - Triphala churna, Triphala ghrita.

KUTAJA

Botanical name – *Holarrhena antidysenterica* Linn.

Family - Apocynaceae.

Synonyms - Kutaja, Vatsaka, Girimallika, Kalinga.

Botanical description - The tree is 7-9 meters tall. The bark is pale or brownish in colour. The inner wood is pale and soft. Leaves are appearing similar to those of kadamba- always green and shiny. Flowers are white, fragrant and similar to jasmine flowers. Fruits - two pods arise on the same stalk. They are long and hard.

a) Macroscopic

Small recurved pieces of varying sizes and thickness, outer surface buff to brownish longitudinally wrinkled and bearing horizontal lenticels, inner surface brownish, rough and scaly fracture short and granular, taste, acrid and bitter.

b) Microscopic

Transverse section of dried stem bark shows cork consisting of 4-12 rows of tangentially elongated cells, radial 15- 45 μ tangential 30-60 μ cork cambium consists of a row of thin walled tangentially elongated cells, secondary cortex usually wide, parenchymatous, interspersed with strands of stone cells, stone cell rectangular to oval, with numerous pits often containing prismatic crystals of calcium oxalate, non-lignified pericyclic fibres upto 52 mm thick, present in bark, secondary phloem wide consisting of sieve-tubes, companion cells, phloem parenchyma and stone cells, stone cells arranged in tangential rows in concentric manner associated with crystal sheath containing prisms of calcium oxalate, medullary rays mostly bi or triseriate rarely uniseriate becoming wide toward, outer part and consist of thin-walled, radially elongated, parenchymatous cells, medullary ray cells near stone cells become sclerosed.

Varieties -Its two varieties are mentioned - 1) Male kutaja 2) Female kutaja.

Habitat - Throughout India but mainly in the jungles of Saharanpur.

Properties - Rasa- tikta, kashaya Guna - ruksha Veerya- sheeta Vipaka - katu.

Doshakarma - Kaphapittashamaka.

Action and uses - The bark and seed are bitter, constipating, astringent, powerful antidysenteric, anthelmintic, antiperiodic, digestive and tonic. It is useful in amoebic dysentery, diarrhoea, asthma, hepatic and gastric disorder, flatulence, malaria, vomiting and leprosy.

Part used - Stem, bark, seeds.

Dose - Drug for decoction - 20 - 30 ml, powder - 3 - 6 g.

Main preparation - Kutajarishta, Kutajavaleha.

DURVA

Botanical name - *Cynodon dactylon* Linn.

Family - Gramineae.

Synonyms - Shataveerya, Harit, Shambhavi.

Botanical description - Durva is a creeper which spreads on the ground. Its stem bears numerous knots from which roots appear and help it in spreading. Leaves - 2.5 to 10 cm long and narrow. Flower is 2.5 to 5 cm in length. Seeds are very minute.

a) Macroscopic

Roots fibrous, cylindrical, upto 4 mm thick, minute hair-like roots arise from the main roots; cream coloured.

b) Microscopic

Mature root shows epiblema or piliferous layer composed of single layered, thin-walled, radially elongated to cubical cells; hypodermis composed of 1-2 layered, thin-walled, tangentially elongated to irregular shaped cells; cortex differentiated into two zones, 1 or 2 layers of smaller, thin-walled, polygonal, lignified sclerenchymatous and 4-6 layers of thin-walled, elongated parenchymatous cells being larger; endodermis quite distinct being single layered, thick-walled, tangentially elongated cells; pericycle 1-2 layers composed of thin-walled sclerenchymatous cells; vascular bundles consisting of xylem and phloem, arranged in a ring on different radials; xylem exarch, having usual elements; centre occupied by wide pith, composed of oval to rounded thick-walled parenchymatous cells containing numerous simple, round to oval or angular starch grains measuring 4-16 µ in dia., and compound starch grains having 2-4 components.

Powder - Cream coloured; fragments of xylem vessels with pitted walls, thick-walled lignified sclerenchymatous cells and numerous simple round to oval or angular starch grains measuring 4-16 μ in dia., and compound starch grains having 2-4 components.

Chemical constituents - Phenolic Phytotoxins and Flavonoids.

Habitat - It is found in all over the India.

Varieties - There are three varieties.

1. White 2. Blue3. Ganda durva.

Properties - Guna -lagu Rasa - madhur, kashaya, tikta Vipaka - madhur Veerya - sheeta.

Doshakarma – Tridoshashamaka (specialy kaphapittashamaka).

Action and Uses - It acts as a astringent, wound healer and reduce burning in cuts. So its juice and paste is used in conjunctivitis, burning sensation and skin diseases. As it has a positive effect on central nervous system. It is useful in epilepsy and schizophrenia. It is useful in diarrhoea, dysentery, and haemoptysis.

Parts used – Whole plant.

Dosage - Juice - 10 to 20 ml, Paste -1 to 3 g, Decoction - 50 to 100 ml.

Main preparation - Durvadi kwath, Durvadi ghrita.

ARJUNA

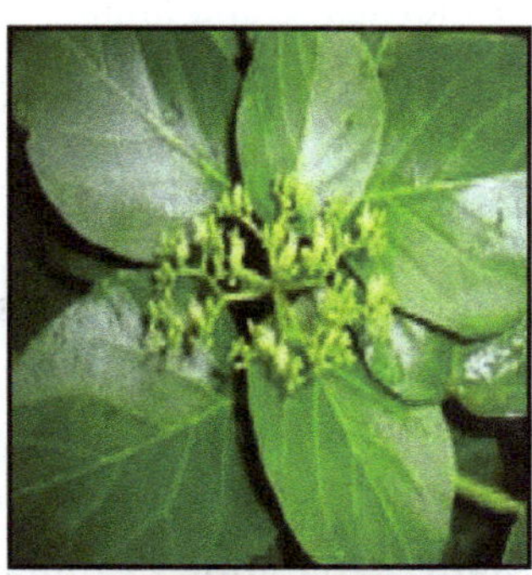

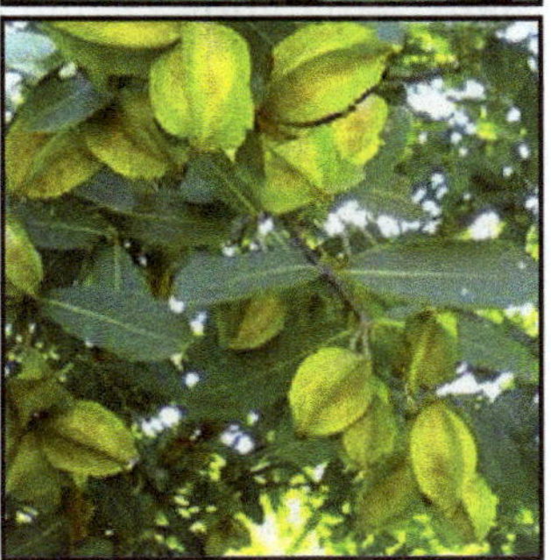

Botanical name- *Terminalia arjuna* Wight & Avn.

Family - Combretaceae

Synonyms - Arjuna, Dhawal, Nadisarja

Botanical description - This is a large tree growing up to a height of 20 to 26 mtrs. Trunk is straight pale white and smooth. Leaves are subopposite, similar to Gvava leaves, 10-15 pairs of leaflets at the end. Leaflets- tounge shaped. Flowers- white or yellow occur all round the stalk. Fruit has five ridges on its surface.

a) Macroscopic

Bark available in pieces, flat, curved, recurved, channelled to half quilled, 0.2-1.5 cm thick, market samples upto 10 cm in length and upto 7 cm in width, outer surface somewhat smooth and grey, inner surface somewhat fibrous and pinkish, transversely cut smoothened bark shows pinkish surface, fracture, short in inner and laminated in outer part; taste, bitter and astringent.

b) Microscopic

Stem Bark -Mature bark shows cork consisting of 9-10 layers of tangentially elongated cells, a few outer layers filled with brown colouring matter; cork cambium and secondary cortex not distinct and medullary rays observed traversing almost upto outer bark; secondary phloem occupies a wide zone, consisting of sieve tubes, companion cells, phloem parenchyma and phloem fibres, traversed by phloem rays, usually uniseriate but biseriate rays also occasionally seen; in the middle and outer phloem region, sieve tubes get collapsed and form ceratenchyma; phloem fibres distributed in rows and present in groups of 2-10; rosette crystals of calcium oxalate measuring 80-180 µ in dia., present in most of the phloem parenchyma, alternating with fibres; idioblasts consisting of large cells having aggregates of prismatic and rhomboidal crystals of calcium oxalate in row throughout the zone, measuring 260-600 μ in dia., starch grains, mostly simple, compound of 2-3 components, sometimes upto 5 components, round to oval, elliptical, measuring 5-13 μ in dia., distributed throughout the tissue (absent in *T. alata*); in a tangential section the uniseriate phloem rays 2-10 cells high and biseriate, 4-12 cells high; in longitudinal section rosette crystals of calcium oxalate found in the form of strands in phloem parenchyma.

Powder - Reddish-brown; shows fragments of cork cells, uniseriate phloem rays, fibres, a number of rosette crystals of calcium oxalate, a few rhomboidal crystals, starch grains simple and compound, round to oval, elliptic, having 2-3 components with concentric striations and small narrow hilum, measuring 5-13 μ in diameter.

Habitat - It is found in lower part of Himalaya, Bengal, Bihar and Madhya Pradesh.

Chemical Constituents - Tannins.

Properties- Guna-laghu, ruksha Rasa - kashaya Vipaka- katu Veerya - sheeta.

Doshakarma - Kaphapittashamaka.

Action and uses - It acts as an astringent, haemostatic, cardiac tonic and antipyretic. The bark paste is locally applied on wounds, ulcer and specially used in promoting the union of fractures. Being haemostatic it controles bleeding in dysentery and haemorrhoids. It gives strength to the cardiac muscles and improves cardiac function and rhythm.

Parts to be used - Leaves and root bark.

Dosage - Decoction in milk- 6 to 12 ml, Powder - 1 to 3 g, Decoction in water- 60-100 ml.

Main preparation - Arjunaristha, Arjuna churna.

Omitection time -In sharad ritu.

ARDRAK

Botanical name – *Zinziber officinale* Roscoe.

Family – Zingiberaceae.

Synonyms – Vishva, Nagar, Vishvabheshaja.

Botanical description– Plants grow up to 1 to 1.5 meter hight. Leaves are 13-30 cms. long, broad and tapering at the top. Flower with stalk is dark violet in colour.

a) Macroscopic

Rhizome, laterally compressed bearing short, flattish, ovate, oblique, branches on upper side each having at its apex a depressed scar, pieces about 5-15 cm long, 1.5-6.5 cm wide (usually 3-4 cm) and 1-1.5 cm thick, externally buff coloured showing longitudinal striations and occasional loose fibres, fracture short, smooth, transverse surface exhibiting narrow cortex (about one-third of radius), a well-marked endodermis and a wide stele showing numerous scattered fibro-vascular bundles and yellow secreting cells, odour agreeable and aromatic, taste, agreeable and pungent.

b) Microscopic

Transverse section of rhizome shows cortex of isodiametric thin-walled parenchyma with scattered vascular strands and numerous isodiametric idioblasts, about 40-80 µ In diameter containing a yellowish to reddish-brown oleo-resin, endodermis slightly thick walled, free from starch immediately inside endodermis a row of nearly continuous collateral bundles usually without fibres stele of thin-walled, parenchyma cells, arranged radially around numerous scattered, collateral vascular bundles, each consisting of a few unlignified, reticulate or spiral vessels upto about 70 µ in diameter, a group of phloem cells, unlignified, thin-walled, septate fibres upto about 30 μ wide and 600 μ long with small oblique slit, like pits, present, numerous scattered idioblasts, similar those of cortex, and associated with vascular bundles, also present, idioblasts about 8-20 μ wide and up to 130 μ long with dark reddish-brown contents: in single or in axial rows, adjacent to vessels, present, parenchyma of cortex and stele packed with flattened, rectangular, ovate, starch grains, mostly 5-15 μ - 30-60 μ long about 25 μ wide and 7 μ thick, marked by five transverse striations.

Habitat– Hot and damp climate like Kerala, Kochin, Bengal & Madras.

Chemical Constituents – Essential oil, pungent constituents (gingerol and shogaol), resinous matter and starch.

Varieties–According to habitat and processing, there are many varieties– (1) Indian (2) Kochi (3) Jamayaka. Wet tubers are known as Adrak and dry as Shunthi.

Properties– Guna-laghu, snigdha Rasa-katu Veerya-ushna Vipaka-madhur.

Doshakarma–Kaphavatashamaka.

Action and uses– By its anti-inflammatory and analgesic action, It is used as local application in swollen joints and rheumatoid arthritis (Joint disorder). It is the best medicine for all vata disorders. It stimulates nerves, improves impulse transmission and relieves pain. Ginger is an excellent appetizer, digestive & antispasmodic. Being appetizer ginger along with salt is given before meals; it is used in nausea, vomiting, loss of appetite, indigestion, flatulence, Jaundice and piles.

Part used– Rhizome.

Dosage– Ginger juice – 5-10 ml, Powder – 1 to 2 g.

Main preparation– Ardrak-khanda,Vyoshadighrt, Panchasama Churna.

DHATAKI

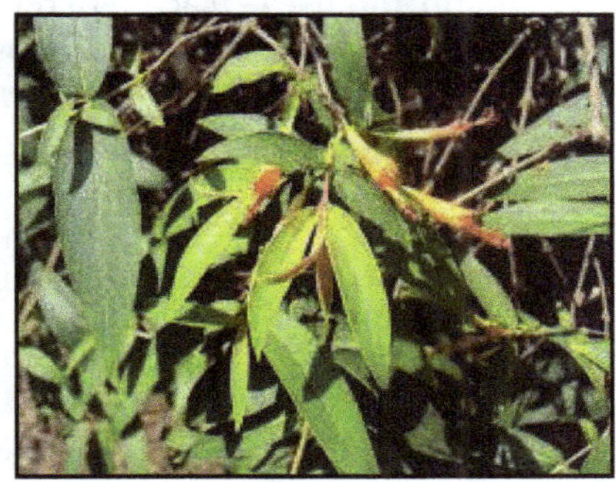

Botanical name – *Woodfordia floribunda* Kurz.

Family – Lythraceae

Synonyms – Dhataki, Tamrapushpi, Dahani.

Botanical description– It is a shrub with many branches. Leaves-sessile; resemble pomegranate leaves, 5 to 10 cm long, stalkless. The leaf is hairy from beneath because of which it looks whitish. Flower-shiny red coloured, seed-grey and slimy. The plant bear flowers in winter and fruits in rainy season.

a) Macroscopic

Flower, about 1.2 cm long, occurs as single or in bunches of 2-15, calyx 1.0-1.6 cm long, ridged and glabrous, bright red when fresh but fades on drying, with campanulate base and oblique apex having 6 triangular and acute teeth, each tooth being, 2-2.5 mm long, 6, very minute accessory sepals attached outside at the juncture of calyx tooth and deeper in colour, petals 6, attached inside the mouth of calyx-tube, shightly longer than calyx tooth, alternating with calyx-tooth pale rose or whitish, thin, papery, lanceolate, acuminate, stamens 12, united at the base, about 1.5-2 cm long, filament filiform, curved at the apex, keeping anthers inside calyx-tube , anthers dorsifixed brown, almost rounded or broadly ovate, carpels 2, united, ovary superior, style filiform, longer, than ovary and stamens, taste, astringent.

b) Microscopic

Transverse section of sepal shows, single layered cuticularised epidermis, provided with both glandular and covering trichomes ;glandular trichomes, multicellular, long, consisting of a stalk and a globose, thin-walled, multicellular head, covering trichomes, unicellular thick-walled broad at base and pointed at the apex, ground tissue consisting of thin-walled, parenchymatous cells surface view of petal shows thin-walled, parenchymatous cells, provided with very few sparsely distributed covering trichomes, transverse section of filament shows, epidermis consisting of single layered tangentially elongated cells, covered with a very thick-cuticle, ground tissue consisting of thin walled parenchymatous cells with intercellular spaces, surrounding a central. vascular cylinder of spirally thickened vessels, transverse section of anther shows, single layered epidermis, covered with cuticle followed by several layers of thickened cells, surrounding both the pollen-sacs having numerous pollen grains, pollen grains roughly tetrahedral with three pores, measuring 12-16 μ approximately , central region consisting of thin-walled cells emboding vascular bundles.

Chemical constituents - Tannin and glucose.

Habitat– All over India mainly in hilly region.

Properties– Guna-laghu, ruksha Rasa-kashaya, katu Vipak-katu Veerya-sheeta.

Doshakarma– Kaphapittashamaka.

Action and uses– It acts as a refrigerant, haemostatic, wound healer, astringent, kusthaghna. It is used externally over bleeds and wounds. Usefull in pervaginal discharges. In leucorrhoea, it is given either with rice water.

Part to be used– Flowers, leaves.

Dosages– Powder 1 to 3 g.

Main preparation– Dhatakyadi churna, Dhatakyadi tail.

DRAKSHA

Botanical name – *Vitis vinifera* Linn.

Family – Vitiaceae

Synonyms – Charuphala, Krishna.

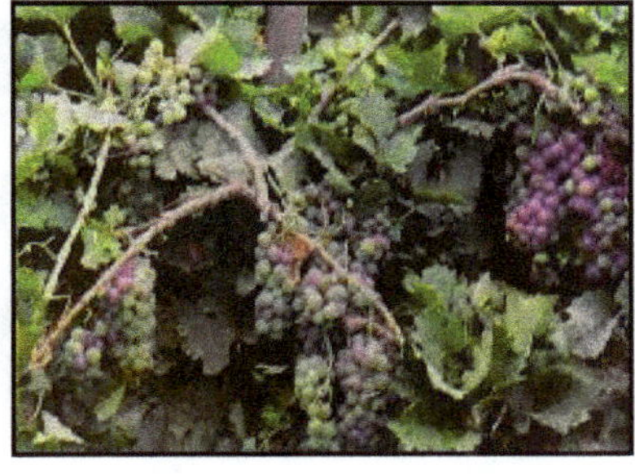

Botanical Description– It is an ascending creeper. Leaves-resemble, bitter guard but have fine hair on them. Flower– greenish, fragrant and in clusters. Fruits are oblong, having 4-5 seeds.

a) Macroscopic

Fruit a berry, sticky and pulpy, dark brown to black; oblong or oval, sometimes spherical; 1.5 -2.5 cm long and 0.5-1.5 cm wide; outer skin irregularly wrinkled forming ridges and furrows; usually contain 1-4 seeds, 4-7 mm long, ovoid rounded to triangular or simply ovoid, brown to black; odour, sweetish and pleasant; taste, sweet.

b) Microscopic

A single layered epidermis cells filled with reddish-brown contents; mesocarp pulpy, made up of thin-walled, irregular cells containing prismatic crystals of calcium oxalate, measuring 13.75 -41 µ in dia.; some fibro-vascular bundles also present in this region; seeds composed of testa and endosperm; testa composed of thick-walled yellowish cells; endosperm composed of angular parenchymatous cells containing oil globules and cluster crystals of calcium oxalate, measuring 11-16 µ in diameter.

Chemical constituents - Malic, Tartaric & Oxalic acids, Carbohydrates and Tannins.

Habitat– North-west India, Punjab, Kashmir, Maharastra.

Properties– Guna-snigdha, guru Rasa-madhur, Vipaka-madhur Veerya-sheeta.

Doshakarma–Vatapittashamaka.

Action and uses– It is useful in thirst, bleeding disorders, burning, gout, pleurisy, tuberculosis, cough, asthma, burning micturation.

Part to be used– Fruit.

Dosage– According to digestive power.

Main preparation– Draksharista, Draksha grita.

KIRATATIKTA

Botanical name – *Swertia chirata* Roxb.ex Flem.

Family – Gentianaceae

Synonyms– Kiratikta, Jwarantak.

Botanical description– This small plant is 0.75 to 1 meter hight, with slightly thick trunk, round shaped but quadrangular at the top. Leaves are unequal of 5-8 cm. length, 1-2 cm wide and pointed at the tip.

a) Macroscopic

Drug consists of whole plant, a peculiar shining yellowish tinge all over the herb in fresh sample, stem upto 1 m long and 6 mm in diameter, glabrous, yellowish-brown to purplish, slightly quadrangular above and cylindrical below, large, continuous, easily separable yellow pith, leaf, opposite, cauline, broad at base, ovate or lanceolate, entire, acuminate, glabrous, usually with 5-7 prominent lateral veins, branching from the axils of the leaves which ramify further into paniculate inflorescence, flower, tetramerous, 2-3 mm wide, ovoid, with two glandular depressions near the base of each of corolla lobes, ovary, superior, bicarpellary, unilocular, ovoid and pointed fruit a capsule with numerous, minute reticulated seed, 0.25-0.55 mm long, 0.16-0.45 mm broad irregularly ovoid.

b) Microscopic

Root-transverse section of root shows, 2-4 layers of cork, secondary cortex represente by 4-12 layers of thick-walled, parenchymataous cells, some showing radial wall formation, tangentially elongated with sinuous walls, secondary phloem composed of thin-walled strands of sieve tubes, companion cells and phloem parenchyma, secondary xylem composed of vessels, tracheids parenchyma and xylem fibres, all elements lignified and thick-walled, in older roots, centre of wood more or less spongy and hollow in most cases, outer woody ring remaining strongly lignified, vessels show scalariform thickening and also simple and bordered pits, tracheids similar in thickening as the vessels, fibres have simple pits, mucilage present in secondary cortical cells, minute acicular crystals present in abundance in secondary cortex and phloem region, resin also present as dark brown mass in secondary cortex cells.

Stem-Transverse section of stem shows single layered epidermis, externally covered with a thick striated cuticle present in young stem, in older epidermis remains intact but cells flattened and tangentially elongated, four ribs also consists of an epidermis and parenchymatous cortical cells, endodermis distinct, showing anticlinal or periclinal walls, followed by single layered pericycle consisting of thin walled cells, stem possesses an amphiphloic siphonostele, external phloem represented by usual elements, cambium between external phloem and xylem composed of a thin strip of tangentially elongated cells, internal phloem similar in structure as that of external phloem excepting that sieve tube strand is more widely separated, xylem continuous and composed mostly of tracheids, a few xylem vessels present singly or rarely in groups of two while tracheids and fibres present in abundance, vessels and fibre tracheids have mostly simple and bordered pits and fibres with simple pits on the walls, medullary rays absent, central part of the stem occupied by a pith consisting of rounded and isodiametric cells with prominent intercellular spaces mucilage present in cortical cells, minute acicular crystals also present in abundance, cortical cells, in resin present as dark brown mass in some cortical cells along with oil droplets.

Leaf- Transverse section of leaf shows very little differentiation of mesophyll tissues, epidermis single layered covered with a thick, striated cuticle, more strongly developed on the upper surface than the lower, stomata of anisocytic type, palisade tissue single layered, cells at places become wider and less elongated particularly in bigger veins, spongy messophyll represented by 4-7 layers of somewhat loosely arranged, tangentially elongated cells, some epidermal cells prominently arched outside at the margin, mucilage present in epidermal and mesophyll cell while minute acicular crystal also present in abundance in mesophyll cells, in leaf parenchymas oil droplets also present.

Habitat– It is found in Himalayan range from Kashmir to Bhutan.

Varieties–There are two verities– 1. Bitter 2. Semibitter.

Properties– Guna-laghu, ruksha, Rasa-tikta, Vipaka-katu Veerya-sheeta.

Doshakarma– Kaphapittashamaka.

Action and Uses– It acts as appetizer, amapachan, cholagogue and laxative, anthelmintic, haemostatic.

It is uses for – generalized itching, dysfunction of liver, constipation, bleeding piles, jaundice and anaemia. Kiratatikta is more famous for its use in fever as an antipyretic.

Parts to be used– Whole plant.

Dosage– Powder-2-6 g, Decoction-60-100 ml.

Main preparation– Kiratadi kwath, Mahasudarshan churna.

KUTAKI

Botanical name – *Picrorrhiza kurroa* Royle ex. benth.

Family – Scrophulariaceae

Synonyms – Tikta (due to bitter Taste), Katurohini, Chakrangi.

Botanical description– Roots are like raddish, trunk hard. Leaves serrated, 5-10 cm long with rounded tip. Flowers stalk is hard; pointing upward and with flowers in bunches. Root is of a finger thickness.

a) Macroscopic

Rhizome - 2.5-8 cm long and 4-8 mm thick, subcylindrical, straight or slightly curved, externally greyish-brown, surface rough due to longitudinal wrinkles, circular scars of roots and bud scales and sometimes roots attached, tip ends in a growing bud surrounded by tufted crown of leaves, at places cork exfoliates exposing dark cortex; fracture, short; odour, pleasant; taste, bitter.

Root - Thin, cylindrical, 5-10 cm long, 0.05-0.1 cm in diameter, straight or slightly curved with a few longitudinal wrinkles and dotted scars, mostly attached with rhizomes, dusty grey, fracture, short, inner surface black with whitish xylem; odour, pleasant; taste, bitter.

b) Microscopic

Rhizome - Shows 20-25 layers of cork consisting of tangentially elongated, suberised cells; cork cambium 1-2 layered; cortex single layered or absent, primary cortex persists in some cases, one or two small vascular bundles present in cortex;

vascular bundles surrounded by single layered endodermis of thick-walled cells; secondary phloem composed of phloem parenchyma and a few scattered fibres; cambium 2-4 layered; secondary xylem consists of vessels, tracheids, xylem fibres and xylem parenchyma, vessels vary in shape and size having transverse oblique articulation; tracheids long, thick-walled, lignified, more or less cylindrical with blunt tapering ends; xylem parenchyma thin-walled and polygonal in shape; centre occupied by a small pith consisting of thin-walled cells; simple round to oval, starch grains, measuring 25-104 µ in dia., abundantly found in all cells.

Root -Young root shows single layered epidermis, some epidermal cells elongate forming unicellular hairs; hypodermis single layered; cortex 8-14 layered; consisting of oval to polygonal, thick-walled, parenchymatous cells; primary stele tetrach to heptarch, enclosed by single layered pericycle and single layered, thick-walled cells of endodermis; mature root shows 4-15 layers of cork, 1-2 layers of cork cambium; secondary phloem poorly developed; secondary xylem consisting of vessels, tracheids, parenchyma and fibres; vessels have varying shape and size, some cylindrical with tail-like, tapering ends, some drum shaped with perforation on end walls or lateral walls; tracheids cylindrical with tapering pointed ends; fibres aseptate, thick-walled, lignified with tapering blunt chisel-like pointed ends.

Powder - Dusty grey; shows groups of fragments of cork cells, thick-walled, parenchyma, pitted vessels and aseptate fibres, simple round to oval, starch grains, measuring 25 - 104 µ in diameter.

Habitat – Grows in the Himalayas from Kashmir to Sikkim at the height of 2 to 4 thousand mtrs.

Properties– Guna-laghu, ruksha Rasa-tikta Vipaka-katu Veerya-sheeta.

Doshakarma– Kaphapittashodhaka.

Action & uses– it acts as appetizer, liver stimulant and cholagogue. So it uses in low appetite, liver disorders and jaundice, useful in fever, pittashodhan and increases flow of bile, reduces burning in fever, it acts a diuretic in diabetes.

Parts to be used– Root.

Dosage– 3-6 g for purgative and fever.

Main preparation– Arogyavardhani, Katukadyavaleha.

ELA

Botanical name – *Elettaria cardemomum* Maton

Family – Zinziberaceae

Synonyms – Chandrabhaga, Bahula, Gandhaphalika.

Botanical description– It is a leafy perinial herb. Leaves are subsessile 20-40 cm long fragrant. Flowers are many, having long stalks. Fruits are many and within the fruit there are several black seeds with pungent odour.

a) Macroscopic

Fruit - 1-2 cm long ovoid or oblong and more or less three sided with rounded, angles, greenish to pale-buff or yellowish in colour, base rounded or with the remains of pedicle, apex shortly beaked, surface almost smooth or with slight longitudinal striations, small trilocular fruit, each containing about 15-20 seeds in a row of doubles, adhering together to form compact mass.

Seed - Dark brown to black, about 4 mm long and 3 mm broad, irregularly angular, transverscly wrinkled but not pitted, with a longitudinal channel containing raphe, enclosed in a colourless, membranous aril, odour, strongly aromatic, taste, characteristic.

b) Microscopic

Transverse section of seed shows flattened, aril, thin-walled parenchymatous cells, testa with outer epidermis of thick-walled, narrow, elongated cells, followed by a layer of collapsed parenchyma, becoming 2 or 3 layered in the region of raphe, composed of large, thin-walled rectangular cells containing volatile oil, a band of 2 or 3 layers of parenchyma and an inner epidermis of thin-walled, flattened cells, inner integument 2 layered, an outer palisade sclerenchyma with yellow to reddish-brown beaker shaped cells, 20 μ long in radial direction and 12 μ wide, thickened on inner and anticlinal walls, each cell with a small bowl shaped lumen containing a warty nodule of silica and an inner epidermis of flattened cells, peri sperm cells thin-walled, packed with minute rounded polyhedral starch grains, about 1-2 to 4-6 μ in diameter and containing 1-7 small prismatic crystals of calcium oxalate, about 10-20 μ long, endosperm of thin-walled parenchyma containing protein as a granular hyaline mass in each cell, embryo, of small thin-walled cells containing aleurone grains, starch absent in endosperm land embryo, fibres sclerenchymatous, large vessels present in pericarp.

Habitat– Gujarat, Mysore, Kerala.

Chemical Composition - Essential oil.

Varieties – These have three varieties – (1) Madhara-those which are ready in Feb-March (2) Kanti - those which are ready in Oct - Nov (3) Neel – black and long variety.

Properties– Guna - laghu, ruksha Rasa - katu, madhur Vipaka - madhur Veerya - sheeta.

Doshakarma– Tridoshahara.

Action & uses– It acts as a mouth fresher, deodorant and antiseptic. Seeds and oil is appetizer, digestive and laxative. It is useful in oral disorders, abdominal pain and piles. Seeds are diuretic, useful in dysuria.

Parts to be used– Fruits.

Dosage– 0-5 to 1 g.

Main preparation– Eladichurna, Eladigutika.

SHANKHAPUSHPI

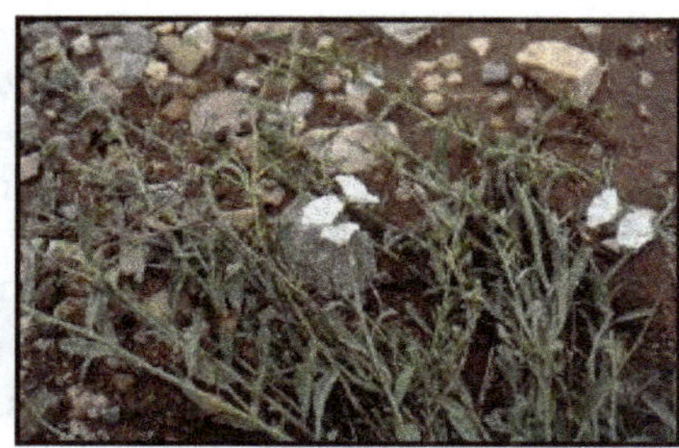

Botanical name – *Convolvulus pluricaulis* chois.

Family – Convolvulaceae

Synonyms–Kshirpushpi,Kusum,Supushpa.

Botanical description– This is a perennial creeper with many branches. Stem is slightly quadrangular. Leaves are thin, long, with three veins, without stalk. Flower - Bell shaped, white. Fruit is small.

a) Macroscopic

Root - Usually branched, cylindrical, ribbed having some rough stem nodules and small secondary roots, 1-5 cm long, 0.1-0.4 cm thick, yellowish-brown to light brown.

Stem - Slender, cylindrical, about 0.1 cm or less in thickness with clear hairy nodes and internodes; light green.

Leaf - Shortly petiolate, linear-lanceolate, acute, hairy on both surfaces; 0.5-2 cm long and 0.1-0.5 cm broad; light green.

Flower - White or pinkish; solitary or in pairs sessile or sub-sessile in the leaf axis; sepals narrowly, linear-lanceolate, sparsely hairy; corolla shortly discoid; stamen 5, free, epipetalous, alternate with the petals, inserted deep in the corolla tube; ovary superior and bicarpellary.

Fruit - Capsule, oblong globose with coriaceous, pale brown pericarp.

Seed - Brown; minutely puberulous.

b) Microscopic

Root - Appears nearly circular in outline; cork composed of 10-15 layers of tangentially elongated, thick-walled cells; cortex composed of 6-10 layers of oval to elongated, elliptical, parenchymatous cells and yellowish-brown, tanniniferous, secretory cells present in this region; phloem composed of sieve elements, phloem parenchyma and phloem rays; xylem consisting of usual elements; vessels solitary or in groups of two with simple pits; fibres and tracheids aseptate and pitted; medullary rays 1-3 cells wide and multicellular in length; starch grains solitary or in groups, simple and composed of 2-3 components, round to oval in shape, measuring 3-8 µ in dia., present in cortex, phloem, xylem rays and parenchyma.

Stem - Shows single layered epidermis, covered with thick cuticle; at places unicellular hairs present; cortex differentiated in two zones, 2-3 upper collenchymatous and 1-2 lower parenchymatous layers, both having round to oval, elongated, thin- walled cells; endoderrnis single layered; pericycle present in the form of single strand of fibres; phloem a narrow zone, mostly composed of sieve elements and parenchyma; xylem consists of vessels, fibres and parenchyma; medullary rays and tracheids not distinct, vessels mostly solitary with spiral thickening; fibres aseptate having pointed ends and narrow lumen; strand of internal phloem present around the slightly lignified pith.

Leaf

Midrib - Appears convex in lower and concave in upper side; epidermis single layered, covered with thick cuticle; lower epidermis followed by 2-3 layers of chlorenchymatous cells; vascular bundle bicollateral, composed of usual elements of phloem and xylem; rest of tissue between chlorenchyrna and vascular bundles composed of 4-5 layers of parenchymatous cells.

Lamina - Shows epidermis on both surfaces covered with thick cuticle; hairs unicellular, present on both surfaces, palisade two layered, spongy parenchyma 4-5 layered; a few bicollateral vascular bundles present in spongy parenchyma; palisade ratio 6-9; vein islet number 21-25 per sq. mm. , stomatal index in lower surface 17-20 and in upper surface, 13.8-17.0; stomatal number in lower surface 184-248, and in upper surface 202-238 per sq. mm.

Powder - Light yellowish-green; shows groups of vessels with sprial thickening and simple pits, fibres and tracheids, simple and compound starch grains, measuring 3 - 8 µ in dia., unicellular hairs, mesophyll cells and gives positive test for tannin.

Chemical constituents - Alkaloid.

Habitat – All over India-specially in rocky and open hand.

Varieties– Three-white, reddish and bluish, but only white one is used in medicine.

Properties– Guna-snigdha, pichchil, guru Rasa-kashaya, katu tikta, Vipaka-katu, Veerya-ushna.

Doshakarma– Tridoshahar, especially vatapittashamak.

Action and uses– It is a brain tonic, sedative and therefore it is used in epilepsy, insomnia and giddiness. In schizophrenia it reduces the severity of attacks. It is a cardiotonic and haemostatic. So it is useful in heart disease, haemoptysis and hypertension.

Parts to be used– Panchang (whole plant).

Dosage– Decoction – 20 to 30 ml , Powder – 3 to 6 g, cold infusion – 12 to 50 ml.

Main Preparation– Shankhapuspi syrup, Amritadi rasayan.

VIDANG

Botanical name – *Embelia ribes* Burm. F.

Family – Myrsinaceae

Synonyms– Vidang, Chitratandul, Jantughna.

Botanical description – It is a big tree, leaves are oval, pointed, flower - grows in cluster, Fruit-contains greyish coloured seed and pulp. On external surface, there are small white spots.

a) Macroscopic

Fruit brownish-black, globular 2-4 mm in diameter, warty surface with a beak like projection at apex, often short, thin pedicel and persistant calyx with usually 3 or 5 sepals present, pericarp brittle enclosing a single seed covered by a thin membrane, entire seed, reddish and covered with yellowish spots (chitra tandula), odour slightly aromatic, taste, astringent.

b) Microscopic

Transverse section of fruit shows epicarp consisting of single row of tabular cells of epidermis, usually obliterated, in surface view cells rounded with wrinkled cuticle, mesocarp consists of a number of layers of reddish-brown coloured cells and numerous fibrovascular bundles and rarely a few prismatic crystals of calcium oxalate, inner part of mesocarp and endodermis composed of stone cells, endodermis consisting of single layered, thick-walled, large, palisade-like stone cells, seed coat composed of 2-3 layered reddish-brown coloured cells, endosperm cells irregular in shape, thick-walled, containing fixed oil and proteinous masses, embryo small when present otherwise most of the seeds sterile.

Powder-Reddish, under microscope shows reddish parenchyma and stone cells.

Identification

(I) Shake 1 g of the powdered seeds with 20ml of Solvent Ether for five minutes and filter. To a portion of the filtrate add 5 per cent *vlv* solution of Sodium Hydroxide, a deep violet colour is developed in the aqueous layer. To the other portion add 2 drops of Dilute Ammonia solution, a bluish violet precipitate is obtained.

(II) Boil 5 g of the powdered seeds :with 25 ml alcohol and filter. Divide the deep red coloured filtrate into two portions. To one portion, add solution of lead Acetate, a dirty green precipitate is produced. To the other portion add solution of ferric chloride a reddish-brown precipitate is produced.

Habitat – All over India in mountainous areas.

Properties – Guna- laghu, ruksha Rasa-katu, Vipaka-katu Veerya-ushna.

Doshakarma– Vatakaphanashak.

Action and uses– Bactericidal, kusthaghna, and shirovirachana, A fine powder of vidanga is used for nasal administration (Pradhaman hasya) in chronic rhinitis, jaundice, it act as a deepan, pachan, anuloman but specially vermincidal. Vidang along with worm water or buttermilk is used in indigestion, vomiting, abdominal colic, flatulence, constipation and piles. In round worm & tape worm infestation 10gms of vidang powder is given on an empty stomach and then a purgative is given.

Parts to be used– Fruit (1 year old).

Doses– 1 to 2 gm and for worm – 10 g.

Main preparation– Vidangadi churna, Vidangarishta, Vidangadi loha.

MANJISTHA

Botanical name – *Rubia cordifolia* Linn.

Family – Rubiaceae

Synonyms– Manjistha, Rakta, Tamramoola.

Botanical description– It is a climber, whose branches spread for a long distance. Trunk square shaped and redish. Leaves- heart shaped pointed, rough on upper surface. Flower- small, yellowish white, Root- reddish, thick & long.

a) Macroscopic

Stem slender, more or less cylindrical, slightly flattened, wiry, about 0.5 cm thick, brown to purple coloured; surface scabrous, stiff and grooved with longitudinal cracks; prickles present in the immature stem; nodes distinct having two leaf scars, one on either side; fracture, short.

b) Microscopic

Mature stem shows exfoliating cork, ruptured at places, forming dome-shaped structure, consisting of 3-12 or more layered radially arranged, squarish and tangentially elongated, thin-walled cells, appearing polygonal in surface view; secondary cortex 3-5 layered consisting of tangentially elongated, thin-walled cells, some of which contain acicular crystals of calcium oxalate as isolated or in bundles; a few cells contain sandy crystals as black granular masses; secondary phloem, a wide zone of reddish colour, composed of sieve elements and phloem parenchyma, fibres absent; phloem parenchyma smaller towards inner side gradually becoming larger and tangentially elongated towards periphery, a few cells contain sandy crystals of calcium oxalate; secondary xylem forms a continuous cylinder of reddish colour, composed of vessels, tracheids, fibres and xylem parenchyma; vessels numerous, distributed uniformly throughout xylem, larger towards outer side and smaller towards centre; in macerated preparation, vessels show great variation in shape and size having lignified walls and pitted thickening; xylem fibres thick-walled, long and short, longer ones have narrow lumen while shorter ones have wide lumen with pitted thickenings; xylem parenchyma also vary in shape and size having pitted or reticulate thickening; centre occupied by narrow pith consisting of thin-walled, parenchymatous cells, a few cells contain sandy crystals of calcium oxalate.

Powder - Pink; shows numerous fragments of cork, lignified xylem vessels, tracheids, and fibres with pitted and reticulate xylem parenchyma having red coloured contents; acicular and sandy crystals as black granular masses.

Chemical constituents - Glycosides.

Habitat– Himalaya, Nilgiri, Hilly areas.

Varieties– Mangistha available in market is of 4 types. 1. Nepali 2. Irani 3. Afghani 4. Hindustani.

Properties– Guna-guru, ruksha Rasa-kashaya, tikta, madhur Vipaka-katu Veerya-ushna.

Doshakarma– Kaphapittashamaka.

Action and uses– It is a good appetizer (deepan) and has good digestive, it has astringent property that reduce molility. It is used in loss of appetite, amadosha, diarrhoea, dysentery associated with bleeding. Being haemostatic & blood purifier (Raktashodhak), it is used in many blood disorder & skin disorders. It increases menstrual flow and it purifies breast milk.

Parts to be used– Root.

Dosage– Decoction – 60 – 120 ml, Powder – 1 – 3 g.

Main preparation– Manjishthadi kadha, Manjistithadyark.

MARICHA

Botanical name – *Piper nigrum* Linn.

Family – Piperaceae.

Synonyms– Maricha, kol, Krimihara.

Botanical description– It is a parasite which grows mostly on coconut and betel nut plants. The roots grow out of the nodes by which it creeps on the host. Leaves resemble betel leaves, on the dorsal aspect they are five veined. Fruit - round, grows in long clusters. When tender it is green in colour, turns red on ripening and black on drying.

a) Macroscopic

Fruits greyish-black to black, hard, wrinkled, 0.4-0.5 cm in dia.; odour, aromatic; taste, pungent.

b) Microscopic

Fruit consists of a thick pericarp for about one third of fruit and an inner mass of perisperm, enclosing a small embryo; pericarp consists of epicarp, mesocarp and endocarp; epicarp composed of single layered, slightly sinuous, tabular cells forming epidermis, below which, are present 1 or 2 layers of radially elongated, lignified stone cells adjacent to group of cells of parenchyma; mesocarp wide, composed of band of tangentially elongated parenchymatous cells having a few isolated, tangentially elongated oil cells present in outer region and a few fibro-vascular bundles, a single row of oil cells in the inner region of mesocarp; endocarp composed of a row of beaker-shaped stone cells; testa single layered, yellow coloured, thick-walled sclerenchymatous cells; perisperm contains parenchymatous cells having a few oil globules and packed with abundant, oval to round, simple and compound starch grains measuring 5.5-11.0 μ in dia.; having 2-3 components and a few minute aleurone grains.

Powder - Blackish-grey; shows debris with a characteristic, in groups, more or less isodiametric or slightly elongated stone cells, interspersed with thin-walled, polygonal hypodermal cells; beaker-shaped stone cells from endocarp and abundant polyhedral, elongated cells from peri sperm, packed tightly with masses of minute compound and single, oval to round, starch grains measuring 5.5-11.0 μ in dia.; having 2-3 component and a few aleurone grains and oil globules.

Chemical constituents - Alkaloids (Piperine, Chavicine, Piperidine, Piperetine) and essential oil.

Habitat– Malaya, Singapore, Assam, Kerala, Konkan.

Varieties– Pepper is of two kinds based on where it grows– (1) Oriental 2. Western.

Properties– Guna-laghu, tikshna Rasa - katu Vipaka-katu Veerya - ushna.

Doshakarma– Vatakaphashamak.

Action and uses– It is a liver stimulant, stimulant to urenlatory system and release digestive juices. So it is used in loss of appetite, indigestion, liver dysfunction and colic (abdominalcolic). The blockage in small capillaries is gradually removed on giving a very fine powder of black pepper in water. There is no better substance than pepper to reverse sluggishness of pranava srotas (lungs) and reduce the mucus secretion.

Parts to be used– Fruit, leave.

Dosage– 0.25 to 0.5 g.

Main Preparation– Marichadi gutika, Marichadi taila.

MADANAPHALA

Botanical name – *Randia dumentorum* Lam.

Family – Rubiaceae.

Synonyms – Madan, Pindiphala, Shalyak, Vishpushpaka.

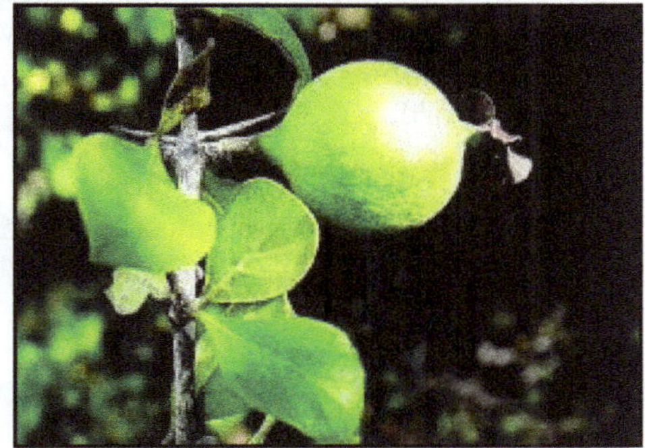

Botanical description– It is thorny medium sized tree. Leaves-like apamarga slightly round and long sharp thorns are situated on either side of leaf axis. Flower-yellowish white. Fruits-like naspati, fruits are round dusky coloured.

a) Macroscopic

Fruit, 1.8-4.5 cm long, globose or broadly ovoid, longitudinally ribbed or smooth yellowish-brown, crowned with persistent calyx-limb, fruit, contains numerous seeds, 0.4-0.6 cm long, compressed, smooth, brown and very hard.

b) Microscopic

Fruit-Trasnverse section shows epicarp consisting of single layered

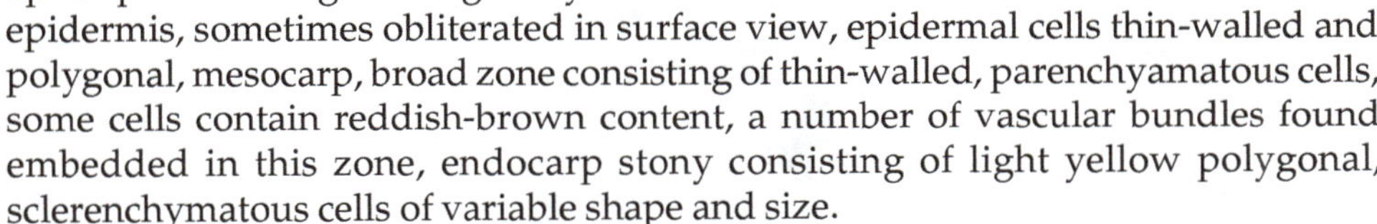

epidermis, sometimes obliterated in surface view, epidermal cells thin-walled and polygonal, mesocarp, broad zone consisting of thin-walled, parenchyamatous cells, some cells contain reddish-brown content, a number of vascular bundles found embedded in this zone, endocarp stony consisting of light yellow polygonal, sclerenchymatous cells of variable shape and size.

Seed- Transverse section shows a seed coat, consisting of single layered, rounded to oval epidermal cells, a few layers of yellowish-brown pigmented cells, endosperm forms bulk of seed consisting of large oval and irregular shaped parenchymatous cells, albumen horny, transluscent, cells of outermost layer smaller in size.

Powder-Reddish brown, under microscope shows numerous, large, irregular, reddish brown cells sclereids of variable shape and size, pieces of xylem vessels with reticulate thickenings, thin- walled, crushed parenchymatous cells and yellow-orange pieces of seed coat

Chemical constituents - Essential oil, saponin, tannin and resin.

Habitat– Foot hills of Himalaya, Sindh, South India & Maharastra.

Properties– Guna-laghu, ruksha Rasa-madhur, tikta, kashaya Vipaka-katu Veerya-ushna.

Doshakarma– Kaphavatashodhak.

Action and uses– Madanphala has been used so extensively for inducing emesis in shodhan chikitsa that its name has become synonymous with emesis. It is extremely suitable for emesis as it does not cause any complications. The oil is used for massage in vata disorders. The paste of fruit is applied over painful and swollen joints in rheumatoid arthritis.

Parts used– Fruit, seed rind.

Dosage– Powder 1-3 g, for emesis decoction of fruit rind 15-30 tola.

Main preparation– Madanadi lepa.

JAIPHAL

Botanical name – *Myristica fragrans* Houtt

Family – Myristicaceae.

Synonyms– Jatiphala (fragrand fruit), jatikash (having fragrant lobule), malatiphala, shalooka.

Botanical description– It is a lofty tree with slender branches. Leaves - like jamun leaves but small, dorsal side of the leaf is dark green while ventrally it is pale yellow. Flower - small, fragrant, yellow in colour. Fruit is a drupe surrounded by arrilus and flesh. Rind - pale yellow and thick. Its hard outer shell is yellow. On drying, it becomes coral coloured. It is called as mace (Jatipatri/jaipatri). A seed with hard shell which is inclosed in this mace is called as nutmeg (Jaiphal).

a) Macroscopic

Seed ellipsoid, 20-30 mm long and about 20 mm broad, externally greenish-brown sometimes marked with small irregular dark brown patches or minute dark points and lines slightly furrowed reticulately, a small light-coloured area at one end indicating the position of the radicle a groove running along the line of raphe to the darker chalaza at the opposite end, surrounded by a thin layer of peri sperm with infoldings appearing as dark ruminations in the abundant greyish-brown endosperm, embryo, in an irregular cavity, small with two widely spreading crumpled cotyledons and a small radicle odour, strong and aromatic, taste, pungent and aromatic.

b) Microscopic

Transverse section of endosperm shows peripheral perisperm, of several layers of strongly, flattened polyhederal cells with brown contents, or containing prismatic crystals, inner layer of perisperm of thin-walled parenchyma about 40 µ thick, infolding into the tissue of the endosperm to form the ruminations containing numerous, very large oil cells with brown cell walls, vascular strands, in the peripheral region, numerous small spiral vessels, large celled, endosperm, parenchymatous With occasional tannin idioblasts with thin brown walls, containing numerous simple, rounded and compound starch grains, with upto about 10 components usually 2-8 individual grains, upto 20 µ in diameter present, most of the cells with crystalline fat and often a large aleurone grain in each cell, containing a rhombic protein crystal upto 12 µ and small aleurone grains with less regular crystalloids, embryo, of shrivelled and collapsed parenchyma.

Chemical constituents - Essential oil and fixed oil.

Habitat– Srilanka and South India.

Varieties– It has another variety which is without smell. It is known as raiphal and its thin covering is called as raipatri.

Properties– Guna-laghu, snigdha, tikshna Rasa-katu, tikta, kashaya Vipaka-katu Veerya-ushna.

Doshakarma–Vatakaphashamaka.

Action and Uses– It is analgesic, stimulant, deodorant, anticonvulsant, vatashamak, expectorant and aphrodisiac. Its paste is used in headache, arthralgia, ointment is used as deodorant and an analgesic in skin diseases. It is used in indigestion, liver disorder, diarrhoea, colitis and helminthiasis.

Part used– Nutmeg, mace.

Dosage– 5 to 10 mg, Oil - 7 to 15 drops.

Main Preparation– Jatiphaladi churna, Jatiphaladigutika.

BABBUL

Botanical name – *Acacia arabica* Wild.

Family – Leguminosae.

Synonyms– Babbul, Kantalu, Tikshnakantak, Sukshmapatrak.

Botanical description– It is a thorny tree 8-10 mtrs High. Trunk is grey in colour. Branches are straight, hanging downward, tender branches are used for brushing teeth. Leaflets are 10 to 20 pairs. Flowers - yellow. Brownish white gum exudate from the trunk, which is available in the market.

a) Macroscopic

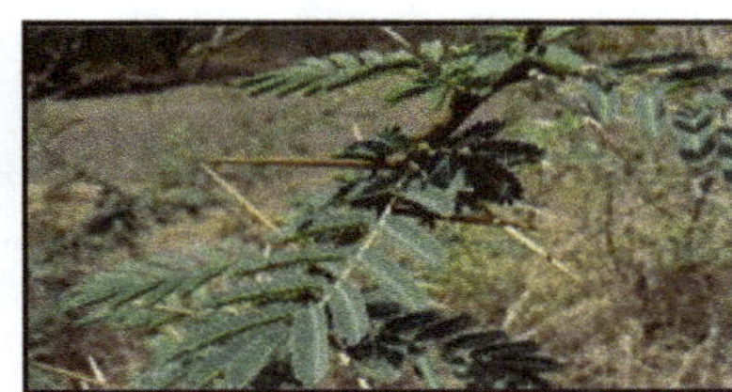

Bark hard, dark brown or black, deeply fissured transversely and longitudinally, inner surface reddish brown, longitudinally striated and fibrous, breaks with difficulty and exhibits a fibrous fracture, taste, astringent.

b) Microscopic

Transverse section of mature bark shows, 15-25 layered, thin-walled, slightly flattened mostly rectangular, brown coloured cork cells, a few lenticels formed by rupturing of cork cells, secondary cortical cells ovate to elongated, many tanniferous stone cel1s, variable in shape and size present in large groups, secondary phloem consists of sieve tubes, companion cells, fibres, crystal fibres and phloem parenchyma phloem fibres in many groups and thick-walled, phloem tissues filled with reddish or brown contents present, crystal fibres thick-walled, elongated, divided by transverse septa into segments, each contain a prismatic crystal of calcium oxalate, medullary rays uni to-multi- seriate run almost straight, ray cells elongated to polygonal, 20-24 cells high and 2-5 cells wide, crystals of calcium oxalate found scattered amongst the stone cell"cel1s of secondary cortex and phloem parenchyma.

Powder-Powder as such reddish brown coloured, under microscope many prismatic crystals of calcium oxalate, stone cells, both with narrow and wide lumen and striations and crystal fibres seen.

Chemical constituents - Tannins and gum.

Habitat– All over India.

Properties– Guna - guru, ruksha Rasa - kashaya Vipaka - katu Veerya - sheeta.

Doshakarma– Kaphapittashamaka.

Action and uses– It acts as blood purifier, haemostatic, vasoconstricting and healing property. Its powder is sprinkled on burnt injuries and bleeds. Decoction of bark is used for enema through vagina (Uttarbasti) in leucorrhoea.Useful in diarrhoea, dysentery, piles, bleeding disorder, cough and urinary disorders.

Past used– Bark, gum.

Dose– Bark decoction – 30 to 80 ml.

Powder of legumes – 3 to 6 g.

Gum – 3 to 6 g.

Main preparation– Babbularishta, Lavangadi vati.

YAVANI

Botanical name – *Trachyspermum ammi* Linn.

Family – Umbelliferae.

Synonyms– Yavani, Deepyaka, Deepya, Vatari.

Botanical description– A small shrub of height 3 to 4 ft. leaves - appear like blunt shape. Flowers – whitish and umbrella shaped. Seed - small, yellowish, rust coloured.

a) Macroscopic

Seeds irregularly reniform or sub-quadrate, slightly over a mm in size, dark grey, surface concave, odour pleasantly aromatic, taste bitter, mucilaginous and pungent, aromatic.

b) Microscopic

Transverse section of seed shows the presence of thick cuticle, testa with two layers, outer one with a row of osteosclereids size ranging from 50 to 80 μ, inner one with crushed parenchyma, endosperm cells thin walled, containing oil globules, embryo coiled; starch absent.

Powder - Dark brown aromatic smell, bitter mucilagenous taste and an oily texture; a number of flask-shaped or dumb-bell shaped osteosclereids seen; fragments of testa in surface view, showing cells with sinuous walls; powder when treated with Sudan IV and mounted in glycerine shows the presence of oil globules which turn orange red; powder cleared with dilute nitric acid shows surface view of sculpturing on testa.

Chemical constituents - Tropane alkaloids hyoscyamine, (its racemic mixture and atropine) and hyoscine.

Habitat – Found all over India.

Properties– Guna-laghu, ruksha, tikshna Rasa-katu, tikta Vipaka-katu Veerya-ushna

Doshakarma– Kaphavatashamaka.

Action & uses– It is used as analgesic, anti inflammatory, laxative, antibacterial, complexion enhancer and on digestive system it is appetizer, deepan and pain killer, anthelmintic. It is used in anorexia, loss of appetite, indigestion, flatulence, abdominal pain, spleen enlargement and intestinal worm. Its external application reduces oedema and pain, skin diseases. It is a uterine stimulant and hence used in dysmenorrhoea and post-portum condition.

Part used– Seed.

Doses– Powder– 1-3 g, Oil– 15-30 drops, Decoction– 15-30 ml.

Main Preparation– Yavanikadi churna, Yavanikadi Kwatha.

GUNJA

Botanical name – *Abrus precatorius* Linn.

Family – Leguminoseae.

Synonyms– Gunja, choodamani, Raktaphalika, Kankatika.

Botanical description– It is a creeper with many branches. Leaves resemble tamarind leaves having 20-40 leaflets. Flowers are pink, bluish and appear in cluster. Legumes are 1.5-3.5 cm. long containing red, white and black coloured seeds. Red coloured seeds have black spot on their tips.

a) Macroscopic

Characterised by smooth, glossy surface and bright scarlet colour with black patch hilum, ovoid or sub-globular, 5-8 mm long, 4-5 mm broad.

a) Microscopic

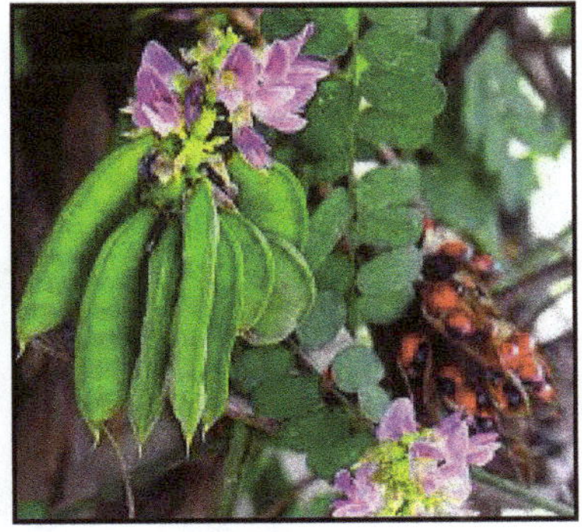

Transverse section of seed shows testa about 75 µ thick, greater parts being formed by epidermis, composed of radially, much elongated cells, arranged irregularly and measure 45-50 µ in length, Inner region of thin testa consists of collapsed cells forming a hyaline layer about 25 µ thick, endodermis composed of thick-walled cellulosic parenchyma, isodiametric cells larger towards inside, walls mainly of hemicellulose and swell considerably in water, outer one or two layers of cells of endodermis (pseudoepidermis) formed of rather smaller cells, walls of which swell to less extent in water.

Chemical constituents - An albuminous substance (abrine and abralin).

Habitat– This creeper grows in India, Srilanka.

Varieties– It is of three varieties. 1. White 2. Red 3. Black.

Properties– Seed properties – Guna - laghu, ruksha, tikshna Rasa - tikta, kashaya Vipaka - katu Veerya – ushna, Root - madhur & snigdha.

Doshakarma– Seed- Kaphavatashamaka, Root- Tridoshshamaka.

Action and uses– Seeds grinded in a paste form are applied over skin disorder, chronic ulcers and alopecia (loss of hair). Paste of the leaves is applied over inflammations and wounds. Oil boiled with the leaves is used for massage in vata disorder. Milk boiled with the leaves is given in horsness of voice, impotency, dysuria and general debility.

Part used– Seed, leaves.

Doses– Seed powder– 75 mg to 175 mg, root powder– 1 to 3 g.

Main preparation– Gunjabhadra ras, Gunja churna.

LAVANG

Botanical name – *Syzygium aromaticum* merr perry.

Family – Myrtaceae.

Synonyms– Devkusum, Chandanpushpa, Shri.

Botanical description– Tree full of green leaves, 10-13 mtrs. Trunk is yellowish white and tender. Flower – brown externally, four triangular petals. Fruit - fleshy 3 cm long. Flowering season is summer and fruiting in pre-monsoon.

a) Macroscopic

Flower bud measuring 10-17.5 mm in length, dark brown or dusty red, consisting of a sub-cylindrical, slightly flattened, four sided hypanthium, readily exuding oil when pressed hypanthium containing in its upper portion a two celled inferior ovary with numerous ovules attached to a axile placenta, surrounded by four thick, divergent sepals and covered by unopened corolla consisting of four membranous imbricate petals, frequently detached, enclosing numerous incurved stamens and one erect-style, odour, strongly aromatic, taste, pungent, aromatic followed by slight tingling of the tongue.

b) Microscopic

Transverse section of hypanthium shows epidermis and calyx teeth composed of straight walled cells, With thick cuticle having large anomocytic stomata, hypanthium tissue spongy, clusters of calcium oxalate crystals varying in size from 6-20 μ in diameter, small number of stone cells and prismatic crystals of calcium oxalate present in stalk, stamens, each with an oil gland in the apex of the connective, triangularly centricular pollen grains, 15-20 µ in diameter anther walls showing a typical fibrous layer, schizolysigenous glands found in all parts of clove, occasional isolate pericyclic fibres present.

Powder-Dark brown, fragments of parenchyma showing large oval, schizolysigenous oil cavities, spiral tracheids and a few rather thick-walled, spindle shaped fibres, calcium oxalate crystals in rosette aggregates, 10-15 µ in diameter, fragments of anther walls with characteristic reticulated cells pollen grains numerous, tetrahedral, 15-20µ. in diameter.

Chemical constituents - Essential oils (eugenalacetate and caryophyllene).

Habitat– Originally from Malaya-sailibius Island. At present it is being cultivated in Sumatra, Brazil, Westindies and Travancore in India.

Properties– Guna-laghu, tikshna, snigdha Rasa-katu, tikta Vipaka - katu Veerya - sheeta.

Doshakarma– Kaphapittashamaka.

Action and uses– It is an appetizer and digestive, liver stimulant, antimicrobial. It is usefull in anorexia, loss of appetite, abdominal colic, hyperacidity, nausea and liver disorders. Clove oil is used for massage in rheumatoid arthritis, sciatica, backache. It is usefull in respiratory disorders namely in cough, foul smell, breathlessness and stimulates respiratory mucosa causing expectoration.

Part used– Buds (clove).

Doses– Churna - 0.5 to 1.5 g, Oil – 4 to 6 drops.

Main preparation - Lavangadivati.

HARITAKI

Botanical name – *Terminalia chebula* Retz.

Family – Combretaceae.

Synonyms– Haritaki, Abhaya, Kayastha, Vayastha.

Botanical description– It is a big tree 25 to 30 mtrs Height. Its wood is hard and bulky. Leaves are 10-30 cm in length and are pointed. The inferior aspect of the leaves show two small nodules wears its attachment with the stalk. Flowers dull white or yellowish with a strong offensive smell, in spikes fruits - ovoid or ellipsoidal.

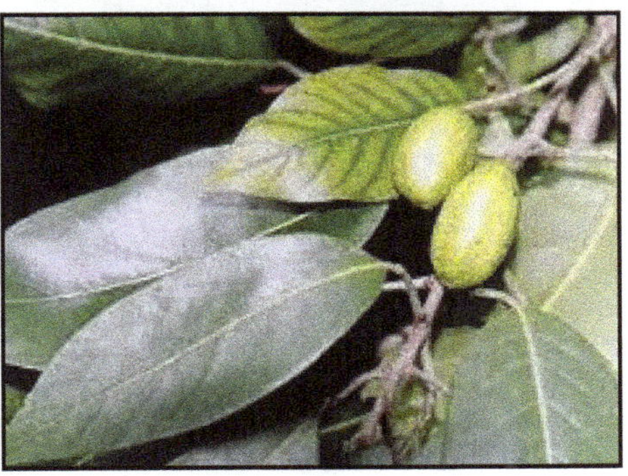

a) Macroscopic

Intact fruit yellowish-brown, ovoid, 20-35 mm long, 13-25 mm wide, wrinkled and ribbed longitudinally, pericarp fibrous, 3-4 mm thick, non-adherent to the seed, taste, astringent.

b) Microscopic

Transverse section of pericarp shows epicarp consisting of one layer of epidermal cells inner tangential and upper portions of radial wall thick, mesocarp, 2-3 layers of collenchyma, followed by a broad zone of parenchyma in which fibres and Sclereids in group and vascular bundles scattered, fibres with peg like out growth and simple pitted walls, sclereids of various shapes and sizes but mostly elongated, tannins and raphides in parenchyma, endocarp consists of thick-walled sclereids of various shapes and sizes, mostly elongated, epidermal surface view reveal polygonal cells, uniformly thick-walled, several of them divided into two by a thin septa, starch grains simple rounded or oval in shape, measuring 2-7 μ in diameter, found in plenty in almost all cells of mesocarp.

Powder- Brownish in colour, under microscope shows a few fibres, vessels with simple pits and groups of sclereids.

Chemical constituents - Tannins, anthraquinones and polyphenolic compounds.

Habitat– Abundant in Northern India also occurs in Bihar, West Bengal, Assam and Central India.

Varieties– Two varieties are found- 1. Laghu 2. Brihat.

Properties– Guna- laghu, ruksha Rasa - pancharasa, Kasaya Mainly Virya-ushna Vipaka-madhur.

Karma– Tridoshahara (Tridosha shamaka).

Action and uses– It acts as anti-inflammatory, antidiabetic, antileprotic, wound healer, anti-emetic, cardio-protective. It is usefull in conjuctivitis, loss of appetite, pain in the abdomen, constipation, ascitis, haemorrhoides, hepatomegally & splenomegaly. It is usefull in dysurea, retention of urine, calculus.

Part used– Fruit.

Dosage– Powder– 3 to 6 g.

Main Preparation– Haritaki churna, Triphala, Agastya Haritaki.

KARKATASRINGI

Botanical name – *Pistacia integerrima* Stewart ex. Brandis

Family – Anacardiaceae

Synonyms–Ajashringi, Kuliravisanika, Vakra, Sringi.

Botanical description– It is a glabrous tree growing upto 16 m, dark grey or blakish bark. Leaves- 4-5 pairs leaflets, with or without terminal leaflet. Flowers are in lateral panicles, fruits - drupe, globose, grey, seeds with a membranous testa.

a) Macroscopic

Dried galls hard, hollow, horn-like, thin-walled, generally cylindrical, tapering at both the ends, greyish brown externally and reddish brown internally, size varies from 2.5-30.0 cm or more, each gall contains numerous dead insects, odour, terebinthine, taste of powdered galls, strongly astringent and slightly bitter.

b) Microscopic

Transverse section of gall shows the collapsed epidermis on both the sides, epidermal cells thin-walled, tangentially elongated, ground tissues thin-walled and oval or circular, the outer two layers tangentially elongated while between vascular bundles radially elongated, outer few layers and some of cells of ground tissue filled with yellowish brown contents, vascular bundle scattered throughout the ground tissues in two rows, consist of phloem accompanied by a large tannin sac in each vascular bundle.

Powder-Powder greyish brown, under microscope, shows orange yellow colour isolated or associated fragments of xylem vessels and ground tissues.

Chemical constituents - Essential oil, tannins and resinous matters

Habitat– North-West Himalayas at 350-2500 m.

Properties– Guna-laghu, ruksha Rasa - kashya, tikta Veerya-ushna Vipaka - katu

Doshakarma– Kaphavatashamaka.

Action & uses– It is an appetizer, carminative, astringent, expectorant & reguvenator. It is usefull in anorexia, dyspepsia, diarrhoea, dysentery. It is indicated in hiccoughs, cough and dyspnoea.

Parts used– Hornlike Cocoons on the leaves (Galls) .

Dosages– Powder 1-3 g.

Main preparation– Shringyadi churna, Balachaturbhadra churna.

GAMBHARI (KASHMARI)

Botanical name – *Gmelina arborea* Linn.

Family – Verbinaceae.

Synonyms– Kashmari, Pitarohini, Madhuparni

Botanical description– A large decidous tree 20 mtr Hight. Leaves are long & broad. Petiole - long like pippal leaf. Leaves are heart shaped. Flowers - yellow fruit - oval shaped.

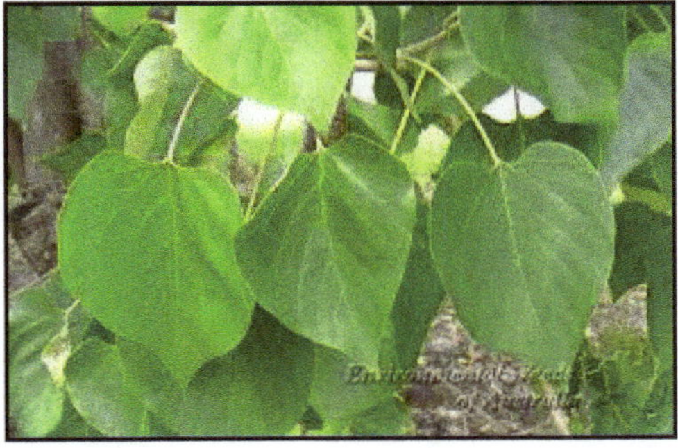

a) Macroscopic

Root - Occurs in pieces with secondary and tertiary branches, root pieces nearly cylindrical with uneven surface, greyish brown, fracture somewhat tough in bark, brittle and predominant in woody portion.

Root bark- Mature root bark when fresh, yellowish in colour, dry pieces curved and channelled, thinner ones forming single quills, external surface rugged due to presence of vertical cracks, ridges, fissures and numerous lenticels, fracture short and granular, taste, mucilaginous, sweetish with slight bitterness.

b) Microscopic

Root-transverse section of root shows 6-8 layers of cork cells, secondary cortex, including primary and secondary phloem about two third consisting of wood, cork brownish, cells arranged in tangential direction and broken at places towards upper layers, cortex characterised by the presence of thin-walled parenchymatous cells with starch grains , resin ducts present in abundance throughout cortex, scattered stone cells fibre like or elongated common, fibres present, occurring mostly in singles, cells of cortex also contain rosette crystals of calcium oxalate and oil globules, primary phloem characterised by the presence of sieve tubes with companion cells, phloem parenchyma, soft bast fibres and ray cells, phloem fibres occur singly and scattered cortical cells 40-70 µ by 25-35 µ and bast fibres, 300-1000 µ by 10-15 µ development of cork takes place in second or third layer of primary cortex, wood consists of simple pitted wood parenchyma and medullary rays, wood cells mainly composed of vessels and tracheids and inner wood consists of a major portion of fibres together with a few vessels, vessels numerous and form almost a ring near the periphery of xylem cylinder and somewhat spares, being scattered in groups or singly nearer the central region, lumen of vessels somewhat large, dimensions of vessels 130-250 µ by 50-100 µ and those of the tracheids 175-300 µ by 30-50 µ wood fibres abundant and with simple pits , cambium distinct, medullary rays generally 1-2 celled thick with abundant starch grains cells oblong to rectangular.

Chemical constituents - Alkaloids and lignans (arboreal, isoarboreal and related lignans).

Habitat– The tree grows in the hilly areas of the Himalaya.

Properties– Guna – guru Rasa-tikta, kashaya Veerya - ushna Vipaka - katu.

Doshakarma– Vatapittashamaka

Action and uses– Fruits and leaves are diuretic, the bask of the root is anti inflammatory and appetizer and digestive. It is indicated in diarrhoea, constipation and haemorrhoides, Leaves relieve burning & pain. Decoction of root bark is given in postpartum.

Parts used– Root & fruits.

Dosage– Root bark decoction – 50 to 100ml, Fruit powder – 1 to 3 g.

Main preparation– Brihatpanchamoladi kwath, Shriparnyadi kwath.

SARSHAPA

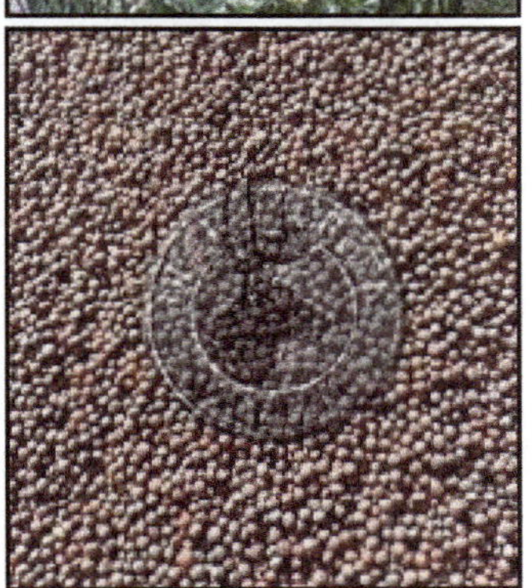

Botanical name – *Brassica compestris* Linn.

Family – Cruciferae.

Synonyms– Bhutaghna, Grahaghna, Kadamba.

Botanical description– It is an annual or biennial, herb 30-90 cms hight, leaves- 30 to 40 cms long. Flowers - yellow Fruits- small containing yellow or red seeds.

a) Macroscopic

Seeds small, slightly oblong, pale or reddish-brown, bright, smooth, 1.2- 1.5 mm in dia.; under magnifying glass it is seen to be minutely reticulated; taste, bitter and sharp.

b) Microscopic

Seed shows single layered colourless testa followed by 3-5 layered, non-lignified, hexagonal, thick-walled cells filled with yellowish-brown contents; embryo and endosperm consists of hexagonal, thin-walled parenchymatous cells containing oil globules.

Powder - Yellow in colour with brown particles and oily, slightly bitter and sharp in taste; shows frequently thick-walled, fragments of reddish-brown cells of hypodermis, yellowish hyaline masses.

Chemical constituents - Fixed Oil.

Habitat – All over India.

Properties – Guna - tikshna, ruksha Rasa - katu, tikta Vipaka – katu Veerya - ushna.

Doshakarma– Kaphavatashamaka.

Action and uses– Seeds are anthelmintic, diuretic, laxative and appetizer. In pyorrhoea, gargles of its oil mixed with salt are useful. It is one of the best medicines for spleenomegaly.

Parts Used– Seed, leaf.

Dosage – Seed powder – 2-4 gm Paste – 0.5 to 1 g.

Main preparation– Marichyadi tail, Sarsapadi pralepa.

PODINA

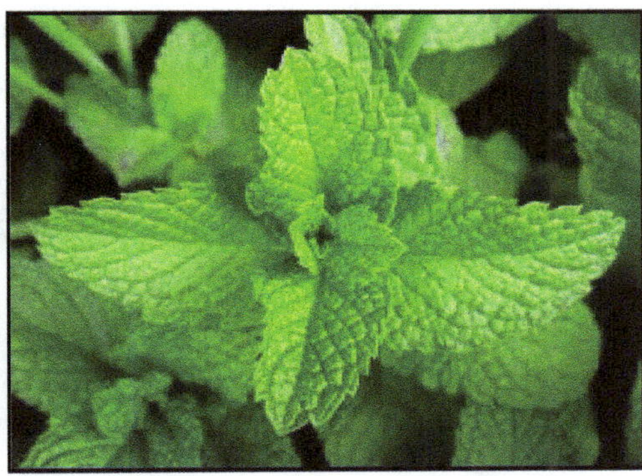

Botanical name – *Mentha spicata* Linn. Emend.

Family – Labiate.

Synonyms– Podina, Rochani, Podinak.

Botanical description– It is perineal herb up to 1-2 feet height. Leaves are smooth, without pedicle, sharp and dentate with glands on lower surface. Flower- Violet coloured inflorescence.

a) Macroscopic

Drug consists of small chopped twigs; leaves opposite, decussate, shortly petiolate, petioles 2-mm long; mature leaves 2.5 to 3.5 cm long and 1.5 to 2.0 cm broad, very minutely hairy, ovate, apex acute, coarsely dentate, comparatively smoother and darker upper surface; stem square, minutely hairy, light brown to brown; flowers in loose cylindrical, slender spikes; awl like, throat of calyx naked, corolla smooth; seeds small, mucilaginous; aromatic odour and slightly pungent taste.

b) Microscopic

Stem - T.S. shows quadrangular outline with corner ridges and thin cuticle; epidermal cells tabular, multicellular uniserate trichomes present, cortex 8 to 9 cells deep below ridges, while 2 to 3 cells deep elsewhere, variable in size; endodermis single layer; pericycle broken, consisting of sclerenchymatous cells; phloem 2 to 4 cells deep and made up of irregular shaped cells; xylem vessels 26 to 46 μ in dia; pith present.

Leaf

Midrib- T.S. shows protruded mid rib towards the lower surface; compact parenchymatous cells enclose a crescent-shaped vascular bundle; collenchymatous cells are absent.

Lamina- Dorsiventral, epidermal cell walls of both the surfaces in the surface view are wavy, stomata diacytic; covering trichomes present on the lower surface, uniseriate, 1 to 4 cells long, 42 to 350 μ in size with pointed apex; glandular trichomes 64 to 80 μ in diam. with a single basal cell and a head of 8 cells, found in depression of the epidermis; a single row of palisade cells towards the upper side followed by spongy parenchyma 3 to 4 cells deep; palisade ratio 6 to 8; vein islet number 18 to 20; stomatal index for upper epidermis 10 to 20, lower epidermis 15 to 30.

Powder - Blackish-brown, fibrous, free flowing, characterized by the presence of uniseriate non-glandular hairs (112 to 350 μ), glandular trichomes 64 to 80 μ in diam, diacytic stomata, epidermal cell walls wavy.

Chemical constituents - Essential oil (0.2 to 0.8 percent) containing terpene such as carvone (60%) and limonene (10%) as major constituents.

Habitat– Throughout all over India.

Varieties– Two varieties are available 1. *M. longifolia* 2. *M. arvensis.*

Properties– Guna- laghu, ruksha, tikshna Rasa- katu Vipaka- katu Veerya-ushna.

Action & Uses– It act as carminative, appetizer, antimicrobial and analgesic, diuretic, uterine-contractor, sweatpromoting, expectorant & cardiac stimulant. It is used in indigestion, vomiting, distension, diarrhoea, cardiac weakness, cough, dysurea and in amenorrhoea diseases.

Parts to be used– Leaf & tail.

Dosage– Leave Juice – 5 to 10 ml, tail – 1 to 3 drops.

Main preparation– Arkapodina.

AMLIKA (CHINCHA)

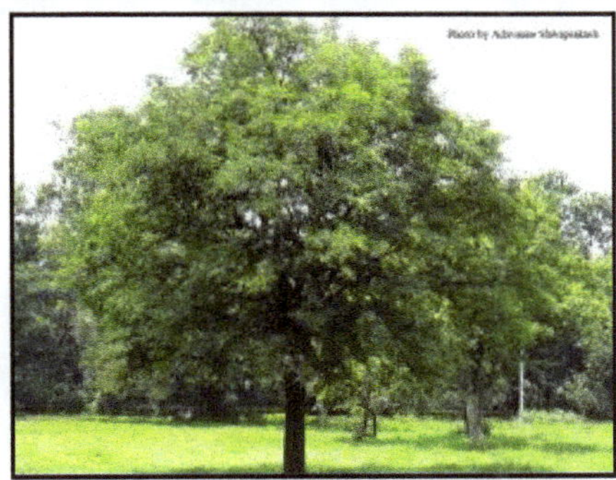

Botanical name – *Tamarindus indica* Linn.

Family – Leguminosae.

Synonyms– Chincha, Tintidik.

Botanical description–It is a evergreen tree height upto 75 feet & 20-22 feet widge. Leave-compound leave, Flower-yellow coloured as inflorescence. Legume–3-7 inch long.

a) Macroscopic

Fruit pulp occurs as a reddish-brown, moist, sticky mass, in which yellowish-brown fibres are readily seen; odour, pleasant; taste, sweetish and acidic.

b) Microscopic

Fruit pulp consists of thin-walled, elongated to polygonal, parenchymatous cells of considerable size, traversed by a number of long fibro-vascular bundles and having a very few small starch granules, and numerous prismatic crystals of calcium oxalate.

Chemical constituents - Inorganic acids, sugars, saponin and bitter principle – Tamarindinca.

Habitat– Throughout all over India especially in South India.

Properties– Green amlika– Guna-guru Rasa-amla Vipaka-amla, Veerya-ushna, ripened amlika-madhur, amla.

Doshakarma– Green-Vatashamaka, ripened– Kaphavatashamaka.

Action & uses– Ripened fruit (Legume)–is appetizer, and purgative. It is used in indigestion.

Part Used– Fruit (legume).

Dose– Panak– 50-100 ml.

Main Preparation– Chincha Panak.

ASHWATTHA

Botanical name – *Ficus religiosa* Linn.

Family – Moraceae.

Synonyms– Ashwattha, Pippal, chalapatra, Bodhidru.

Botanical description– It is a large tree, having white-grey coloured bark. Leaves-thin, smooth, 5-7 veins top shaped, fruit-small rounded, green in early stages and after ripen it's become violet or black colour. It is religious plant in Hindu Dharma.

a) Macroscopic

Bark occurs in flat or slightly curved pieces, varying from 1.0-2.5 cm or more in thickness, outer surface brown or ash coloured, surface uneven due to exfoliation of cork, inner surface smooth and somewhat brownish, fracture, fibrous, taste, astringent.

b) Microscopic

Transverse section of bark shows compressed rectangular to cubical, thick-walled cork cells and dead elements of secondary cortex, consisting of masses of stone cells, cork cambium distinct with 3-4 rows of newly formed secondary cortex, mostly composed of stone cells towards periphery, stone cells found scattered in large groups, rarely isolated, most of parenchymatous cells of secondary cortex contain numerous starch grains and few prismatic crystals of calcium oxalate, secondary phloem a wide zone, consisting of sieve elements, phloem fibres in singles or in groups of 2 to many and non-lignified, numerous crystal fibres also present, in outer region sieve elements mostly collapsed while in inner region intact, phloem parenchyma mostly thick-walled, stone cells present in single or in small groups similar to those in secondary cortex, a number of ray-cells and phloem parenchyma filled with brown pigments, prismatic crystals of calcium oxalate and starch grains present in a number of parenchymatous cells, medullary rays uni to multiseriate, wider towards outer periphery composed of thick-walled cells with simple pits, in tangential section ray cells circular to oval in shape, cambium when present, consists of 2-4 layers of thin-walled rectangular cells.

Chemical constituents - Tannins.

Habitat– Throughout All over India.

Properties– Guna- guru, ruksha Rasa- kashaya, madhur Vipaka- katu Veerya- sheeta.

Doshakarma– Kaphapittashamaka.

Action and uses– It is acts as analgesic, wound healer, anti-inflammatory, blood-purifier & anti-diuretic. Milk is uses in inflammation, pain & bleeding. In wound swelling, fistula & stomatitis its barks is uses. In vomiting, diarrhoea & dysentery its bark is given. In abdominal pain & constipation its ripped fruit is used.

Parts to be used– Bark, fruit, milk.

Dosage– Kwath- 50 to 100 ml.

Main preparation– Ashwattha kashaya.

SANAYA

Botanical name – *Cassia angustifolia* Vahl.

Family – Leguminosae.

Synonyms– Swarnapatri, Kalyani, Hemapatri, Rechani.

Botanical description– It is erect shrub, height up to 2-3 feet. Leaves-compound leave, Fruit-legume 1.4-2.8 inch long brown coloured, in which 5-7 top shaped brown coloured smooth seed are present.

a) Macroscopic

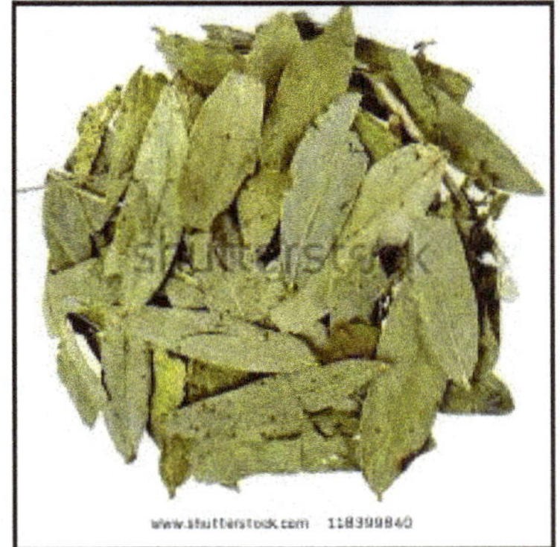

Leaflets, 2.5-6 cm long and 7-15 mm wide at centre, pale yellowish-green, elongated lanceolate, slightly asymmetric at base, margins entire, fiat apex acute with a sharp spine, both surfaces smooth with sparse trichomes, odour, faint but distinctive, taste mucilagenous and disagreeable but not distinctly bitter.

b) Microscopic

Transverse section of leaflet through midrib shows an isobilateral structure, epidermal cells, straight walled containing mucilage, both surfaces bear scattered, unicellular hair, often conical, curved near base, thick-walled, non-lignified, warty cuticle, stomata, paracytic, numerous on both surfaces, mesophyll consists of upper and lower palisade layers with spongy layer in between, palisade cells of upper surface longer than those of lower surface the latter having wavy anticlinal walls, prismaatic crystals of calcium oxalate present on larger veins and clusters of calcium oxalate crystals distributed throughout the palisade and spongy tissues, midrib biconvex, bundles of midrib and larger veins, incompletely surrounded by a zone pericyclic fibres and a crystal sheath of parenchymatous cells containing prismatic crystals of calcium oxalate.

Chemical constituents - Anthraquinone, glucoside, flavonoids, steroids and resin.

Habitat– It is a native of Africa and Arab country. In India it is cultivated in South India especially in Chennai, and Mysore.

Properties– Guna-laghu, ruksha, tikshna Rasa-katu, tikta, madhur Vipaka-katu Veerya-ushna.

Doshakarma– Pittashodhak and Vatanulomaka (Passout the Abdominal gases through anal route).

Action & uses– It is act as a liver stimulant, purgative, blood purifier, scraping of doshas. So it is uses in skin diseases, constipation, distension of abdomca, fever, rheumatoid arthritis, gout & in worms.

Parts to be used– Leave & fruit.

Dosage– Churna – 0.6 - 2 g.

Main preparation– Shatasakar churna, Yashtyadi churna.

VARUNA

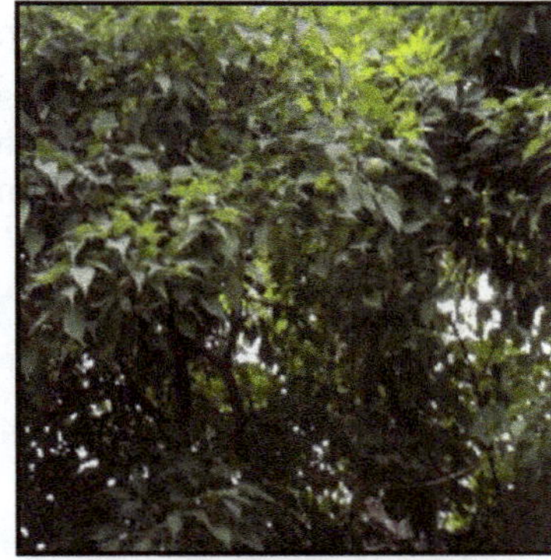

Botanical name – Crataeva nurvala Buch-Ham.

Family – Capparidaceae.

Synonyms– Tiktashaka, Varana, Setu.

Botanical Description– A medium sized tree. Leaves digitately tri-foliate, long petioted, leaflets ovate, shining above and pale beneath. Flowers white & pale yellow, on filiform pedicles. Fruits-barrie's ovoid, woody, 3-5 cm in diameter, many seeded. Seeds are 6 mm long, reniform immersed in yellow pulp.

a) Macroscopic

Thickness of bark varies, usually 1-1.5 cm according to the age and portion of the plant from where the bark is removed, outer surface, greyish to greyish-brown with ash- grey patches, at places, surface rough due to a number of lenticels, shallow fissures and a few vertical or longitudinal ridges, inner most surface smooth and cream white in colour, fracture tough and short, odour, indistinct, taste, slightly bitter.

b) Microscopic

Transverse section of mature stem bark shows, an outer cork composed of thin-walled, rectangular and tangentially elongated cells, phellogen single layered, thin-walled, tangentially elongated cells followed by a wide secondary cortex, consisting of thin-walled, polygonal to tangentially elongated cells with a number of starch grains, starch grains mostly simple, occasionally compound with 2-3 components also present', large number of stone cells in groups of two or more, found scattered in secondary cortex, single stone cells not very common, stone cells vary in size and shape, being circular to rectangular or elongated with pits and striations on their walls, stone cells distributed somewhat in concentric bands in phloem region except in inner region of phloem which is devoid of stone cells, secondary phloem comparatively a wide zone, consisting of sieve tubes, companion cells, parenchyma and groups of stone cells, alternating with medullary rays, sieve elements found compressed forming ceratenchyma in outer phloem region, whereas in inner region of phloem, intact, medullary rays mostly multiseriate composed of thin-walled, radially elongated cells, tangentially elongated towards outer periphery, a number of starch grains similar to secondary cortex also present in phloem and ray cells, few rhomboidal crystals of calcium oxalate also found in this region, inner most layer is cambium.

Chemical constituents - Saponin and tannin.

Habitat– It is a deciduous tree, found throughout India, either wild or cultivated.

Properties– Guna-laghu, ruksha Rasa-kashaya, tikta Vipaka- katu Veerya-ushna.

Doshakarma– Kaphavatashamaka.

Action & uses– Bark is bitter, astringent acrid, thermogenic, carminative, digestive, diuretic, stomachic and antilithiatic. It is uses in dyspepsia, loss of appetite, abdominal tumours, hepatomegally, abdominal distention, renal & vesicle calculi, fever, cough, bronchitis & gouty arthritis.

Part used– Stembark, leaf.

Dosages– Decoction– 50 to 100 ml.

Main preparation– Varunadi Kwath, Varunashigru Kwath.

VACHA

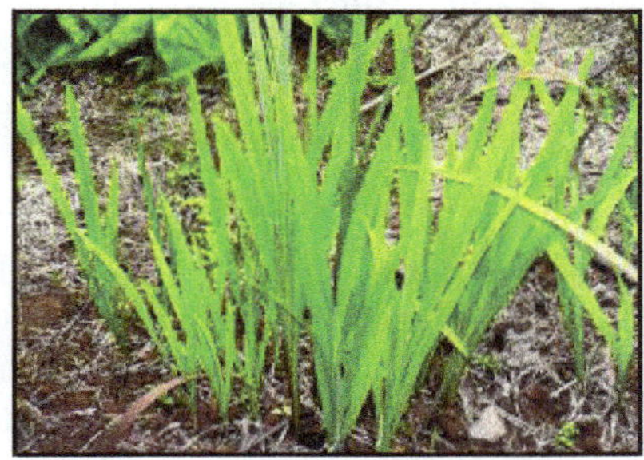

Botanical name – *Acorus calamus* Linn.

Family – Araceae.

Synonyms– Ugragandha, Shadagrantha, Golomi.

Botanical description– It is a semiaquatic, perenial, aromatic herb. Rhizomes – horizontal, rounded, spongy Leaves–grass like a sword shaped, longer. Flower - small, yellow green in a spadix. Fruit- berries green angular.

a) Macroscopic

Drug occurs in simple or rarely with thumb-like branches at nodes; sub cylindrical to slightly flattened, somewhat tortuous or rarely straight, cut pieces of 1-5 cm long, and 0.5-1.5 cm thick; upper side marked with alternately arranged, large, broadly, triangular, transverse leaf scars which almost encircle the rhizome; at nodes leaf sheath mostly having an appearence present; lower side shows elevated tubercular spots of root scars; light-brown with reddish-tinge to pinkish externally, buff coloured intemally; fracture, short; odour, aromatic; taste, pungent and bitter.

b) Microscopic

Rhizome - Shows single layered epidermis; cortex composed of spherical to oblong, thin-walled cells of various sizes, cells towards periphery, smaller, somewhat collenchymatous, more or less closely arranged cells towards inner side, rounded and form a network of chains of single row of cells, enclosing large air spaces, fibro-vascular bundles and secretory cells having light yellowish-brown contents, present in this region; endodermis distinct; stele composed of round, parenchymatous cells enclosing large air spaces similar to those of cortex and several concentric vascular bundles arranged in a ring towards endodermis, a few vascular bundles scattered in ground tissues; starch grains simple, spherical, measuring 3-6 μ in dia., present in cortex and ground tissue.

Powder - Buff coloured; shows fibres, reticulate, annular vessels and simple spherical starch grains, measuring 3-6 μ in diameter.

Observation of powder and its extracts on exposure under UV light :-

a. Powder as such: - Yellowish-cream

b. Extracts in

    i. Petroleum ether-No change

    ii. Chloroform-Light green

    iii. Methanol-Yellowish green

    iv. Benzene-No change

Habitat – Throughout India upto an altitude of 2200 mtr in Himalayas.

Properties– Guna-laghu, teekshna Rasa-katu, tikta Vipaka-katu Veerya-ushna.

Doshakarma– Kaphavatashamaka.

Action & uses– The rhizome of vacha is acrid, bitter, thermogenic, aromatic, intellect promoting, emetic, laxative, carminative, stomachic, diurectic and anodyne. Usesfull in hoarshness of voice, abdominal colic, flatulence, dyspepsia, amenorrhoea, nephropathy, calculi, cough, bronchitis and are in odontalgia.

Part used– Rhizome.

Dose– 125 m to 500 m, for vamana karma – 1-2 g.

Main preparation– Saraswat churna, Medhya rasayan.

PALASHA

Botanical name – *Butea monosperma* Lam & Kuntze.

Family – Leguminosae.

Synonyms– Kinshuka, Parna, Ksharshreshtha.

Botanical description– A deciduous tree, upto 15m hight, leaves - long petioled, 3-foliate. Flower-bright orange red; Pods silky tomentose, 10-13 cm long pendulous, thickened at the suture containing a single seeds. Seeds flat, reniform.

a) Macroscopic

Mature stem bark, 0.5 - 1 cm thick, greyish to pale brown, curved, rough due to presence of rhytidoma, and scattered dark brown spots of exudate; rhytidoma 0.2 cm thick usually peels off, exposing light brown surface, exfoliation of cork and presence of shallow longitudinal and transverse fissures; fracture, laminated in outer part and fibrous in inner part; internal surface rough, pale brown; taste, slightly astringent.

b) Microscopic

Stem Bark -Mature bark shows rhytidoma consisting of alternating layers of cork, secondary cortex and phloem tissue; cork cells, thin-walled, 5-10 or more layered, rectangular, dark-brown; secondary cortical cells round and irregular in outline, dark brown, moderately thick-walled; tanniniferous cells, often in groups, having brown colour, sometimes containing mucilage and other materials found scattered in this zone; beneath this zone regular cork consisting of 4-12 rows of radially arranged, rectangular cells followed by a zone of 2 - 4 layers of sclereids; secondary phloem consisting of sieve tubes, companion cells, phloem parenchyma, phloem fibres, crystal fibres, traversed by phloem rays; in outer and middle phloem regions phloem tissues get crushed and form tangential bands of ceratenchyma; phloem fibres arranged in tangential bands alternating with sieve tubes and phloem parenchyma; most of fibre groups contain prismatic crystals of calcium oxalate forming crystal sheath; in macerated preparation phloem fibres appear thick-walled lignified elongated with tapering or bifurcated ends; crystal fibres divided into a number of chambers containing a prismatic crystal of calcium oxalate in each chamber; phloem rays multiseriate 4 - 12 cells wide, 7 - 50 cells in height, straight; prismatic crystals of calcium oxalate found scattered in the secondary phloem tissues and phloem rays; starch grains simple or compound having 2 - 3 components, measuring 2.75 - 13.75 μ in dia., found scattered in phloem parenchyma and phloem ray cells abundantly; tanniniferous cells and secretory cavities also occur in secondary phloem.

Powder - Reddish-brown; shows numerous prismatic crystals of calcium oxalate, starch grains simple and compound with 2 - 3 components measuring 3-14 μ in dia., dark brown coloured cells, sclereids mostly in groups, thin-walled cork cells, numerous crystal fibres in group or singles.

Habitat– Throughout India, except in very arid part.

Properties– Guna - laghu ruksha Rasa - katu, tikta, kashaya Vipaka - katu Veerya – ushna.

Doshakarma– Kaphavatashamaka.

Action and uses– The bark is acrid, bitter, astringent, thermogenic, aphrodisiac, appetizer, digestive, anthelmintic and tonic. It is uses in anorexia (low appetite), dyspepsia, diarrhoea, dysentery, haemorrhoids, intestinal worms, bone fracture, rectal diseases, ulcers, hepatapathy and diabetes.

Part used– Bark, flower, exudate.

Dosage- Decoction bark – 50 to 100 ml, flower powder – 3 to 6 g, gum – 1 to 3 g.

Main preparations– Palashbijadi churna, Palashkshar ghrita.

NIMBUK

Botanical name – *Citrus lemon* Linn.

Family – Rutaceae.

Synonyms– Jambir, Dantashath.

Botonical description– Medium size tree 10-12 feet long, branches small, thick and spine present on it. Leaf-oval shaped fruit-round in shape, on ripping it turned to yellow color.

a) Macroscopic

Small creeping tree with slender stem, rooting at nodes giving rise to thin, brownish-grey, roots of about 2.5 to 6.0 cm in length; leaves 1 to 3 from each node, orbicular-reniform, crenate, base cordate, petioles channelled with adnate stipules; flowers fascicled umbels each carrying 3 or 4 flowers, short stalked; fruits cremocarp, ovoid, with laterally compressed seeds.

b) Microscopic

Root - Shows wavy outline, consisting of 3 to 5 layered, rectangular, cork cells having exfoliated cells, followed by 3 or 4 layers of parenchyma cells containing oval to round, simple, starch grains measuring 8 to 16 µ in dia., having centric hilum and microsphenoidal crystals of calcium oxalate; secondary cortex composed of thin-walled, oval to polygonal, parenchymatous cells; secretory cells present, scattered towards periphery region; secondary phloem and secondary xylem consisting of usual elements; vessels lignified with reticulate and spiral thickening; pith nearly obliterated.

Stem - More or less concave-convex outline, shows single layered epidermis composed of round to cubical cells covered by striated cuticle; below this 2 or 3 layers of collenchymatous cells, followed by 6 to 8 layers of thin-walled, isodiametric, parenchymatous cells with intercellular spaces present; vascular bundles collateral, open, arranged in a ring, capped by patches of sclerenchyma and traversed by wide medullary rays; vessels with spiral thickening present, resin duct present in parenchymatous cells of cortex and generally one in between vascular bundles; pith of isodiametric, parenchyma with intercellular spaces.

Leaf-

Petiole - Shows a characteristic outline due to two projections adjacent to ventral groove; epidermis single layered, cells cubical covered by a thick cuticle; inner walls of epidermal cells adjoining the cortex much thickened; hairs absent; collenchyma 2 or 3 layered, absent on the projections, a broad zone of more or less rounded parenchyma cells present with intercellular spaces, and a few containing rosette crystals of calcium oxalate; resin canal present on dorsal side of each vascular bundle except in the vascular bundles occurring projecting arms; vascular bundles seven in number, two of which less developed and present in projections.

Midrib - Show a single layered epidermis, 2 or 3 layered collenchyma on both surfaces, 4 or 5 layered parenchyma, mostly devoid of chloroplasts; central zone occupied by vascular bundles differentiated into xylem towards ventral side and phloem towards dorsal side; phloem consisting of sieve tubes, companion cells and phloem parenchyma; xylem consisting of radial rows of vessels with xylem parenchyma in between.

Lamina- Shows an epidermis of tangentially elongated cells on both surfaces, larger on the upper surface, covered by striated cuticle; mesophyll differentiated into 2 or 3 layers of palisade cells, 5 to 7 layers of loosely arranged, somewhat isodiametric spongy parenchyma; rosette crystals of calcium oxalate present in a few cells; stomata more on the lower surface, anisocytic in general, but anomocytic type also occurs on both surfaces, palisade ratio 3 to 5, stomatal index on upper surface, 9 to 12, and lower surface 11 to 17.

Fruit - Shows several ridges in outline; epicarp consists of single layered epidermis covered externally with thick cuticle; mesocarp consists of polygonal, thin walled parenchymatous cells having patches of sclerenchymatous cells on both lateral side; each ridge having a vittae and patch of sclerenchyma; endocarp consists of columnar shaped sclereids arranged in wavy layers; endosperm and embryo composed of oval to polygonal, thin-walled parenchymatous cells.

Powder - Green to greenish-brown, shows fragments of epidermal cells polygonal in surface view with stomata, palisade cells, vessels with spiral, reticulate and annular thickening; microsphenoidal and rosette crystals of calcium oxalate; simple, oval to round starch grains measuring 8 to 16 µ in dia.

Chemical constituents - Glycosides – Saponin, Glycosides.

Habitat– Throughout all over India especially wildly in Assam, Sikkim.

Properties– Guna- guru, tikshna Rasa- amla Vipaka- amla Veerya – ushna.

Doshashamak karma– Kaphavatashamaka.

Action & uses– It is act as rochana (increase to taste), deepana (carminative), anuloman (passout air), trishnanigrahara (Antithirst), liver stimulant, hridya (cardio tonic) and krimighna (anti-microbial). It is used in thirst, vomiting, indigestion, abdominal calic, constipation, liver diseases.

Part used– Fruit, leaf, bark.

Dosage– Fruit juice – 10-20 ml, Bark decoction – 40-80 ml.

NIMBA

Botanical name- *Azadirachta indica* Linn.

Family- Meliaceae.

Synonyms- Pichumard, Hingu niryas, Sheeta, Peetsarak, Arishtaphala, Netrasukhpriya.

Botanical description-Tree measuring 8-10mts in hight.Trunk is straight with branches in all directions. Bark is thick black, rough, from which secretion is obtained. Leaves are compound, equidistant eye shaped 6to 14, paired, foliated bilateral on the stalk. Flowers – small white, scented. Fruits are green and hard on ripening it turns yellow and soft. The native term for such fruits is nimboli. Fruit contains sweet, slightly pungent sticky pulp and single seed. Oil is extracted from the seed .Leaves fall in winter, and in autumn new redish tender leaves appears first followed by flowering and then fruit at the begning of rainy season.

a) Macroscopic

Leaves - Compound, alternate, rachis 15-25 cm long, 0.1 cm thick; leaflets with oblique base, opposite, exstipulate, lanceolate, acute, serrate, 7-8.5 cm long and 1.0-1.7 cm wide, slightly yellowish-green; odour, indistinct; taste, bitter.

b) Microscopic

Leaf

Midrib -Leaflet through midrib shows a biconvex outline; epidermis on either side covered externally with thick cuticle; below epidermis 4-5 layered collenchyma present; stele composed of one crescent-shaped vascular bundle towards lower and two to three smaller bundle towards upper surface; rest of tissues composed of thin-walled, parenchymatous cells having secretory cells and rosette crystals of calcium oxalate; phloem surrounded by non-lignified fibre strand; crystals also present in phloem region.

Lamina - Shows dorsiventral structure; epidermis on either surface, composed of thin walled, tangentially elongated cells, covered externally with thick cuticle; anomocytic stomata present on lower surface only; palisade single layered; spongy parenchyma composed of 5-6 layered, thin-walled cells, traversed by a number of veins; rosette crystals of calcium oxalate present in a few cells; palisade ratio 3.0-4.5; stomatal index 13.0-14.5 on lower surface and 8.0-11.5 on upper surface.

Powder - Green; shows vessels, fibres, rosette crystals of calcium oxalate, fragments of spongy and palisade parenchyma.

Chemical constituents - Triterpenoids and Sterols.

Habitat- All over India, does not grow and survive in heavy rain fall.

Properties: Guna- laghu, Rasa- tikta, kashaya Vipaka- katu Veerya- sheeta.

Action and uses – Anthelmintic, antiseptic, bitter, diuretic, emmenagogue and febrifuge. Used in blood disorder, eye disease, intermittent fever, as well as persistant low fever, leprosy, skin disease, ulcers and wounds.

Part Used- Flower, Leaves, seed and oil.

Dose- Bark powder -1 to 3 g, Leaf juice- 10 to 20 cc, oil- 4 to 10 drops.

Main preparation- Nimb tail, Nimbadi churna.

NIRGUNDI

Botanical name – *Vitex negundo* Linn.

Family – Verbenaceae.

Synonyms– Sinduvar, Nilpuspi, Shephalee.

Botanical description– An aromatic large shrub of about 3m in height with quadrangular branches, bark grey, leaves opposite, stipulate, long petioted and digitally 3-5 foliate. All leaflets with petiole the middle one longer. Flowers are small, bluish purple; fruits ovoid four seeded drupe, black when ripe.

a) Macroscopic

Leaves palmately compound, petiole 2.5 - 3.8 cm long; mostly trifoliate, occasionally pentafoliate; in trifoliate leaf, leaflet lanceolate or narrowly lanceolate, middle leaflet 5- 10 cm long and 1.6 -3.2 cm broad, with 1- 1.3 cm long petiole, remaining two sub-sessile; in pentafoliate leaf inner three leaflets have petiolule and remaining two sub-sessile; surface glabrous above and tomentose beneath; texture, leathery.

b) Microscopic

Petiole - Shows single layered epidermis having a number of unicellular, bicellular and uniseriate multicellular covering trichomes and also glandular trichomes with uni to tricellular stalk and uni to bicellular head; cortex composed of outer collenchymatous tissue and inner 6 - 8 layers of parenchymatous tissue; collenchyma well developed in basal region and gradually decreases in middle and apical regions; pericyclic fibres absent in basal region of petiole and present in the form of a discontinuous ring in apical region surrounding central horse shoe-shaped vascular bundle; a few smaller vascular bundles present ventrally between arms of central vascular bundle and two, or rarely three, bundles situated outside the arms.

Lamina - Shows single layered epidermis having mostly unicellular hairs, bi and multicellular and glandular trichomes being rare; hypodermis 1 - 3 layered interrupted at places by 4- 8 palisade layers containing chlorophyll; a large number of veins enclosed by bundle sheath traverse mesophyll; stomata present only on the ventral surface, covered densely with trichomes; vein-islet and vein termination number of leaf are 23-25 and 5-7 respectively.

Powder - Shows number of pieces or whole, uni-bi and multicellular covering trichomes, glandular trichomes, palisade tissues with hypodermis, and upper and lower epidermis, xylem vessels with pitted walls.

Chemical constituents - Alkaloids and Essential Oil.

Habitat– Throughout India.

Properties– Guna-laghu, ruksha Rasa-katu, tikta Vipaka-katu Veerya-ushna.

Doshakarma– Kaphavatashamaka.

Action & uses– It is act as a analgesic, anti-inflammatory, wound healer, hair tonic, antimicrobial, braintonic, deepan (carminative), digestive, liver stimulant, cough supresant & antipyretic. It is uses in arthritis, fever, asthma, lung diseases, spleen enlargement, sciatica, worm, diarrhoea, cholera.

Part used– Panchanga.

Dosage– Leaf juice – 10-20 ml, Root bark Powder – 3-6 g, Seed Powder – 3-6 g.

Main preparation– Nirgundi tail, Nirgundi kalp.

HARIDRA

Botanical name – *Curcuma longa* Linn.

Family – Zingiberaceae.

Synonyms– Rogaghni, Nisha, Gauri, Yoshitpriya (female favourate).

Botanical description– It is a perennial herb 2-3 feet hight, small rhizomer tuber (a modification of root), tuber cylendric, leaves upto 50x8 cm, oblong lanecolate, base tapering, flowering bracts pale green.

a) Macroscopic

Rhizomes ovate, oblong or pyriform (round turmeric) or cylindrical, often short branched (long turmeric), former about half as broad as long, latter 2-5 cm long and about 1-1.8 cm thick, externally yellowish to yellowish-brown with root scars and annulations of leaf bases, fracture horny, fractured surface orange to reddish brown, central cylinder twice as broad as cortex: odour and taste characteristic.

b) Microscopic

Transverse section of rhizome shows epidermis with thick-walled, cubical cells of various dimensions, cortex characterised by the presence of mostly thin-walled rounded parenchyma cells scattered collateral vascular bundles, a few layers of cork developed under epidermis and scattered oleo-resin cells with brownish contents; cork generally composed of 4-6 layers of thin-walled, brick-shaped parenchyma, cells of ground tissue contain starch grains of 4-15 µ in diameter, oil cell with suberised walls containing either orange-yellow globules of volatile oil or amorphous resinous matter, vessels mainly spirally thickened, a few reticulate and annular.

Identification

1) On the addition of Concentrated Sulphuric acid or a mixture of Concentrated Sulphuric acid and alcohol to the powdered drug, a deep crimson colour is produced.

2) A piece of filter paper is impregnated with an alcoholic extract of the powder, dried, and then moistened with a solution of Boric acid slightly acidified with Hydrochloric acid, dried again, the filter paper assumes a pink or brownish red colour which becomes deep blue or greenish-black on the addition of alkali.

Chemical constituents- Essential oil and a colouring matter (curcumin).

Habitat– Plant is a native of South Asia and is cultivated throughout warmer parts of the world including India.

Properties– Guna- ruksha, laghu Rasa-tikta, katu Vipaka-katu Veerya-ushna.

Doshakarma– Tridoshashamaka.

Action & uses– The rhizomes are bitter, acrid, thermogenic, anti-inflammatory, antiallergic, antiseptic, diuretic. It is use in various diseases liprosy, pruritus allergic condition, discolouration of skin, dyspepsia, asthma, bronchitis, hiccough, anaemia, haemorrhage, hepatomegally, spleenomegally, conjunctivitis, general debility and diabetes.

Part used– Rhizome.

Dosages– Juice– 10 to 12 ml of fresh rhizome, Powder– 1 to 3 gm of dry drug.

KAMPILLAKA

Botanical name – *Mallotus philippinensis* Muell Arg.

Family – Euphorbiaceae.

Synonyms– Kampillaka, Karkash, Raktang, Rechanak.

Botanical description– It is medium sized tree, 8 to 10 mtr hight. Leaves are like udumber and have 2 nodes at the base of the leaf stalk as well as on the vental surface, it has 3 red vains. Flowers are small whitish- yellow coloured. Fruit small, having 3 compartments and appear in bunches. On the ripe fruit there is red powder which falls of when fruits are rubbed. It is called as 'Kabila'.

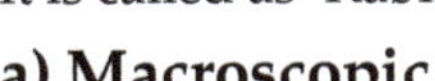

a) Macroscopic

Fine, granular powder, dull-red or madder-red coloured, floating on water.

b) Microscopic

Under microscope glands appear depressed and globular, containing deep-red coloured resin, secreted by many club shaped cell radiating from a common centre, a number of stellate trichomes present, trichomes thick-walled, branching lignified with smooth margins, yellow coloured, arranged in small radiating groups.

Chemical constituents - Resinous colouring matter (rottlerin).

Habitat – All over India, in outer part of Himalaya region and in Srilanka, Malaysia, Singapore.

Properties– Guna- laghu, ruksha, tikshna Rasa - katu Vipaka - katu Veerya - ushna.

Doshakarma– Kaphavatashamaka.

Action & uses– It is use as wound cleaner. In ulcerated and pyogenic wounds it is apllied externally mixed with oil and honey. It is laxative and wormicidal so kills and expels the worms. It acts especially on tape worm.

Part used– Phalaraj (Powder on the fruit).

Dosages– 1 to 3 g.

Main preparation- Krimighatini gutika.

MANDOOKAPARNI

Botanical name – *Centella asiatica* Linn.

Family – Umbelliferae.

Synonyms– Mandookaparni, Mandooki, Brahmi.

Botanical description– It is a perinial herb (creeper) growing in wet places.

a) Macroscopic

Small creeping herb with slender stem, rooting at nodes giving rise to thin, brownish-grey, roots of about 2.5 to 6.0 cm in length; leaves 1 to 3 from each node, orbicular-reniform, crenate, base cordate, petioles channelled with adnate stipules; flowers fascicled umbels each carrying 3 or 4 flowers, short stalked; fruits cremocarp, ovoid, with laterally compressed seeds.

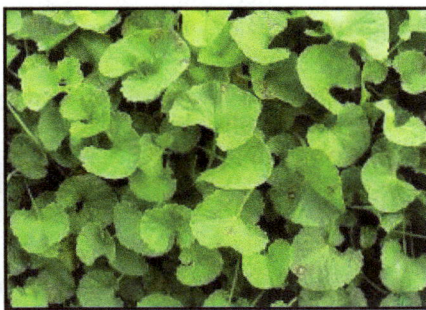

b) Microscopic

Root - Shows wavy outline, consisting of 3 to 5 layered, rectangular, cork cells having exfoliated cells, followed by 3 or 4 layers of parenchyma cells containing oval to round, simple, starch grains measuring 8 to 16 μ in dia., having centric hilum and microsphenoidal crystals of calcium oxalate; secondary cortex composed of thin-walled, oval to polygonal, parenchymatous cells; secretory cells present, scattered towards periphery region; secondary phloem and secondary xylem consisting of usual elements; vessels lignified with reticulate and spiral thickening; pith nearly obliterated. **Stem** - More or less concave-convex outline, shows single layered epidermis composed of round to cubical cells covered by striated cuticle; below this 2 or 3 layers of collenchymatous cells, followed by 6 to 8 layers of thin-walled, isodiametric, parenchymatous cells with intercellular spaces present; vascular bundles collateral, open, arranged in a ring, capped by patches of sclerenchyma and traversed by wide medullary rays; vessels with spiral thickening present, resin duct present in parenchymatous cells of cortex and generally one in between vascular bundles; pith of isodiametric, parenchyma with intercellular spaces.

Leaf

Petiole - Shows a characteristic outline due to two projections adjacent to ventral groove; epidermis single layered, cells cubical covered by a thick cuticle; inner walls of epidermal cells adjoining the cortex much thickened; hairs absent; collenchyma 2 or 3 layered, absent on the projections, a broad zone of more or less rounded parenchyma cells present with intercellular spaces, and a few containing rosette crystals of calcium oxalate; resin canal present on dorsal side of each vascular bundle except in the vascular bundles occurring projecting arms; vascular bundles seven in number, two of which less developed and present in projections.

Midrib - Show a single layered epidermis, 2 or 3 layered collenchyma on both surfaces, 4 or 5 layered parenchyma, mostly devoid of chloroplasts; central zone occupied by vascular bundles differentiated into xylem towards ventral side and phloem towards dorsal side; phloem consisting of sieve tubes, companion cells and phloem parenchyma; xylem consisting of radial rows of vessels with xylem parenchyma in between.

Lamina - Shows an epidermis of tangentially elongated cells on both surfaces, larger on the upper surface, covered by striated cuticle; mesophyll differentiated into 2 or 3 layers of palisade cells, 5 to 7 layers of loosely arranged, somewhat isodiametric spongy parenchyma; rosette crystals of calcium oxalate present in a few cells; stomata more on the lower surface, anisocytic in general, but anomocytic type also occurs on both surfaces, palisade ratio 3 to 5, stomatal index on upper surface, 9 to 12, and lower surface 11 to 17.

Fruit - Shows several ridges in outline; epicarp consists of single layered epidermis covered externally with thick cuticle; mesocarp consists of polygonal, thin walled parenchymatous cells having patches of sclerenchymatous cells on both lateral side; each ridge having a vittae and patch of sclerenchyma; endocarp consists of columnar shaped sclereids arranged in wavy layers; endosperm and embryo composed of oval to polygonal, thin-walled parenchymatous cells.

Powder - Green to greenish-brown, shows fragments of epidermal cells polygonal in surface view with stomata, palisade cells, vessels with spiral, reticulate and annular thickening; microsphenoidal and rosette crystals of calcium oxalate; simple, oval to round starch grains measuring 8 to 16 μ in dia.

Chemical constituents - Glycosides - Saponin Glycosides.

Habitat– All over India and Sri Lanka, Specially on the bank of river, ponds, dams and mainly in rainy season.

Properties – Guna- laghu Rasa – tikta, kashaya, madhur Vipaka - madhur Veerya - sheeta.

Doshakarma – Kapha pitta shamaka.

Action & uses – Its external application in wound and skin diseases. The drug is orally given in psychological disorders epilepsy, hysteria and memory loss. It is a good brain tonic.

Part used – Wholed plant.

Dosages– Powder-2 to 5 gm, Swarasa- 10 to 20 ml.

Main preparation- Saraswataristh, Bramhighrita.

JYOTISHMATI

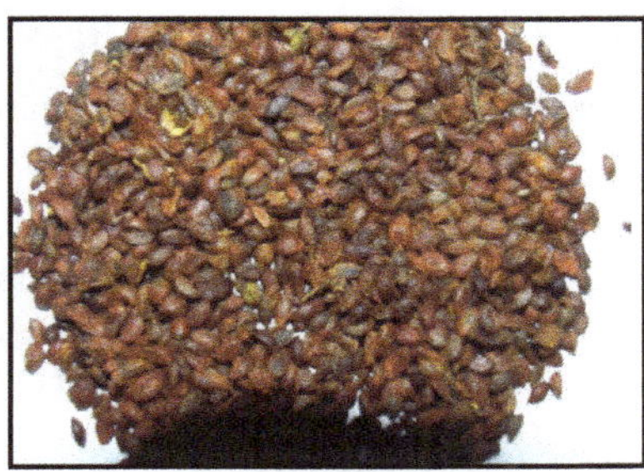

Botanical name – *Celastrus paniculatus* Willed.

Family – Celastraceae.

Synonyms – Parawatpadi, Kakandaki, Kangunika, Peetataila.

Botanical description – It is large creeper growing vertically attaining a large height. Leaves are egg shaped. Flowers are greenish and sweet in taste. Fruits are round like pea, yellow and trilobular. Each of the parts is saffron coloured, triangular single seed, fruiting in rainy season.

a) Macroscopic

Dried ripe seeds more or less covered by orange-red crusty aril, seed without aril also prescent, measuring 5-6 mm in length and 2.5-3.35 mm in breadth, a few roughly three sided being convex on the sides and a few two sided with one convex and other more or less flat side, one edge of many seeds show a faint ridge or raphe on the whole margin; surface generally smooth and- hard; colour, light to dark brown; odour, unpleasant; taste, bitter.

b) Microscopic

Seed - Shows single layered epidermis covered externally with thick cuticle and filled with tannin, followed by 4-6 layers of thin-walled, collapsed, parenchymatous cells and layer of radially elongated stone cells; parenchyma of top one or two layers longer than of the below with triangular intercellular spaces; inner most layer of parenchyma containing prismatic crystals of calcium oxalate; beneath stone cells layer quadrangular to octagonal, tangentially elongated cells filled with brownish contents; endosperm composed of polygonal, thin-walled, parenchymatous cells having oil gloubles and aleurone grains; embryo spathulate in fleshy endosperm containing oil globules and aleurone grains.

Powder - Oily, dark brown; under microscope shows groups of endospermic parenchyma, stone cells, oil globules and aleurone grains and shows fluorescence under U.V. light as following :-

Powder as such	:	Grenish -brown
Powder + 1 N NaOH in Methanol	:	Light green
Powder + Nitrocellulose in Amyl Acetate	:	Yellowish-green

Chemical constituents - Alkaloids, Oil and Tannins.

Habitat – All over India, especially in hilly areas Punjab, Kashmir.

Properties – Guna- tikshna Rasa – katu, tikta Vipaka - katu Veerya - ushna

Doshakarma – Kaphavatashamaka

Action & uses – Its oil is vatanashaka, vedananashaka due to katu vipaka and ushna veerya. It is used in paralysis, backache, arthritis and sciatica. It improves memory by medhya guna. Jyotishmati oil with cow's ghee is used as a memory enhancer. It is a medhya rasayana. Due to its katu, tikta and ushna guna, it is used in loss of appetite, stimulates heart and improves cardiac out put. Black oil of jyotishmati is usefull in beriberi.

Part used – Seed and oil.

Dosages– Seed-1 to 2 gm, Oil- 5 to 15 dops.

Main preparation- Jyotishmati tail.

RASNA

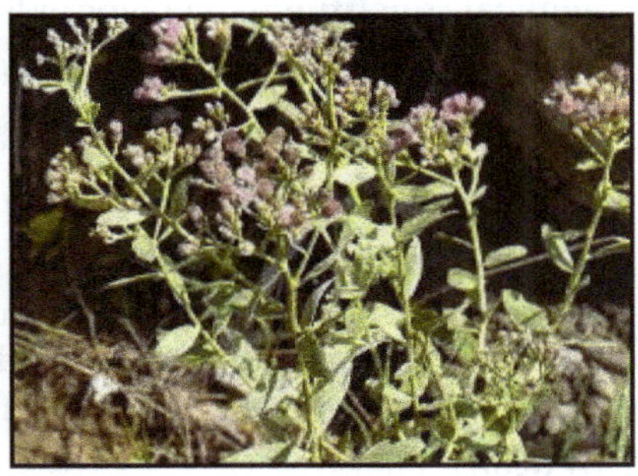

Botanical name – *Pluchea lanceolata* C.B Clarke.

Family – Compositae.

Synonyms – Uktarasa, Elaparni, Surasa, Sugandha.

Botanical description – It is a herb 1-4 feet height, with various branches and sub-branches. It has brown hair on stem. Leaves are 1-2 inch long like sanaya patra (Cassia angustifolia). Flowers are violet coloured.

a) Macroscopic

Leaves simple, 3-5 cm long, 0.6-2 cm broad; sessile, obtuse, lanceolate to ovate-lanceolate; margin entire or toothed around the apex, unequal at base; both surfaces pubescent, distinct small hairs more prominent near veins; texture, brittle, papery; odour, characteristic; taste, astringent and slightly bitter.

b) Microscopic

Leaf

Midrib - Shows single layered epidermis covered by thick, striated cuticle; collenchyma 2-5 layered towards xylem, 1-3 layered towards phloem; beneath collenchyma 2-5 layers of parenchyma present on both sides; central portion occupied by a large vascular bundle, xylem facing towards upper and phloem towards lower epidermis; vascular bundle surrounded by sclerenchymatous sheath appearing as a cap above and below; vascular bundle consists of wide phloem, a thin cambium and xylem; phloem consists of phloem parenchyma and a few phloem fibres; xylem consists of tracheids, vessels and xylem parenchyma; vessels arranged radially; parenchyma and palisade cells of leaf contain oil globules, scattered rosette crystals of calcium oxalate are both in lamina and midrib.

Lamina - Shows isobilateral structure with palisade occurring in upper and lower mesophyll regions; epidermal cells tangentially elongated, covered by thick, striated cuticle; uniseriate, unbranched covering trichomes 2-3 cells long, present on both surfaces, basal cell short and slightly swollen, apical cells long; stomata, anisocytic and anomocytic present on both surfaces but more on lower surface; palisade tissue 2 or 3 layered on both sides, composed of radially elongated, thin-walled cells; spongy parenchyma composed of thin-walled, circular to elliptical, parenchymatous cells containing abundant chloroplasts with prominent intercellular spaces; a number of small veins, surrounded by a sclerenchymatous sheath present in mesophyll; vascular tissue much reduced and represented by a few phloem and xylem elements; average value of stomatal index on upper surface 14-24 and on lower surface 20-24; palisade ratio not more than 5; average value of vein islet number 27.

Powder - Light green; shows fragments of parenchyma, palisade cells, pointed 2-5 celled trichomes, a few oil globules and rosette crystals of calcium oxalate.

Chemical constituents - Flavonoids - Quercetin and Isorhamnetin.

Habitat – Upper part of Gangetic region, Bengal, Bihar, Punjab and Rajasthan.

Properties – Guna- guru Rasa – tikta Vipaka - katu Veerya - ushna.

Doshakarma – Kaphavatashamaka.

Action & uses – It is a digestive, analgesic, elevates kapha and uterine contractor. Its decoction is used in abdominal pain, indigestion, amenorrhea, desmenorhea, lungs disorder like asthma, bronchitis and pleuritis as well as in rheumatoid arthritis and in vatic disordes. It is the best drug for vatashamana.

Part used – Leaf.

Dosages– Decoction- 50 to 100 ml.

Main preparation- Rasnadi kwath, Maharasnadi kwath.

TAGAR

Botanical name – *Valeriana wallichi* DC.

Family – Valerianaceae.

English name- Indian valerian.

Synonyms – Vakra, Kutil, Nata, Nihush, Kalanusarya.

Botanical description – It is a perinial hairy herb. Stem is 15-45 cm small in bunches. Leaves are cardiac or top shaped, dentate and sharp on margins. Flowers are white or pink, flowering in July and fruting in September- October. Fruits have also hair on them. Root is 4-8 cm long and 5-10 mm thick, fragile, dull yellowish brown coloured and have strong odour.

a) Macroscopic

Rhizome, of about 4-8 cm long and 4-10 mm thick pieces, dull yellowish-brow. sub-cylindrica1 and dorsiventrally somewhat flattened, rough, slightly curved and unbranched, upper surface marked with raised encircling leaf scars, under surface bearing numerous, small, circular prominent, root scars and a few stout rootlets, crown bearing remains of aerial stems with scale leaves, fracture short and horny, stolon connecting rhizomes stout, 1-5 mm long and 2-4 mm thick, yellowish-grey in colour, longitudinally wrinkled, usually with nodes and internodes and bearing adventitious roots, occasionally thin stolons 1-2 mm thick, root, yellowish-brown, 3-5 cm long and 1 mm thick, odour, strong and reminiscent of isovaleric acid, taste, bitter and somewhat camphoraceous.

b) Microscopic

Rhizome - Transverse section of rhizome shows cork, consisting of 4-14 layers of lignified, cells occasionally containing oil globules, cortex parenchymatous containing numerous starch grain oil globules and yellowish-brown substance, outer 2 or 3 layers of cortex, collenchymatous occasional root traces appear as paler strands, endodermis single layered, pericycle, parechymatous .and within it 12-18 collateral vascular bundles, separated by dark medullary ray present, pith large, parenchymatous, lacunar, containing starch grams, starch occurs as single or occasional compound grains of two components, individual grains being 7-30 μ mostly, 10-25 μ in diameter calcium oxalate crystals absent.

Stolon- Transverse section of stolon shows cork, consisting of 2-5 layers, cortex upto 25 layers, pareachymatous, followed by 20 collateral vascular bundles, which in young stolons separated by cellulosic parenchymatous medullary rays and in older stolons become lignified, pith wide and lacunar, root traces absent.

Root- Transverse section of root shows small, central parenchymatous pith, surroundod by tetrach to polyarch xylem and a wide parenchymatous bark.

Chemical constituents - Essential oil

Habitat – At the altitude of 10 thousand feet in the Himalaya region- from Kashmir to Bhutan.

Properties – Guna- laghu, snigdha Rasa – tikta, katu, kahsaya Vipaka - katu Veerya - ushna.

Doshakarma – Kaphavatashamaka.

Action & uses – Due to its vatahar properties it is useful in hemiplagia, epilepsy, osteoarthritis, rheumatoid arthritis and gout. It reduces pain, convulsions and nourishes the central nervous system. It is an appetizer, digestive, antispasmodic, laxative and hepatostimulant used in loss of appetite, colic, flatulence, hepatomegaly, jaundice and spleenomegally.

Part used – Root.

Dosage – Root powder- 1 to 3 g.

DEVADARU

Botanical name – *Cedrus deodara* Roxb-Loud.

Family – Conifereae.

English name- Deodara.

Synonyms – Deodara, Bhadradaru, Suradaru, Drukilim, Shambhava, Snehadaru.

Botanical Description – It is a big tree height upto 160 – 180 feet. The branches of the stem are big erected. Leaves are green, triangular, in bunches on small branches with tapering end. Flowers are green yellow appear in clusture. Fruits are black on ripened. The oil of deodaru is known as turpentine oil.

a) Macroscopic

Fruit pulp occurs as a reddish-brown, moist, sticky mass, in which yellowish-brown fibres are readily seen; odour, pleasant; taste, sweetish and acidic.

b) Microscopic

Fruit pulp consists of thin-walled, elongated to polygonal, parenchymatous cells of considerable size, traversed by a number of long fibro-vascular bundles and having a very few small starch granules, and numerous prismatic crystals of calcium oxalate.

Chemical constituents - Inorganic acids, sugars, saponin and bitter principle – tamarindinca.

Habitat – In Himalaya region at the height of 2000 to 3000 thousand feet.

Properties – Guna- laghu, snigdha Rasa – tikta Vipaka - katu Veerya - ushna

Doshakarma – Kaphavatashamaka.

Action & uses – Its local application and oil are used in arthritis. Its oil is a very good wound cleaner and healer. Due to vatashamaka properties it is a good analgesic and used in so many vatic disorders like chronic osteoarthritis, rheumatoid arthritis, sciatica and headache. It is an appetizer, digestant and laxative. Oil is useful in mumps, filaria and chancre. Deodar is a very good medicine for chronic cough and cold.

Part used – Heart wood and Oil.

Dosage – Powder 2-6 g, Oil- 20 – 40 ml.

Main preparation – Deodarvyadi kwath, Deodarvyadi churna.

VATSANABH

Botanical name – *Aconitum ferox* Wall.

Family – Rannunculaceae.

English name- Monk's hood.

Synonyms – Vatsanabh, Visha, Amrit, Mahaushadha.

Botanical description – It is shrub 30 to 40 cm hight. Leaves are like water melon, short and stalk with hairy structure. Flower is blue and resembling the flower of pea. Fruit is thorny like hurhur. Seeds are black and winged. Root is 9 cm in length. In external appearance it is brown in colour but inside it is white, oily and shiny.

a) Macroscopic

Roots paired, occasionally separated due to breakage, ovoid, conical, small portions of stem sometimes attached, tapering downwards to a point, 2-4.5 cm, rarely 5 cm long, 0.4 - 1.8 cm thick, gradually decrease in thickness towards tapering end; wrinkled longitudinally and transversely, rough due to root scars; dark brown to blackish-brown; fracture, cartilaginous, hard and white within the cambium ring and brownish outside cambium; odour indistinct, taste, slightly bitter followed by a strong tingling sensation, poisonous.

b) Microscopic

Root -Shows epidermis 1-3 layered, suberised, papillose on outside, primary cortex consisting of 8-10 layers of oval to tangentially elongated, thin-walled, parenchymatous cells, without or with a few intercellular spaces, a few rectangular or triangular stone cells in singles found scattered in this zone; primary cortex separated by distinct endodermis; inner bark parenchymatous, consisting of round to oval cells, containing a few groups of phloem strands, occupying more than half the radius; cambium having 6 - 10 angles; xylem vessels arranged almost in a ring, some scattered, often forming 'V' shaped ring, enclosing xylem parenchyma in older portions; bundles compact often wedge-shaped having acute apex; xylem exarch, metaxylem vessels met in centre; starch grains simple measuring 6-18 µ in dia. and compound grains consisting of 2-5 components with hilum in centre, present in cortical cells, phloem parenchyma and xylem parenchyma.

Powder - Light grey; shows vessels, a few aseptate fibres, and numerous simple and compound starch grains having hilum in the centre, single grain measuring 6-18 µ in dia.

T.L.C.

T.L.C. of alcoholic extract of the drug on Silica gel 'G' plate using Chloroform: Methanol (90:10) shows six spots at Rf. 0.10, 0.20, 0.39, 0.56, 0.74 and 0.96 (all yellow) on exposure to Iodine vapour. On spraying with Dragendorff reagent two spots appear at Rf. 0.39 and 0.96 (both orange).

Chemical constituents - Alkaloids

Habitat – It is found in Himalaya region from Sikkim to Garhwal at a height of 3000 to 5000 meters.

Properties – Guna- laghu, ruksha, teekshna, vyavayi and vikasi Rasa – madhur Vipaka - madhur Veerya - ushna.

Doshakarma – especially kaphavatashamaka.

Action & uses – On external application it reduces pain and inflammation. In therapeutic doses it acts as an appetizer (deepan), pachan and reduce pain. Purified vatsanabh acts as a cardiostimulant due to its vyavayi and vikasi properties where as impure vatsanabh has a depressant effect on heart. It is best antipyretic due to it sweating property.

Part used – Rhizome.

Dosage – 1.5 mg.

Main preparation – Mrityunjay ras, Hinguleshwer ras, Tribhuvankirti ras.

DANTI

Botanical name – *Balliospermum montanum* Arg.

Family – Euphorbiaceae.

Synonyms – Erandfala, Vishodhini, Akhuparni, Sheeghra.

Botanical description – It is a small tree of height 1-2 meters. Leaves are defferent size and shape from bottom to top. Lower leaves are broad. Flowers are pale green colour. Fruits are divided in to 3 campartments having 3 coloured seeds. Growth of danti needs the shade of other trees.

a) Macroscopic

Root pieces almost cylindrical, straight or ribbed with secondary and tertiary roots, 0.2-1 cm thick and upto 10 cm or more in length, tapering at one end, tough, externally brown; surface, rough due to longitudinal striations, transverse cracks and scars of rootlets; internally cream-coloured; transversely smoothened root shows thin, brown bark and yellowish-white central core; taste, bitter.

b) Microscopic

Shows 5-18 layered cork, consisting of brown coloured, suberised or lignified brick-shaped cells, a few cells containing tannin and red colouring matter; secondary cortex consists of 2-7 layers of oval to elliptical, tangentially elongated cells, a few cortical fibres are also present in this region; secondary phloem consists of usual elements, traversed by uni to biseriate phloem rays; secondary xylem consists of usual elements; vessels and tracheids, bordered pits, a few having reticulate thickening; fibres slightly thick-walled, narrow lumen and blunt tips; xylem rays 1 or 2 cells wide; rosette crystals of calcium oxalate and starch grains, present only in secondary cortex and phloem; starch grains solitary and in groups, simple, round to oval measuring 6-17 µ in dia.

Powder - Brown; shows fragments of cork more or less rectangular, thick-walled in surface view; rosette crystals of calcium oxalate; numerous phloem fibres with narrow lumen and blunt tips, border pitted- and reticulate vessels, tracheid and tannin cells, round to oval simple starch grains measuring 6-17 µ in diameter, and in groups occasionally.

Chemical constituents - â - Sitosterol and Triterpenoids, Resinous Glycosides, Phorbol Esters.

Habitat – Bengal, Bihar and South India.

Properties – Guna- guru, ruksha, teekshna Rasa – katu Vipaka - katu Veerya - ushna.

Doshakarma – Kaphapittashamaka.

Action & uses – In pain and oedema its root and seed paste is used externally. Internally it is used in various disorders- anorexia, heamorrhoids and helminthiasis due to its appetiezer, liver stimulant, cholagogue and laxative action. It is useful for elimination for doshas in jaundice.

Part used – Root, seeds.

Dosage – Root powder- 1-3 g, seeds- 60-125 m.

Main preparation – Dantiharitaki, Dantyadi churna.

GOJIHVA

Botanical name – *Onosma bracteatum* Wall.

Family – Boraginaceae.

Synonyms – Darvipatra, Kharapatra, Gojihva, Vrushjihva.

Botanical description – It is a small herb spreding on the ground. Leaves are big, fleshy and rough like cow's tongue and have small, round spot on them. Flowers are blue, appear in clusters and become red on drying.

a) Macroscopic

Stem - Cut pieces available in 5-9 cm long and 3.2 to 4.7 cm in dia., flattened, erect, stout; rough due to white, hard, hispid hairs and cicatrices, and longitudinal wrinkles; colour greenish-yellow; fracture, short; odour and taste not characteristic.

Leaf - Lanceolate to ovate-lanceolate, 12-30 cm long, 1.5-3.5 cm broad, acuminate tubercle-based hispid hairs present on both surfaces; greenish to light yellow on top and white beneath.

b) Microscopic

Stem - Shows single-layered epidermis, covered with thick cuticle, some epidermal cells elongate to form long, warty, tubercle-based unicellular hairs, cortex differentiated in two zones, 5-7 layered outer collenchyma, 3-4 layered inner parenchymatous cells, consisting of thin-walled, round to oval cells; phloem composed of usual elements; phloem fibres absent; xylem consisting of usual elements, vessels mostly solitary or rarely 2-3 in groups having spiral thickening, and fibres and tracheids having blunt tips and simple pits; xylem ray not distinct: pith consisting of round, thin-walled, parenchymatous cells.

Leaf -

Midrib - Single layered epidermis with thick cuticle and long warty, tubercle-based unicellular hairs present on both surfaces followed by 5-7 layers of collenchymatous and 3-4 layers parenchymatous cortical cells; vascular bundle situated centrally.

Lamina - Isobilateral, single layered epidermis on either surface covered with thick cuticle, long warty, tubercle-based, simple, unicellular hairs present on both surfaces; palisade 2 layered, spongy parenchyma 8-10 layered, stomata paracytic

Powder - Greenish-brown; shows groups of oval to polygonal, thin-walled straight epidermal cells; spiral vessels; a few fibres entire or in pieces, elongated with blunt tips; long warty, tubercle-based unicellular hairs and a few paracytic stomata.

Chemical constituents - Tannin and sugars.

Habitat – In Himalayas at higher altitude and in Kashmir, Kumaon.

Properties – Guna- laghu, snigdha, Rasa – madhur, tikta Vipaka - madhur Veerya - sheeta.

Doshakarma – Vatapittashamaka and kapha nihsaraka.

Action & uses – Leave bhasma is applied over wound and stomatitis. Its expectorant activity is seen in cold, cough, asthma, pleuritis and throat irritation. It is carminative and mild laxative indicated in jaundice, udavarta and constipation.

Part used – Leaves.

Dosage – Leaves- 4-6 g, kwath- 20-40 ml.

Main preparation – Gojihvadi kwath.

LANGALI

Botanical name – *Gloriosa superba* Linn.

Family – Liliaceae.

English name- Glory lily.

Synonyms – Garbhapatini, Agnishikha, Kalihari, Swarnapuspa.

Botanical description – It is a creeper having pleasant attractive flowers. The stalk near the ground is twisted like plough and contracted distantly. Leaves are stalkless, long with hook like tip and climbs taking its support. Flowers are yellow and red towards periphery. Tuber is round flat and 3 cm in long.

a) Macroscopic

Tuberous roots thick, almost cylindrical or slightly laterally flattened, occurring in pieces of 15-30 em long and 2.5 - 3.8 cm thick, often bifurcated with tapering ends, resembling a plough-share, one arm generally more than double the length of the other; brownish externally and yellowish internally; fracture, short; taste, acrid and bitter.

b) Microscopic

Tuberous root shows single layered epidermis, externally cuticularised, consisting of rectangular cells, followed by ground parenchyma, with scattered small vascular bundles; parenchyma cells large, thin-walled, polygonal to circular, having conspicuous intercellular spaces, most of the cells specially of the outer layers filled with starch grains, simple, round to oblong, or polyhedral, measuring 8-33 µ in dia., showing clear hilum and concentric striations, occasionally compound with 2-3 components, measuring 24-36 µ in dia.; vascular bundles collateral, numerous, scattered throughout ground tissue, consisting of xylem and phloem; each vascular bundle enclosed by sclerenchymatous sheath, xylem composed of vessels, tracheids and parenchyma; vessels having mostly reticulate thickening, smaller ones having spiral thickening, tracheids with reticulate thickening; xylem parenchyma cells usually rectangular; phloem consisting of sieve tubes, companion cells and phloem parenchyma; phloem parenchyma cells very small and thin-walled.

Powder - Brown; shows fragments of parenchyma cells, simple starch grains, round to oblong or polyhedral measuring 8-33 µ dia. showing clear hilum and concentric striations, occasionally compound with 2-3 components, measuring 24-36 µ in dia., sclerenchymatous cells, a few xylem vessels and tracheids.

Chemical constituents – Alkaloids- superbine, gloriojine, starch and resins.

Habitat – South India, Bangladesh, Sri Lanka and in Bengal.

Properties – Guna- laghu, tikshna Rasa – katu, tikta Vipaka - katu Veerya - ushna.

Doshakarma – Kaphavataghna.

Action & uses – Its paste is applied on wounds, inflammation, lymphodenopathy, piles and scorpion sting. Its application on palms, soles, umbilicus and lumber region increases the process of delivery and expulsion of placenta. Its tuber kept in vagina acts as an abortefacient. It is indicated in loss of appetite, worms and general debility internally.

Part used – Tuber.

Dosage – For easy delivery and abortion- 3-6 ratti, normal dose-1-2 g.

Main preparation – Langali rasayana, Kasisadi tail.

PATALA

Botanical name – *Stereospermum suaveolens* DC.

Family – Bignoniaceae.

Synonyms – Krishnvrinta, Madhudooti, Allivallabha.

Botanical description – It is a giant tree of height 10 to 20 meters. Stem is smoke coloured externally with black spot on it and pale yellow internally. It has wing shaped compound leaves. Leaflets are 17-20 cm long and 8-10 cm broad. Flowers are yellow, red and sweet scented. Fruits are 40 cm long and 1 cm broad, hairy, surrounded by four strips. It contains 10 - 20 seeds.

a) Macroscopic

Root occurs in about 6-9 cm long, 1-1.5 cm thick cut pieces, cylindrical, externally brown to creamy, rough due to vertical fissures, cracks, ridges and transverse fine lenticels, internally dark brown, lamellation or stratification due to presence of concentric bands of fibres; fracture tough and fibrous; odour, not distinct; taste, bitter.

b) Microscopic

Root cork consists of 25-35 layers of rectangular cells with 3-5 stratified layers, lignification being more prominent where the stratification starts, arranged with 1-3 tangential rows of narrow cells alternating with 3-5 tangential rows of wider cells; cork cambium composed of 1 -2 layers of tangentially elongated cells; secondary cortex arranged more or less radially, becomes polyhedral to isodiameteric in inner region, a few cells getting converted into stone cells which are regular in shape and show projection; secondary phloem wide, forms cerantenchyma between two obliquely running rays; some rays and phloem cells get converted into irregular, polygonal shaped stone cells, measuring 10- 150 μ in width, phloem parenchyma being intact; medullary rays multiseriate, being 3-4 cells wide, and 8-11-15 cells high; fibres tapering, pointed or slightly blunt, with a small peg-like projection at both ends; sieve tube gets collapsed in outer region forming strips of ceratenchyma; a few small microsphenoidal crystals of calcium oxalate present in phloem parenchyma and rays; secondary xylem wide having usual elements; vessels simple, pitted, lignified; fibres large, pointed, aseptate; rays multiseriate, 2- 3 cells wide.

Powder - Dark brown; shows fragments of rectangular cork and phloem parenchyma cells; groups of single, thick- walled, cubical to rectangular, lignified stone cells having striations and wide lumen; a number of microsphenoidal crystals of calcium oxalate, intact and scattered outside.

Chemical constituents - Bitter Substances, Sterols, Glycosides and Glyco-Alkaloids.

Habitat – Terai regions of Himalaya, Bihar and Bengal.

Properties – Guna- laghu, ruksha Rasa – tikta, kashaya Vipaka - katu Veerya - ushna.

Doshakarma – Kaphapittashamaka.

Action & uses – Externally it has analgesic and wound healing properties. It is astringent, antidyspeptic and liver stimulant so indicated in acid-peptic disease to reduce gastric acidity. Flowers are used in so many disorders- cardiac disease, semen debility, general weakness. Patalakshar is used in dysuria and calculi.

Part used – Root bark, flowers.

Dosage – Bark powder- 0.25-1.5 g, flower juice- 10-20 ml.

Main preparation – Dashmoolaristh, Brihatpanchmooladi kwath.

SHALMALI

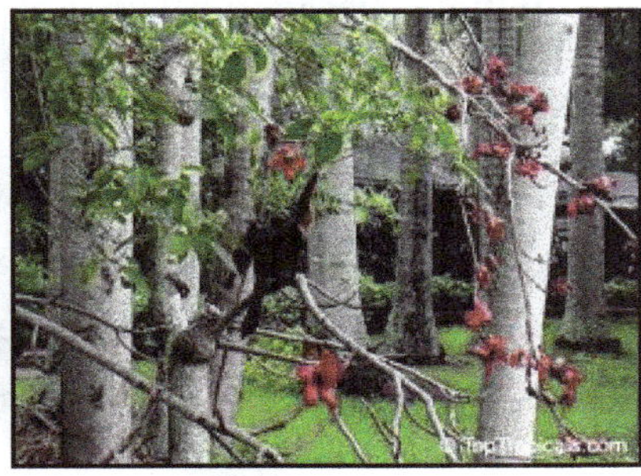

Botanical name – *Salmalia malabarica* Linn.

Family – Malvaceae.

English name- Silk cotton tree.

Synonyms – Raktapushpa, Sthirayu, Mocha, Toolini.

Botanical description – It is big and tall tree having stout thorn. Leaves are like fingers of hand about 10-15 cm in length. Flowers are red having big petals. Fruits are 13-15 cm long like banana, has 5 ovaries within cotton and many black seeds. Flowering season is the end of winter and fruiting occurs in summer. The latex is oozes out from its bark are called as mocharasa.

a) Macroscopic

Bark 0.5-1 cm thick, pale-ashy to silvery-grey externally, brownish internally, external surface rough with vertical and transverse cracks, mucilaginous on chewing; fracture, fibrous.

b) Microscopic

Stem bark shows 10-15 layered, transversely elongated, radially arranged, thin-walled, cork cells with a few outer layers having brown coloured contents; rhytidoma present at certain places interrupting the cork; secondary cortex con- sists of moderately thick-walled, parenchymatous cells containing orange brown contents; stone cells in singles or in groups, thick-walled, oval to irregular, and tangential bands of stone cells having striations with narrow lumen, measuring 13-33 µ in dia., occur throughout the secondary cortex; secondary phloem consists of usual elements traversed by phloem rays, elements in the outer region form tangential bands of ceratenchyma; a number of concentric bands of fibres alternating with groups of sieve elements also present; fibres lignified having narrow lumen and pointed tips; phloem rays numerous and wavy, 1-6 seriate, cells being radially elongated and moderately thick-walled; rosette crystals of calcium oxalate scattered throughout the secondary cortex, phloem parenchyma and ray cells; mucilage canals and tannin cells present in the parenchymatous cells of cortex.

Powder - Reddish-brown; shows fragments of cork cells, parenchymatous cells, single or groups of thick-walled, oval to irregular, stone cells having striations with narrow lumen, measuring 13-33 μ in dia., rosette crystals of calcium oxalate, phloem fibres and numerous reddish-brown coloured masses and tannin cells.

Chemical constituents - Saponins, Tannins and Gums.

Habitat – All over India.

Properties – Guna- laghu, snigdha Rasa – madhur Vipaka - madhur Veerya - sheeta.

Doshakarma – Vatapittashamaka, Its root, flowers and fruits are used in vata-pitta disorders and mocharas is used in kapha-pitta disorders.

Action & uses – The bark of shalmali is anti-inflamatory and reducing burning sensation. Its fowers are haemostatic. Mocharas is astringent in property, So it is used in diarrhea, dysentery, bloody dysentery, grahani, bleeding piles and other bleeding disorders. The thorns of shalmali are doing lekhan karma. Mocharasa is used in oral ulcers and in tooth powders.

Part used – Bark, flower, mocharasa.

Dosage – Bark powder- 2-3 g, flower juice- 10-20 ml.

Main preparation – Shalmali ghrita, Piccha basti.

SHIREESH

Botanical name – *Albbizia lebbek* Benth.

Family – Leguminoseae.

Synonyms – Shalmali, Shukapriya, Shukrapushpa, Lomapushpak.

Botanical description – It is a big and tall tree has a height of 15 to 20 meters. Leaves are compound, unctuous and hairy. Leaflets are wide and there are 4 to 8 pairs. Flowers are white, fragrant and tender. It has flat legumes containes 6 to 10 seeds. Seeds are brownish in colour, flat and circular.

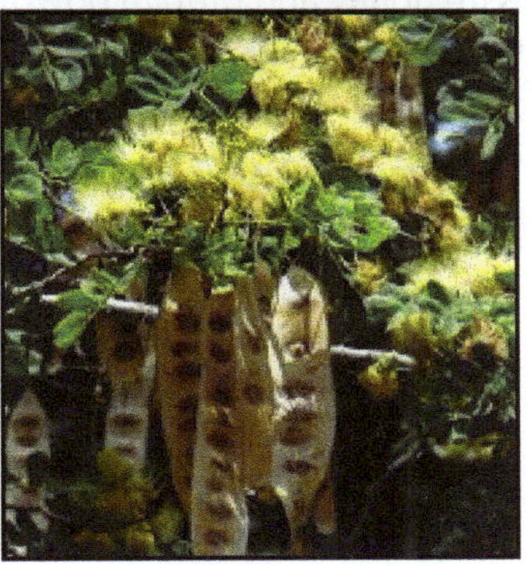

a) Macroscopic

Bark 1.5 - 2.5 cm thick, external surface dark brown, rough due to longitudinal fissures and transverse cracks, rhytidoma forming major part of bark and peeling off in flakes exposing buff coloured surface, middle bark brown, inner bark much fibrous. light yellow to grey; fracture, laminated in outer region and fibrous in inner region; taste, very astringent.

b) Microscopic

Mature bark about 2 cm thick, shows dead tissue of rhytidoma; cork consists of a few layers of thin-walled, transversely elongated and radially arranged cells; secondary cortex wide, composed of radially elongated to squarish, moderately thick-walled cells containing orange to reddish-brown contents; a few of the cells contain prismatic crystals of calcium oxalate; stone cells, variable in shape and size, present in singles or in groups throughout the region; secondary phloem consists of sieve elements, phloem parenchyma, phloem fibres and crystal fibres, traversed by phloem rays; prismatic crystals of calcium oxalate present in most of the phloem parenchyma cells; tangential bands of ceratenchyma present in middle and outer phloem region; phloem fibres. elongated, thick-walled, lignified, present in many concentric strips, mostly enclosed by crystals sheath throughout the middle and inner regions of phloem; crystal fibres having a number of septa, each chamber containing a single prismatic crystal of calcium oxalate; phloem rays numerous, radially elongated, somewhat wavy in outer phloem region and bi to multiseriate in the inner phloem region. being 2 - 5 cells wide and 7 - 25 cells high.

Powder - Greyish-brown; shows large number of stone cells, prismatic crystals of calcium oxalate, crystal fibres and phloem fibres.

Chemical constituents - Saponins and Tannins.

Habitat – All over India.

Properties – Guna- laghu, ruksha, tikshna Rasa – kashaya, tikta Vipaka - katu Veerya – mild ushna.

Doshakarma – Tridoshashamaka.

Action & uses – It is act as anti-inflamatory, analgesic, antitoxic and beneficial to eye. Its bark decoction is used for gargaling in toothache and looseness of tooth. In night blindness its leaf juice is dropped in eye. Its bark decoction is indicated in blood diseases, inflammation and lymphadenopathy. In asthma its flower juice is given with honey and pippali powder. Seed powder with milk is given for improve semen quality. Flowers are used for delaying ejaculation.

Part used – Bark, seeds, flowers.

Dosage – Bark powder- 3-6 g, seed powder- 1 to 2 g.

Main preparation – Shireeshasava, Shireesharishta.

SHYONAK

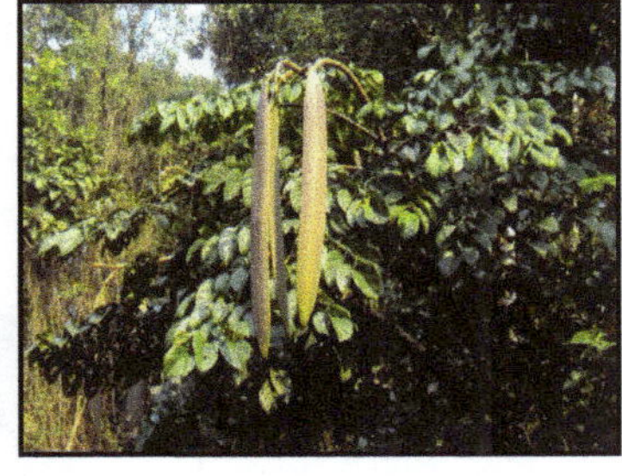

Botanical name – *Oroxylum indicum* Vent.

Family – Bignoniaceae.

English name- Indian trumpet flower.

Synonyms – Shukanasa, Katwanga, Kutannata, Priyajeeva.

Botanical description – It is a big tree 10 to 12 meters height. Leaves are compound and bird shaped. Leaflets are 15 cm long and 9 cm broad, having a serrated margin and pointed at the tips. Flowers are brinjal coloured having bigul shaped 24 cm long stalks. Petals are fleshy with bad smell. It has bearing a sword like legumes up to 1 meter in length.

a) Macroscopic

Drug available in cut pieces, having secondary roots, greyish-brown to light brown, cut surface brownish-cream, cylindrical, ribbed at few places, 5-16 cm long, 1-3 cm thick, external surface rough due to longitudinal and transverse cracks, fracture, short; taste, slightly sweet.

b) Microscopic

Mature root shows 10-30 or more layers of tangentially elongated, radially arranged cork cells filled with reddish-brown content; secondary cortex composed of oval to polygonal, parenchymatous cells; stone cells, thick-walled, lignified of various shapes and sizes with narrow lumen, distinct pits and striations; secondary phloem composed of sieve tubes, parenchyma, fibres and groups of stone cells; groups of fibres traversed by 2-8 cells wide phloem rays; secondary xylem consists of usual elements; xylem vessels of various sizes, occur in singles and groups of 2-5 cells arranged radially having reticulate thickening; xylem rays 2-4 cells wide; fibres having wide lumen and pointed tips, and tracheids present.

Powder - Brownish-cream; shows groups of stone cells, fragments of cork, phloern fibres with wide lumen and pointed tips and reticulate vessels and tracheids.

Chemical constituents - Flavonoids and Tannins.

Habitat – All over India.

Properties – Guna- laghu, ruksha Rasa – tikta, kashaya Vipaka - katu Veerya – sheeta.

Doshakarma – Tridoshashamaka.

Action & uses – It acts as anti-inflammatory, wound healing and analgesic. Its decoction is indicated in rheumatoid arthritis, osteoarthritis and in other painfull and inflammatory diseases. It reduces amadosha in the blood circulation and helps in reducing inflammation and increases appetite, digests the food, stomch is clear as well as takes away the body weakness originated due to amadosha.

Part used – Root bark.

Dosage – Decoction- 20-40 ml, juice-10 to 20 ml.

Main preparation – Bhrihatpanchamool kwath, Shyonak putpak.

AGARU

Botanical name – *Aquilaria agallocha* Roxb.

Family – Thymelaecae.

English name- Eagle wood.

Synonyms – Loha, Krimij, Agnikashtha, Mallikagandha.

Botanical description – It is an evergreen tree and it bark is paper like thin. The wooden bark is whitish, soft and fragrant on cutting. The inner portion of the stem is become black as the plant grows and smells like honey. Leaves are thin and leathery. Flowers are white and in clusters. Fruit is soft in touch. Inner bark of old tree is known as aguru. A small worm grows in the old bark that secretes a type of resin, which gives fragrance to aguru.

a) Macroscopic

Drug available in cut pieces, dark brown to nearly black; fracture, hard; no characteristic smell and taste.

b) Microscopic

Shows mostly uniseriate sometimes biseriate xylem rays; vessels isolated having simple pitted thickening and filled with dark brown contents; xylem fibres short having narrow lumen occupying a major portion of wood; xylem parenchyma less in number and simple pitted; included phloem tissues in pockets partially disorganised, leaving large circular or oval holes, containing collapsed and broken tissues.

Powder - Dark brown; shows numerous aseptate fibres, simple pitted vessels with dark brown contents.Che

Chemical constituents- Essential Oil.

Habitat – Eastern Himalaya, Assam, Chitgaon, Bhutan and Malaysia.

Properties – Guna- laghu, ruksha, tikshna Rasa – katu, tikta Vipaka - katu Veerya – sheeta.

Doshakarma – Kaphavatashamaka.

Action & uses – It is anti-inflammatory and analgesic on external application. It is indicated in skin disorders like wounds, ring worm, chronic ulcers, inflammatory and painful conditions. It is vatahar in nature and indicated in paralytic conditions. In digestive disorders it increases the appetite. In respiratory disorders it reduces the cough in asthma and bronchitis.

Part used – Inner bark and oil.

Dosage – Bark powder- 0.5 – 1.5 g, oil-1-2 drops.

Main preparation – Aguruvadi tail.

MAYAPHALA

Botanical name – *Quercus infectoria* Oliv.

Family – Fagaceae.

English name- Gall-oak, Magic nut .

Synonyms – Mayaphala, Mayuk, Chhidraphala, Mayika.

Botanical description – It is a tree with a greyish trunk height up to 16 feets. Leaves are long, rough with dentate margins. Fruits are small, round and yellowish. On new branches of this plant an insect called Adleria gallae tinctoria sucks the juice from its branches and makes a cocoon like shelter for itself in which it lays its eggs. This is called Mayaphala.

Stem - More or less concave-convex outline, shows single layered epidermis composed of round to cubical cells covered by striated cuticle; below this 2 or 3 layers of collenchymatous cells, followed by 6 to 8 layers of thin-walled, isodiametric, parenchymatous cells with intercellular spaces present; vascular bundles collateral, open, arranged in a ring, capped by patches of sclerenchyma and traversed by wide medullary rays; vessels with spiral thickening present, resin duct present in parenchymatous cells of cortex and generally one in between vascular bundles; pith of isodiametric, parenchyma with intercellular spaces.

Leaf

Petiole - Shows a characteristic outline due to two projections adjacent to ventral groove; epidermis single layered, cells cubical covered by a thick cuticle; inner walls of epidermal cells adjoining the cortex much thickened; hairs absent; collenchyma 2 or 3 layered, absent on the projections, a broad zone of more or less rounded parenchyma cells present with intercellular spaces, and a few containing rosette crystals of calcium oxalate; resin canal present on dorsal side of each vascular bundle except in the vascular bundles occurring projecting arms; vascular bundles seven in number, two of which less developed and present in projections.

Midrib - Show a single layered epidermis, 2 or 3 layered collenchyma on both surfaces, 4 or 5 layered parenchyma, mostly devoid of chloroplasts; central zone occupied by vascular bundles differentiated into xylem towards ventral side and phloem towards dorsal side; phloem consisting of sieve tubes, companion cells and phloem parenchyma; xylem consisting of radial rows of vessels with xylem parenchyma in between.

Lamina - Shows an epidermis of tangentially elongated cells on both surfaces, larger on the upper surface, covered by striated cuticle; mesophyll differentiated into 2 or 3 layers of palisade cells, 5 to 7 layers of loosely arranged, somewhat isodiametric spongy parenchyma; rosette crystals of calcium oxalate present in a few cells; stomata more on the lower surface, anisocytic in general, but anomocytic type also occurs on both surfaces, palisade ratio 3 to 5, stomatal index on upper surface, 9 to 12, and lower surface 11 to 17.

Fruit - Shows several ridges in outline; epicarp consists of single layered epidermis covered externally with thick cuticle; mesocarp consists of polygonal, thin walled parenchymatous cells having patches of sclerenchymatous cells on both lateral side; each ridge having a vittae and patch of sclerenchyma; endocarp consists of columnar shaped sclereids arranged in wavy layers; endosperm and embryo composed of oval to polygonal, thin-walled parenchymatous cells.

Powder - Green to greenish-brown, shows fragments of epidermal cells polygonal in surface view with stomata, palisade cells, vessels with spiral, reticulate and annular thickening; microsphenoidal and rosette crystals of calcium oxalate; simple, oval to round starch grains measuring 8 to 16 µ in dia.

Chemical constituents - Glycosides - Saponin Glycosides.

Habitat – Asia Minor, Syria, Iran and Persia. Alleppo gall is best.

Properties – Guna- laghu, ruksha Rasa – kashayaVipaka - katu Veerya – sheeta.

Doshakarma – Kaphapitta-shamaka.

Action & uses – Externally its powder is used for application in rectal prolapsed, piles and wounds. It stops bleeding from wounds. It is a good anti-diarrhoeal and haemostatic. It is indicated internally for diarrhea, dysentery, bleeding piles, diabetes, pyuria and leucorrhoea.

Part used – Gall.

Dosage – Powder- 1 to 2 g.

Main preparation – Mayaphaladi oientment, Vajradantamanjan.

ATIVISHA

Botanical name – *Aconitum heterophyllum* Wall.

Family – Ranunculaceae.

English name- Aconite.

Synonyms – Ativisha, Shishubhaishajya, Shuklakanda, Pittadivallbha.

Botanical description – It is a herb height of 1 to 3 feet with soft and smooth hairy stem. Leaves are 2 to 4 inch in length, heart shaped. Lower leaves are big and divided into five parts whereas upper leaves are small. Flowers are blue or greenish blue with inner one petal is largest and like a hood. Fruits are embedded with seeds. Roots are modified tubes, which are two in number one from the previous year and one from the current year. Fresh tuber is round shaped brownish on upper surface and whitish on inner surface on transverse section with 4-5 black spots.

a) Macroscopic

Roots, ovoid-conical, tapering downwards to a print, 2.0-7.5 cm long, 0.4-1.6 cm or more thick at its upper extremity, gradually decreasing in thickness towards tapering end, externally light ash-grey, white or grey-brown, while internally starch white, external surface wrinkled marked with scars of fallen rootlet and with a rosette of scaly rudimentary leaves on top: fracture, short, starchy, showing uniform white surface, marked towards centre by 4-7 concentrically arranged yellowish-brown dots, corresponding to end of fibrovascular bundles traversing root longitudinally taste, bitter with no tingling sensation.

b) Microscopic

Transverse section of mature root shows, single layered epidermis consisting of light brown tabular cells rupturing on formation of cork, cork consists of 5-10 rows of tangentially elongated, thin-walled cells, cork cambium single layered consisting of tangentially elongated, thin-walled cells, cortex much wider consisting of tangentially elongated or rounded, thin-walled parenchymatous cells with intercellular spaces, cells fully packed with both simple as well as compound starch grains, compound starch gains composed of 2-4 components of spherical body, endodermis distinct composed of barrel-shaped cells, elements of vascular bundles poorly developed, vascular bundles, arranged in a ring, inter-fascicular cambium present in form of a ring composed of few layered thin-walled cells, central core consisting of thin-walled parenchymatous cells, possessing starch grains similar to those found in cortical cells.

Powder- Ash coloured to light brown, under microscope shows abundant simple and compound starch grains and parenchymatous cells.

Chemical constituents - Alkaloids (atisine, dihydroatisine, hetisined and heteratisine).

Habitat – In Himalaya region at the height of 6 to 15 thousand feet.

Properties – Guna- laghu, ruksha Rasa – tikta Vipaka - katu Veerya – ushna.

Doshakarma – Tridoshahara, especially kaphapittashamaka.

Action & uses – It is a best appetizer, digestive, anti-emetic, anti-diarrhoeal and anti-helminthic. It reduces kapha and clears airways. Indicated in amadosha, indigestion, vomiting, fever associated with diarhoea, cough and breathlessness. It is best and effective medicine for fever, diarrhea and cough in children.

Part used – Tuber.

Dosage – Powder- 1 to 3 g.

Main preparation – Balachaturbhadra churna, Ativishadi churna.

HINGU

Botanical name – *Ferula foetida* Linn.

Family – Umbelliferae.

English name- Asafoetida.

Synonyms – Hingu, Bahlik, Sahasravedhi, Jatuk.

Botanical description – It is a scented perinneal herb of 5-7 feet height. Leaves are ciliated with wings 1-2 feet long, stalk bear a single leaf with broken margin at the tip. Flowers are small yellow coloured. Roots are big and branched. Its oily latex is known as hingu.

a) Macroscopic

Rounded, flattened or masses of agglutinated tears, greyish-white to dull yellow, mostly 12-25 mm in diameter, freshly exposed surface, yellowish and translucent or milky white, opaque, slowly becoming pink, red, finally reddish brown, odour, strong, characteristic and persistent, taste, bitter and acrid.

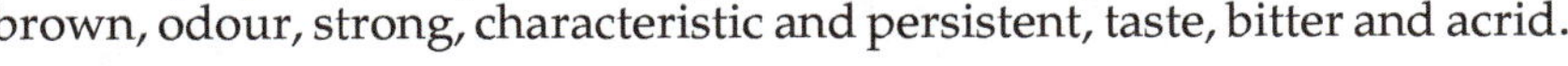

Identification

(I) Freshly broken surface when touched with sulphuric acid a bright red or reddish-brown colour is produced, changing to violet when acid washed off with water.

(II) Boil 0.2 g with 2 ml Hydrochloric acid for about 1 minute, cool, dilute with an equal volume of water, and filter into 3 ml of dilute solution of Ammonia, fluorescence is produced.

Absence of colophony resin-Triturate 1 g with 10 ml of Light Petroleum (b.p. 40°-60°) for 2 minutes, filter into a test tube and add to the filtrate 10 ml of a fresh 0.5 per cent w/v aqueous solution of copper acetate, shake well and allow the liquids to separate, petroleum layer does not show any green colour, indicating absence of colophony resin.

Chemical constituents - Essential oil- This contains rason oil and allyl persulphide, gum and resin.

Habitat – Iran, Afganistan, Punjab and Peshawar.

Properties – Guna- laghu, snigdha, tikshna Rasa – katu Vipaka - katu Veerya – ushna.

Doshakarma – Kaphavatashamaka.

Action & uses – Its topical use is helpful in cough and breathlessness. The paste of hingu is applied on abdomen in flatulence in infant. It is indicated in paralysis, facial palsy, sciatica and epilepsy due to its stimulant, anticonvulsant and analgesic properties. In digestive disorder it is a good appetizer and digestive as well as improves peristalsis, relieve colic pain and intestinal worm. Roasted hingu is indicated in loss of appetite, flatulence, abdominal pain and constipation. In repiratory disorder, it is indicated in cough, pneumonia, chronic cough and whooping cough due to its tikshna guna and kaphanihsarak property.

Part used – Latex.

Dosage – 0.125 to 0.5 g.

Main preparation – Hingwadi vati, Hingwastak churna, Hingukarpur vati.

KUSHTHA

Botanical name – *Saussurea lappa* C.B. Clarke

Family – Compositeae.

English name- Costus.

Synonyms – Shoolahara, Pushkarasya, Padmateerth, Kashmeer.

Botanical description – It is a perrineal shrub 6-7 feet height. Stem is solid like the thickness of little finger. Root leaves are with petiole is 2-3 feet long and stem leaves large and heart shaped. Flowers are round and violet coloured. Roots are thick; perrineal and fresh root emerge from these roots every year. Roots are long and thick; carrot shaped and has strong fragrance.

a) Macroscopic

Drug greyish to dull brown, thick, stout, fusiform to cylindrical, 7-15 cm long, 1.0-5.5 cm broad, thicker roots with collapsed centre, occasionally ridged, wrinkles longitudinal and anastomosed, rootlets rarely present, cut surface shows two regions, outer periderm ring thin, inner porous woody portion lighter in colour showing fine radial striations and often the central portion collapsed, fracture, short, horny, odour, strong, characteristically aromatic, taste, slightly bitter.

b) Microscopic

Transverse section of thin root shows thin periderm, followed by broad zone of phloem and still broader zone of xylem traversed by wide medullary rays, cork, 3-5 layered wide secondary cortical cells polygonal, mostly elongated, secondary phloem consists of mostly storage parenchyma, small groups of sieve tubes and companion cells and often phloem fibres, bast fibres thick-walled, lignified, upto 350 μ in length, with many simple pits associated with fibre, tracheids and parenchyma, wood fibres smaller than bast fibres, with wider lumen and obtusely tapering ends, meduallary rays multi seriate and wider in phloem region, resin canals found throughout as large cavities, some roots possess a central cylinder of sclerenchyma, while others have parenchymatous centre with scattered xylem elements, in older roots, wood parenchyma collapses and takes a spongy appearance in the centre of root, inulin present in storage parenchyma.

Powder-Deep brown or rusty, under microscope irregular bits of yellow, brown or orange-red fragments of resins and oils associated with thin-walled parenchymatous cells, broken bits of xylem vessels with scalariform, reticulate thickening and horizontal end walls.

Chemical constituents - Essential oil, alkaloid (saussurine) and bitter resin.

Habitat – In Himalaya higher altitude, Kashmir.

Properties – Guna- laghu, ruksha, tikshna Rasa- tikta, katu, madhur Vipaka- katu Veerya- ushna.

Doshakarma – Kaphavatashamaka.

Action & uses – It acts as antibacterial, analgesic, varnya, kushthagna and remove bad smell hence indicated in toothache, headache, rheumatoid arthritis and chronic ulcer. It is appetizer, digestive and carminative. It is indicated in loss of appetite, indigestion, colicky pain, diarrhea and cholera. Due to vatahara and antiepileptic action, it is indicated in epilepsy. It is a uterine stimulant, enhance ovulation, breast milk, improve semen quality and thus indicated in amenorrhoea, dysmenorrhoea and as a uterine tonic in post partum condition.

Part used – Root.

Dosage – 0.25 to 1.25 g.

Main preparation – Kusthadi churna, Kusthadi kwath.

PRISHNAPARNI

Botanical name- *Ureria picta* Desv.

Family- Leguminosae.

Synonyms- Kalashi, Deerghaparni,Upchitra, Snigdhaparni, Vedavahni.

Botanical description- The tree is of 0.75 to 1.5 mt. height. Leaves are compound, having different shape, with round base and elongated tapering end. Leaflets - 8cm long, with white coloured veins. Flowers are white bell shaped. Seeds are Kidney –shaped, yellow, 1 to12 in numbers. The plant bears the flowers in rain and seeds in Autumn.

a) Macroscopic

Root - Occur in pieces of varying size, thickness of 1 to 2 cm, gradually tapering, tough, woody and cylindrical; externally light yellow to buff internally pale yellow; surface bearing fine longitudinal striations; fracture, splintery or fibrous; taste, slightly acrid.

Stem - About 8.0 to 16.0 cm long, 0.2 to 0.4 cm in diameter, in cut pieces; cylindrical, branched, pubescent, external surface light yellow to brown; transversely cut and smoothened surface shows buff-white colour, mature stem longitudinally wrinkled, leaf scar present at nodes; fracture, fibrous.

Leaf - Very variable, imparipinnate, upto 20 cm or more long, upto 2 cm wide; leaflets on the upper part of the stem 5 to 7, rigidly sub-coriaceous, linear-oblong, acute, blotched with white; glabrous above, finely reticulately veined and minutely pubescent beneath, base rounded; leaflets on the lower part of the stem 1 to 3, sub-orbicular or oblong.

b) Microscopic

Root - Shows 5 or 6 layers of thin-walled, tabular, regularly arranged cork cells; cork cambium single layered; secondary cortex composed of 4 to 6 layers of oval, tangentially arranged, thin-walled, parenchymatous cells, a few fibres present singly or in groups; secondary phloem composed of sieve elements, parenchyma and fibres traversed by phloem rays; sieve elements somewhat collapsed towards periphery but intact in inner phloem region; phloem parenchyma composed of rounded to somewhat oval cells, larger towards periphery; fibres thick-walled, lignified with narrow lumen and tapering ends;phloem rays 1 to 5 cells wide, their cells being oval or rectangular in the portion nearer the wood but broader towards their distal ends; secondary xylem composed of vessels, tracheids, fibres, crystal fibres and parenchyma traversed by xylem rays; vessel very few, mostly confined to inner and outer part of xylem; fibres similar to those of phloem fibres and arranged in close set concentric bands; in isolated preparation vessels are cylindrical, pitted with transverse to oblique perforation; tracheids possess bordered pits; xylem parenchyma mostly rectangular with simple pits; xylem ray cells isodiametric showing simple pits; starch grains simple, round to oval, measuring 6 to 17μ in dia., distributed throughout parenchymatous cells of secondary cortex, phloem and xylem; prismatic crystals of calcium oxalate present in crystal fibres, as well as in many parenchymatous cells of secondary cortex, phloem and ray cells.

Stem - Shows single layered epidermis covered with cuticle, a few epidermal cells elongate outwards forming papillae; cortex 8 to 10 cells wide, consisting of oval to circular, thin walled, parenchymatous cells; groups of pericyclic fibres present in the form of discontinuous ring; phloem consisting of usual elements except phloem fibres; phloem rays 2 to 4 cells wide; xylem consisting of usual elements; vessels mostly simple pitted; fibres simple with blunt tips; xylem rays 1 to 4 cells wide and 2 to 8 cells in height; pith wide, consisting of thin-walled, round to oval parenchymatous cells.

Leaf

Midrib - Single layered epidermis on either surfaces covered with striated cuticle having a few unicellular or bicellular, hooked or straight and pointed tipped hairs present on bothsurfaces but more on lower surface ; collenchyma 2 or 3 layered, followed by 2 layers of parenchyma cells; single row of pericyclic fibers present on both sides; vascular bundle located centrally.

Lamina - Shows single layered epidermis on either surfaces, a few unicellular or bicellular, hooked or straight, pointed tipped hairs present on lower surface; mesophyll differentiated into single layered palisade and spongy parenchyma; spongy parenchyma cells oval to rounded having small intercellular spaces; numerous paracytic stomata present on lower surface; stomatal index 27 to 36 on lower surface; palisade ratio 4 or 5; vein-islet number 29 to 32 per sq. mm.; vascular bundle present centrally.

Powder - Greenish-yellow; shows simple pitted vessels; fragments of fibres, tracheids, parenchyma cells; pieces of hairs; palisade cells; a few prismatic crystals of calcium oxalate; epidermal cells wavy walled in surface view showing paracytic stomata and starch grains simple, round to oval, measuring 6 to 17 μ in diameter.

Habitat- All over India in sandy regions and in barren land.

Properties: Guna- laghu, snigdha Rasa- madhur, tikta Vipaka- madhur Veerya- ushna.

Action and uses – It is an appetizer, deflatulence and astringent, so it is indicated in dyspepsia, flatulence, diarrhea, bleeding piles and sprue. It is a nervine tonic. It acts as cardiotonic, expectorant, diuretic and nervine tonic, hence used in heart diseases, blood disorder, vatarakta and oedema. It is a good aphrodisiac.

Part used- Panchanga(Whole Plant), root.

Dose- 5-10 ml fresh juice, 15-25 ml decoction along with 15ml honey twice daily.

Main preparation- Dashamoolaristh.

PUSHKARMOOL

Botanical name- *Inula racemosa* Hook-F.

Family- Compositeae.

Synonyms- Poushkara, Hamvati, Kashmir, Sugandhika, Shoolahara.

Botanical description- It is a herb of 1 to 5 feet height. The leaves emerge from the root is long with peduncle and those leaves which originated from shoot is attached with stem and splintered.This plant resembles the plant of kushta. The inflorecense is yellow coloured in bunches. Root of pushkarmool is resemble with kusth.

a) Macroscopic

Root available in cut pieces, upto about 15 cm long and 0.5 to 2.0 cm in dia.; cylindrical, straight or somewhat curved; surface rough due to longitudinal striations and cracks, scars of lateral rootlets and rhytidoma present, externally brownish-grey and internally yellowish-brown; fracture, short and smooth; odour, camphoraceous and aromatic; taste, bitter and camphoraceous.

b) Microscopic

Mature root shows a wavy outline due to development of rhytidoma; cork composed of 8 to 12 layers of thick-walled, tangentially elongated, rectangular cells, some filled with reddish-brown contents; secondary cortex 1 or 2 layers or absent; secondary phloem consists of sieve elements and parenchyma having secretory cavities and traversed by medullary rays; cambium not distinct; wood occupies bulk of root consisting of vessels, tracheids, fibres, parenchyma, secretory cavities and medullary rays; vessel have reticulate thickenings, a few fibres occur in small patches adjacent to vessels and abundant in xylem parenchyma, thin-walled; a few small tracheids; parenchyma in general contain granular, slightly yellowish or colourless inulin granules and also a few yellowish oil globules; starch grains either absent or very rarely seen in cortical and ray cells; yellowish resinous masses present in secretory canals.

Powder - Reddish- brown; under microscope shows fragments of cork cells, vessels, fibres and parenchyma cells containing tannin and inulin.

Chemical constituents – It contains a volatile oil, bitter principale and benzoic acid.

Habitat- It grows at the height of 2.25 thousand meters in Kashmir.

Properties: Gunu- laghu, tikshna Rasa-tikta, katu Vipaka- katu Veerya- ushna.

Doshakarma: Vatakaphaghna.

Action and uses- It stimulates brain and nervous system, hence it is indicated in cerebral impairment. It is a good appetizer, digestive indicated in indigestion, anorexia and flatulence. It is an expectorant by bitter and ushna properties, thus it clear the airways and reduce cough, dyspnoea and hiccough. It is a good cardiac stimulant and blood purifier hence useful in cardiac disability and blood disorder; it is the best medicine for cardiac asthma.

Part used- Root.

Dose- 1 to 3 g with honey twice daily.

Main preparation- Pushkarmooladi churna, Pushkaradi churna.

SHIGRU

Botanical name- *Moringa oleifera* Lam.

Family- Moringaceae.

Synonyms-Shobhanjan, Mochak, Haritshaka, Komalpatrak, Laghupatra.

Botanical description- It is a tree 7 to 8 meter height with soft bark and wood. Leaves are compound, 14 to 24cm.in lenth, Leaflets 6 to 9 in numbers 1.25 to1.5 cm. in lenth and slightly turned to inferior surface. Flowers are blue in clusters. Fruit Legume 15 to 25 cm. long, have 6 striations longitudinally, green–grey in colour. Seeds-have 3 striation and are winged and bitter in taste. These are used as vegetables. It contains vitamins –A, B, E, D.

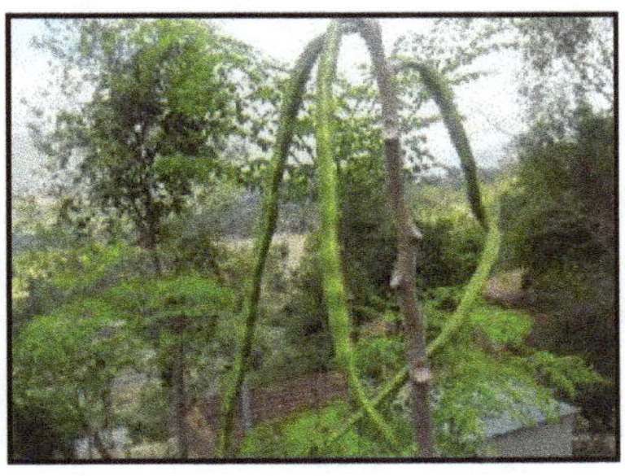

a) Macroscopic

Drug occurs in pieces of variable sizes, external surface, light greyish-brown, rough, reticulated, marked with transverse row of lenticels; outer bark, thin, peeling off in small bits, internal surface, white.

b) Microscopic

Mature bark shows a very wide zone of cork, consisting of 25 or more rows of rectangular cells, arranged radially, a few inner layers, larger and cubicular in shape; secondary cortex composed of rectangular, thin-walled cells, a few containing starch grains and rosette crystals of calcium oxalate and a few others containing oil globules and coloured resinous matter; starch grains mostly simple and rarely compound, composed of 2 or 3 components, round to oval in shape, measuring 6 to 28 µ in dia., groups of stone cells, round to rectangular, of various sizes, present in secondary cortex; mucilagenous cavities found scattered towards inner secondary cortical region;

secondary phloem appreciably wide, consisting mainly of phloem fibres and phloem parenchyma; phloem fibres in large patches, alternating with phloem parenchyma; numerous starch grains and cell contents as described above also present in phloem cells; phloem rays numerous, long, 2 to 4 seriate, consisting of radially elongated, thin-walled cells containing numerous starch grains, similar to those present in secondary cortex.

Powder - Pinkish-brown; shows stone cells, phloem fibres, starch grains, measuring 6 to 28 μ in dia., rosette crystals of calcium oxalate and oil globules.

Chemical constituents- In root, there is a bad smelling volatile oil; seeds contains stable, clean, colourless oil 36.6%. In root bark two alkaloids namely Moringine is found. Pterigospermin an antibiotic is found in root which kills the many bacteria and fungi.

Habitat- In India and Myamar.

Properties: Guna- laghu, ruksha, Rasa- katu Vipaka- katuVeerya- ushna.

Action and uses – The paste of bark and leaf is applied topically on inflammation and abscess. Seed powder is administered nasally in headache and heaviness of the head due to the congestion of cough. It is a nerve stimulant. It is indicated in loss of appetite, pain in abdomen, ascitis, gulm and worms. It stimulates menstruation, hence indicated in dysmenorrhoea and obstructed menstruation.

Part used- Bark, Legume Seed, Oil, Leaves.

Dose-15 to 25 ml of decoction along with two fruits of pepper, twice.

Main preparation- Shigru kwatha, Shobhanjanadi lepa.

TALISHPATRA

Botanical name- *Abies webbiana* Linde.

Family- Pinaceae.

English name- Silver fur.

Synonyms- Talispatra, Patradhya, Shukodara, Dhatripatra.

Botanical description- It is an evergreen tree of 50 to 70 meters height. The branches are tender, thin, greyish and hairy are bent downwards. Leaves are smooth and dark dorsally while ventral side is dusky with central vein. Fruits are elongated violet or blue in colour. Seeds are feathery with 1.5 to 3 cm length.

a) Macroscopic

Leaves flat, 1 to 5.5 cm long, about 2 mm broad; shining, midrib in the upper surface channelled down the middle but raised beneath; with two faint white lines on either side of the midrib beneath, petiole very short, greyish-brown; odour, terebinthine-like; taste, astringent.

b) Microscopic

Mature leaf shows single layered epidermis on either side covered with thick cuticle; upper epidermis followed by single layered sclerenchymatous hypodermis, lower epidermis shows papillate projections at some places followed by 1 or 2 layers sclerenchymatous hypodermis; palisade 2 layered; spongy parenchyma 4-6 layered; vascular bundle single, situated centrally, consisting of xylem and phloem, enclosed by a single layered endodermis; xylem on upper side and phloem on lower side; cambium inconspicuous; secretory cavities two in numbers, located on either side of vascular bundle, stomata sunken type, present only on the lower surface.

Powder - Greenish-brown; shows sclerenchymatous cells, palisade, spongy parenchyma and a few epidermal cells.

Chemical constituents - Essential Oil & Alkaloid.

Habitat- Afganistan, Bhutan and in India Kumaon hill.

Properties: Guna- laghu, tikshna Rasa- tikta Vipaka- madhur Veerya- ushna.

Doshakarma- Vatakaphashamaka.

Action and uses – Externally in headache it is applied due to analgesic properties. Due to its appetizer and carminative action, it is indicated in flatulence, anorexia and loss of appetite. It is a good expectorant, so indicated in cough, hoarseness of voice, asthma and tuberculosis.

Part used- Leaves.

Dose- Powder- 4-8 g.

Main preparation- Talishadi churna.

KUPILU

Botanical name- *Strychnos nux vomica* Linn.

Family- Loganiaceae.

English name- Nux-vomica.

Synonyms- Kupeelu, Vishatinduk, Kunchvruksha, Kakatinduka.

Botanical description-The tree is 12-16 mtrs in height, branches thin but strong. Bark-thin soft, grey in colour, Interior surface of the stem looks white on breaking but becomes pale yellow after some time. Leaves are glabrous and having slite odour. Leaf stalk-thick, leaves are 5-9 cm. long. Flower small, greenish white smell like turmeric. Androceium- five in numbers. Ovary is divided in 2 parts. Fruit –ovoid like guava, ripe fruit is orange colour fruit shell is very hard. Fruit pulp is tender, white and very bitter. Seeds –each fruit has 2 to 5 seeds having a breadth of 2cm. and thikness 0.5mm. Seed is big round like button, concave on one side and convex on the outer side , soft covered with striations. It is hard and whitish dusky in colour. The tree bears flowers in spring and fruits in winter.

a) Macroscopic

Seeds greenish-grey to grey, extremely hard, silky to touch with a satiny sheen; disc-shaped, almost flat, umbonate but a few seeds somewhat irregularly bent, 10 to 30 mm in diameter, 4 to 6 mm thick, margin rounded or depressed; when cut open, endosperm found to be horny, having a central cavity in which the embryo is situated with two small, thin, cordate, leafy cotyledons with 5 to 7 veins and a terete radicle; odourless.

b) Microscopic

Seed shows single layered epidermis, each epidermal cell elongated externally to form closely appresed trichomes, lignified, comprising of pitted bulbous base and a thick-walled narrowly elongated, projection; trichome slightly bent beyond the base, with about ten strongly lignified ribs of thickenings; inner testa composed of collapsed parenchymatous cells with yellowish-brown contents; outermost layer of endosperm consists of palisade-like cells while the inner layers have thick-walled, cellulosic polyhedral cells, showing plasmodesmata; endosperm cells also contain oil, and aleurone grams.

Powder - Greenish-grey; shows narrowly elongated and slightly bent thick-walled, lignified trichomes with bulbous base without ramification, thin-walled, parenchymatous cells filled with yellowish-brown content, oil globules and aleurone grains.

Habitat- It is a wild tree. In India, it appears in the forest of Torrid Zone. Mainly Chennai, Kerala, Konkan, Odisha and Sri Lanka.

Chemical Composition- Nux vomica fruit is available in India .It contains 1.6 to 3% alkaloids. It contains 1.25to1.5 % strychnine. Besides , It also contains brucin , bomisine , Loganine, Glucoside , protid (11%), bile principle, oil gum , White sugar (6%), wax phosphate and ash . Brucin is a major component of fresh bark (3.1%) Leaf and wood contain less quantity.

Properties: Guna- ruksha, laghu, tikshna Rasa- tikta, katu Vipaka- katu Veerya-ushna.

Action and uses – It is a cardiac stimulant and nervine tonic. It is used in disease of bladder, cholera, colic, cough, enteric fever, heart disease, impotence, spermatorrahea and wounds.

Part used- Inner hard button like purified seed.

Dose- 0.5- 1g powder along with 10 ml honey twice daily.

Main preparation- Akangaveer ras.

VRIDHADARU

Botanical name- *Argyria speciosa* Sweet.

Family- Convolvulaceae.

Synonyms- Ambika, Jeernadaru, Jantuka, Sukshmaparna.

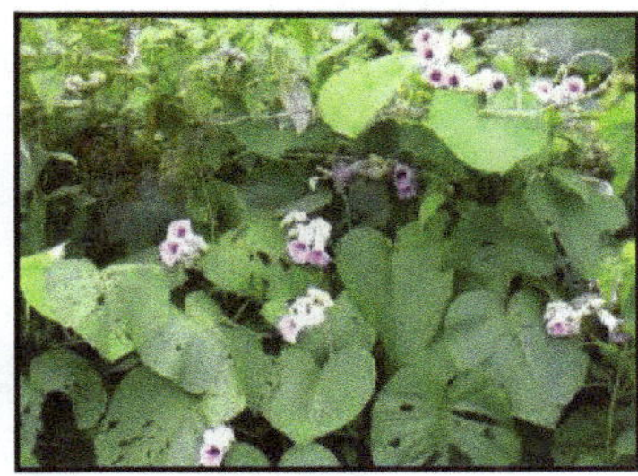

Botanical description-This widely spread creeper has stalk which is hard, thick round and bears cotton wool. Leaves are large 20- 30 cm. oblong pointed , upper surface is oily while the lower surface shows hairy growth. Flowers –bell shaped, white from out side and violet or red with in. Flowers bloom in night and are fragrant. Fruits- oval 3 cm. long green, self breaking on ripening. Flower season -4 months from rainy season , followed by fruiting. The creeper is usually planted in gardens.

a) Macroscopic

Roots of varying sizes and thickness, thin pieces show somewhat smooth brownish exterior, thick pieces tough and woody, light brown in colour, rough, longitudinally striated, lenticellate and with circular root scars; fracture fibrous; rootlets and branches, thin and somewhat fibrous; odour, nil; taste, pungent, bitter and astringent.

b) Microscopic

T.S. comprises of 6 to 9 layers of cork cells, a single layer of phellogen and usually 10 to 12 layers of phelloderm; cortical cells thin walled and tangentially elongated, containing circular starch grains, rosette crystals of calcium oxalate found scattered; a wide zone of secondary phloem consisting of sieve tubes, companion cells and phloem parenchyma present, traversed by medullary rays containing circular starch grains; resin canals present; secondary xylem a wide zone comprising of xylem vessels, tracheids, fibre-tracheids and fibres.

Powder - Creamish brown when fresh turning greyish brown on storage; shows under microscope, cortical cells parenchymatous filled with circular starch grain measuring between 3 to 16 µ in diameter; brown colouring matter and rosette crystals of calcium oxalate present; vessels, tracheids, xylem parenchyma, fibres and fibre tracheids present; vessels, drum shaped, pitted with large end perforations; tracheids, much longer than wide with bordered pits; fibres having pointed ends; fibre tracheids, having blunt ends and a few oblique pits.

Habitat- All over India.

Properties: Guna- snigdha Rasa- katu, tikta, kashaya Vipaka- madhur Veerya- ushna.

Doshakarma- Kaphavatashamaka.

Action and uses – It is used in vata diseases, heart diseases, cough, horseness of voice, diabetes mellitus, sexual weakness and emaciation. It is indicated in impotency, disease of nervous system, piles and rheumatism.

Part used- Root, seeds.

Dose- 3 to 6 g powder twice daily with milk.

Main preparation- Vridhaddaruk churna.

LAJJALU

Botanical name- *Mimosa pudica* Linn.

Family- Leguminoseae.

Synonyms- Shamipatra, Samanga, Namaskari, Sparsharodanika.

Botanical description-This thorny shrub grows 30 to 100 cms. in height. The stalk of the leaf is long and 4 leaves in the form of paw arise from it leaves resemble that of khadir leaves and close up on touching, hence the name lajjalu, namaskari. Flowers are delicate and pink arise from the tip of the flower stalk. They contain four big androceiums. Legume contains 3 to 4 seeds. The plant flowers in winter followed by fruiting.

a) Macroscopic

Root - Cylindrical, tapering, rependant , with secondary and tertiary branches, varying in length, upto 2 cm thick, surface more or less rough or longitudinally wrinkled; greyish brown to brown, cut surface of pieces pale yellow; fracture hard, woody, bark fibrous; odour, distinct; taste, slightly astringent.

Stem - Cylindrical, upto 2.5 cm in dia; sparsely prickly, covered with long,week bristles longitudinally grooved, external surface light brown, internal cut surface grey, bark fibrous; easily separable from wood.

Leaf - Digitately compound with one or two pairs of sessile, hairy pinnae, alternate, petiolate, stipulate, linear lanceolate; leaflets 10-20 pairs, 0.6-1.2 cm long, 0.3-0.4 cm broad, sessile, obliquely narrow or linear oblong; obliquely rounded at base, acute, nearly glabrous; yellowish-green.

Flower - Pink, in globose head, peduncles prickly; calyx very small; corolla pink, lobes 4, ovate oblong; stamens 4, much exserted; ovary sessile; ovules numerous.

Fruit - Lomentum, simple, dry, 1-1.6 cm long, 0.4-0.5 cm broad with indehisced segments and persistent sutures having 2-5 seeds with yellowish, spreading bristle at sutures, 0.3 cm long, glabrous, straw coloured.

Seed - Compressed, oval-elliptic, brown to grey, 0.3 long, 2.5 mm broad having a central ring on each face.

b) Microscopic

Root - Mature root shows cork 5-12 layered, tangentially elongated cells, a few outer layers crushed or exfoliated; secondary cortex consisting of 6-10 layered, tangentially elongated thin-walled cells; secondary phloem compossed of sieves elements, fibres, crystal fibres and phloem parenchyma traversed by phloem rays, phloem fibres single or in groups, arranged in tangential bands; crystal fibres thick-walled, 3-25 chambered, each with single or 2-4 prismatic crystals of calcium oxalate; phloem rays uni to multiseriate, 2-3 seriate more common; secondary xylem consists of usual elements traversed by xylem rays; vessels scattered throughout secondary xylem having bordered pits and reticulate thickenings; crystal fibres containing one or rarely 2-4 prismatic crystals of calcium oxalate in each chamber; parenchyma, thick-walled, scattered throughout secondary xylem; xylem rays uni to bi-seriate, rarely multiseriate, wider towards secondary phloem and narrower towards centre; starch grains, prismatic crystals of calcium oxalate and tannin present in secondary cortex, phloem and xylem rays and parenchyma; starch grains both simple and compound having 2-3 components, rounded to oval measuring 6-20 µ and 16-28 µ in dia. respectively.

Stem - Mature stem shows 4-8 layered, exfoliated cork of tangentially elongated cells filled with reddish-brown contents; secondary cortex wide, consisting of large, moderately thick-walled, tangentially elongated to oval, parenchymatous cells, filled with reddish-brown contents, a few cells containing prismatic crystals of calcium oxalate, a number of lignified, fibres single or in groups, scattered throughout; secondary phloem consisting of usual elements, 2-5 transversely arranged strips of fibres occur alternating with narrow strips of sieve elements and parenchyma, crystal fibres elongated, thick-walled, containing single crystal of calcium oxalate in each chamber; phloem rays thick-walled, radially elongated; secondary xylem composed of usual elements traversed by xylem rays; vessels drum-shaped with spiral thickenings, tracheids pitted with pointed ends, fibres of two types, shorter with wide lumen and longer with narrow lumen; xylem rays radially elongated, thick-walled, 1-6 cells wide and 3-30 cells high; pith consisting of polygonal, parenchymatous cells with intercellular spaces.

Leaf

Petiole - Shows single layered epidermis with thick cuticle; cortex 4-7 layered of thin walled, parenchymatous cells; pericycle arranged in a ring; 4 central vascular bundles present with two smaller vascular bundles arranged laterally, one in each wing.

Midrib - Shows single layered epidermis, covered with thin-cuticle; upper epidermis followed by a single layered palisade, spongy parenchyma single layered, pericycle same as in petiole; vascular bundle single.

Lamina - Shows epidermis on both surfaces, palisade single layered; spongy parenchyma, 3-5 layers consisting of circular cells; rosette crystals and a few veins present in spongy parenchyma.

Fruit - Shows single layered epidermis with a few non-glandular, branched, shaggy hairs; mesocarp of 5-6 layers of thin-walled, parenchymatous cells; some amphicribral vascular bundles found scattered in this region; endocarp of thick-walled, lignified cells followed by single layered, thin-walled, parenchymatous cells

Seed - Shows single layered radially elongated cells; followed by 5-6 layered angular cells filled with dark brown contents; endosperm consists of angular or elongated cells, a few containing prismatic crystals of calcium oxalate; cotyledons consists of thin-walled cells, a few cells containing rosette crystals of calcium oxalate; embryo straight with short and thick radicle.

Powder - Reddish-brown; shows, reticulate, pitted vessels, prismatic and rosette crystals of calcium oxalate, fibres, crystal fibres, yellow or brown parenchymatous cells, palisade cells non glandular, branched, shaggy hairs, single and compound starch grains, measuring 6-25 µ in dia. with 2 - 3 components

Chemical constituent – Alkaloid.

Habitat- It is said that this plant was originally found in Brazil and now it is seen all over India.

Properties: Guna- laghu, ruksha Rasa- kashaya, tikta Vipaka- katu Veerya- sheeta.

Doshakarma- Kaphapittashamaka.

Action and uses – It acts as wound healer, haemostatic and sandhaniya (for destruction) hence indicated in wounds, ulcers and fistula. It is indicated in haemoptysis, menorrhagia and bleeding piles due to its astringent properties. It is used in burning sensation in body, diarrhea, dysentery, haemostatic condition, leucorrhoea, morbid condition of vagina, skin disease and wounds.

Part used- Leaves and seeds.

Dose- Leaf juice- 10 to 20 ml.

Main preparation- Lajjalu sidh tail.

NAGKESHAR

Botanical name- *Mesua ferrea* Benth+Hook.

Family- Guttifereae.

Synonyms- Kinjalika, Champeya, Nagkinjalka, Suvarna.

Botanical description- It is an evergreen medium size, beautiful tree. Bark is reddish. Pale yellow green gum comes out from the bark. Leaves are 5 to 15 cm. long, 4 to 7cm. broad tapering at both sides.

a) Macroscopic

Stamen consists of anther, connective and filament; coppery or golden brown; filament united at base forming a fleshy ring; each stamen 0.9-1.9 cm long; anther about 0.5 cm long, linear, basifixed, containing pollen grains; filament 0.8 - 1.0 cm long; slender, filiform, more or less twisted, soft to touch, quite brittle; connective not visible with naked eye; odour, fragrant; taste, astringent.

b) Microscopic

Androecium - Anther shows golden-brown, longitudinally dehiscent anther wall, consisting of thin-walled, parenchymatous cells, pollen grains numerous in groups or in single, yellowish and thin-walled, many pollen grains having 1-3 minute, distinct protuberances on walls, thick-walled, exine and intine distinct.

Powder - Brown; shows elongated cells of filament, connective and numerous golden yellow pollen grains having 1-3 protuberances.

Chemical constituents - Essential oil and Oleo-resin. Saffron contains two bitter principles.

Habitat- Mainly Nepal, Eastern Himalayas, Bengal, Assam, Konkan, Kerala and Andman.

Properties: Guna- laghu and ruksha Rasa- kashaya tikta Vipaka- katu, Veerya- slightly ushna.

Action and uses – Astringent aromatic, bitter, cooling, expectorant used in blood dysentery, fever, leucorrhea and piles. The phenolic constituents isolated from the seed oil showed bronchodialator effect.

Part used- Stemens.

Dose- 1 to 2 g.

Main preparation- Nagkeshar churna.

ASTHISHRINKHALA

Botanical nane- *Cissus quadrangularis* Linn.

Family-Vitaceae.

Synonyms- Vajravati, Vajrang, Granthiman, Vajri, Vajravallari, Asthisandhanini.

Botanical description- It is creeper having green and quandrangular stalk which have nodes in between. Therefore it looks like a chain. Leaves are small, divided in to 3 to 5 parts and have serrated margins. Flowers – white. Fruits are round and red. If it is bent into the ground it grows.

a) Macroscopic

Drug occurs as pieces of stem of varying lengths; stem quadrangular, 4--winged, internodes constricted at nodes; a tendril occasionally present at nodes; internodes 4-15 cm long and 1-2 cm thick; surface smooth, glabrous, buff coloured with greenish tinge, angular portion reddish-brown; no taste and odour.

b) Microscopic

Mature stem shows squarish outline with prominent projection at each anular point; epidermis single layered, covered externally with thick cuticle; epidermal cells thin-walled, rectangular and tangentially elongated, followed by 2-3 layers of cork and single layered cork cambium; cortex composed of 8-16 layers of thin-walled, circular to oval parenchymatous cells; four patches of collenchymatous cells present in all the four angular points embedded in cortical region like an umbrella arching over large vascular bundles; in the projected portion of angular region cortical cells filled with brown-red contents present; endodermis not distinct; stele consists of a large number of vascular bundles varying in size arranged in the form of a ring separated by rays of parenchyma; 3 -4 vascular bundles larger in size, in each angular region, below collenchymatous patch, while rest of bundles smaller in size; vascular bundles collateral and open type, capped by sc1erenchymatous sheath which is well developed in larger bundles; cambium and interfascicular cambium quite distinct; central region occupied by a wide pith composed of thin-walled, circular to oval parenchymatous cells; idioblasts containing raphides and isolated acicular crystals of calcium oxalate present in the outer region of cortex and also in a number of cells throughout the region; rosette crystals of calcium oxalate also found in most of the cells in cortical region; starch grains present throughout the cortical and the pith regions.

Powder - Brown; shows fragments of vessels, fibres, parenchymatous cells and a few rosette crystals of calcium oxalate, starch grains and idioblast, containing raphides and isolated acicular crystals of calcium oxalate.

Chemical constituents - Calcium oxalate, carotene and ascorbic acid.

Habitat- In south India and Sri Lanka, its stalk is used as a vegetable.

Properties: Guna- laghu, ruksha, Rasa- madhur, Vipaka- amla,Veerya- ushna.

Doshakarma- Kaphavatashamaka and pittavardhaka.

Action and uses – Used in fracture of bones, gastro-interstinal disorders and irregular menstruation. Stem paste is applied over the fracture of bones by traditional healers.

Part used- Whole plant.

Dose- 5 to10 ml fresh juice with 15 ml honey.

Main preparation- Asthishrinkhala churna.

VIDARI

Botanical name- *Pueraria tuberosa* DC.

Family- Leguminoseae.

Synonyms- Swadukanda, Sita, Ikshugandha, Payaswini, Bhukushmandi, Shukla.

Botanical description- It is a spreading creeper. Its stalk has pores. Leaves are trifoliate, resembling the leaves of B. frondosa with 10-15 cms. long leaflets pointed with dense hairs on the lower surface. In florescence -purple. Fruits - 6 to 9 cms. long. Tubers are circular with a diameter of 30 to 40 cms. There is another variety called ksheervidari the tuber of which oozes out a thick latex on cutting.

a) Macroscopic

Roots of varying sizes and thickness, thin pieces show somewhat smooth brownish exterior, thick pieces tough and woody, light brown in colour, rough, longitudinally striated, lenticellate and with circular root scars; fracture fibrous; rootlets and branches, thin and somewhat fibrous; odour, nil; taste, pungent, bitter and astringent.

b) Microscopic

T.S. comprises of 6 to 9 layers of cork cells, a single layer of phellogen and usually 10 to 12 layers of phelloderm; cortical cells thin walled and tangentially elongated, containing circular starch grains, rosette crystals of calcium oxalate found scattered; a wide zone of secondary phloem consisting of sieve tubes, companion cells and phloem parenchyma present, traversed by medullary rays containing circular starch grains; resin canals present; secondary xylem a wide zone comprising of xylem vessels, tracheids, fibre-tracheids and fibres.

Powder - Creamish brown when fresh turning greyish brown on storage; shows under microscope, cortical cells parenchymatous filled with circular starch grain measuring between 3 to 16 µ in diameter; brown colouring matter and rosette crystals of calcium oxalate present; vessels, tracheids, xylem parenchyma, fibres and fibre tracheids present; vessels, drum shaped, pitted with large end perforations; tracheids, much longer than wide with bordered pits; fibres having pointed ends; fibre tracheids, having blunt ends and a few oblique pits.

Habitat- Himalayas, Nepal, Bengal, Assam, Punjab and Konkan region.

Properties: Guna- guru, snigdha Rasa- madhur Vipaka- madhurVeerya- sheeta.

Dosha- Vattapittashamaka.

Action and uses – Aphrodisiac, cardiotonic, demulcent, diuretic, used in emaciation , entricfever and spermatorrhea.In traditional medicine it is considered as restorative of high value. The study indicates 12% proteins and aminoacids in the drug. Significant oestrogenic potentiality was observed with petroleum ether extract.

Part used- Tuber.

Dose-Vidari churna 1 to 3 gms with milk.

Main preparation- Vidarikand churna.

AGNIMANTH

Botanical name- *Premna mucronata* Roxb.

Family-Verbenaceae.

Synonyms- Jaya, Shriparna, Ganikarika, Asavarika, Vataghni, Ketu, Nadeyi.

Botanical description-The tree grows up to a height of 8-10 meters. Bark –smoke coloured. Leaves are opposite ovate tapering and soft, Flowers – white, aromatic and in cluster. Fruits are small like coriander fruit. Flowering season is spring.

a) Macroscopic

Drug pieces 7-15 cm long, 0.2 -3.0 cm thick, occasionally branched, cylindrical, tough, yellowish-brown externally, bark thin, occasionally easily peeled, outer surface rough due to exfoliation, wood light yellow, fracture hard; taste, slightly astringent.

b) Microscopic

Root shows exfoliating cork, consisting of 10-15, occasionally more, rows of tangentially elongated, thin-walled cells; secondary cortex consists of round to oval parenchymatous cells, a few containing rhomboidal crystals of calcium oxalate; secondary phloem consists of isodiametric, thin-walled, parenchymatous cells, a few of them containing rhomboidal crystals of calcium oxalate; phloem rays distinct, consisting of radially elongated cells; secondary xylem shows a wide zone, consisting of usual elements, all being lignified; vessels found in single as well as in groups of 2-3, scattered throughout xylem region; xylem parenchyma simple pitted, squarish wide lumen; xylem rays 1-5 seriate, consisting of radially elongated cells; rhomboidal crystal of calcium oxalate packed in xylem parenchyma and xylem rays; abundant simple, round starch grains measuring 6-17 µ in dia., found scattered throughout.

Powder - Dull yellow; shows fragments of cork cells, small, pointed, aseptate, lignified fibres, simple, pitted vessels, lignified cells packed with rhomboidal crystals of calcium oxalate and numerous simple, round to oval starch grains having narrow hilum, measuring 6-11 μ in diameter.

Chemical constituents - Sterols.

Habitat-Through out India mainly in Gangetic plain, Utter Pradesh, Bihar and Odisha.

Properties: Guna- ruksha, laghu Rasa- tikta, katu, kashaya, madhur Vipaka- katu Veerya - ushna.

Doshakarma- Kaphavattashamaka.

Action and uses – Cardiotonic, carminative, febrifuge, laxative, stomachic and tonic. It is used in abdominal diseases, anasaraka, constipation, fever, heart disease, nervine and neurological disease and rheumatism. Taditionally this drug is highly valued for anti inflammatory property.

Part used- Whole plant.

Dose- 1 to 3 g powder with honey, Agnimanth kwatha- 20-40 ml twice daily.

Main preparation- Dashamool kwath, Brihat panchmool kwath.

BHALLATAKA

Botanical name- *Semecarpus anacardium* Linn.

Family-Anacardiaceae.

Synonyms- Bhallataka, Arushkar, Agnik, Agnimukh.

Botanical description- It is a tree of the height 25 – 40 feet. The bark of the stem is grey, smoke coloured. Leaves are develop from the tip of the branches, having dorsal surface ciliated, broad at the tip and having 15-25 veins on it. Flowers are yellow and homosexual. Fruit is 1 inch long, heart shaped. Unripe fruit is green while ripe fruit is shiny black. Fruit juice is white; on exposure to air it becomes black. Fruit has cap like structure on it. It is red and can be eaten when ripe. Like almond fruit has a kernel which is called as 'Godambi'.

Ideal fruit- Marking nut which sinks in water should be used for medicinal purpose.

a) Macroscopic

Fruit laterally flattened, drupaceous, dark brown, nut 2.5-3 cm long, obliquely ovoid, smooth, shining with residual receptacle.

b) Microscopic

Fruit - Pericarp differentiated into epicarp, mesocarp and endocarp; in longitudinal section pericarp shows outer epicarp consisting of single layer of epidermal cells which are elongated radially and lignified, characteristic glands found in pericarp which exude oil globules and arise as small protuberances in epicarp and due to pressure exerted by cells of mesocarp, some of epidermal cells and cuticle rupture and oil globules exude from oil glands; mesocarp a very broad zone, 30-40 layers thick, composed mostly of parenchymatous cells having lysigenous cavities and fibro-vascular bundles, below epidermis a few outer cells of parenchyma smaller as compared to rest; rosette crystals of calcium oxalate found scattered in parenchymatous cells, some cells get dissolved and form lysigenous cavities which increase in size with maturity of fruit, cavities do not have any special lining and contain an acrid and irritant yellowish oily secretion; endocarp consists of two distinct layers, innermost prismatic, very much elongated radial walls, being highly thickened, outer layer shorter and thinner than prismatic layer but cells similar to the former; number of mesocarp parenchyma contain rosette crystals of calcium oxalate and oil drops in oil glands; lysigenous cavities of mesocarp contain oily vesicating substance, insoluble in water and soluble in alcohol, ether, chloroform.

Powder - Dark-brown; shows rosette crystals of calcium oxalate and oil globules.

Chemical constituents - A Tarry oil containing anacardic acid, non-volatile alcohol (cardol).

Habitat-Through out India mainly torrid zone and lower region of Himalaya, Bengal, Bihar.

Properties: Guna- laghu, snigdha, tikshna Rasa- katu, tikta, kashaya Vipaka- madhur Veerya - ushna.

Doshakarma- Kaphavattashamaka.

Action and uses – It is externally used in heamorrhoids. Heamorrhoids become dry and fall of by an incense of bhallataka.It is indicated in gastrointestinal disorders by its actions like appetizer, digestive, purgative, liver stimulant and anti-helmenthic. It is useful in loss of appetite, constipation, abdominal distention, gulma, ascites, sprue, piles and worms. It promotes virility, stimulates penile nerve, hence it is an aphrodisiac. It is indicated in seminal weakness, impotency and dysmenorrhoea. Bhallataka is the best medicine for dermatoses, vitiligo and vatarakta. It improves angi of each dhatu and also acts as a tonic and rejuvenator.

Part used- Fruit, oil, godambi.

Dose- Fruit powder- 3 to 6 g, oil- 10 to 20 drops.

Main preparation- Bhallataka tail, Amrittbhallataka, Bhallataka kshar.

BRIHATI

Botanical name- *Solanum indicum* Linn.

Family- Solanaceae.

Synonyms- Brihati, Sinhi, Mahavartakini, Mahati, Chhudrabhantaki.

Botanical description- It is a shrub like brinjal 1 to 2 meters height. It has crooked and flattened thorns on its trunk and leaves with many branches and sub-branches. Leaves are 3-6 inch long, 1-4 inch wide, wing shaped with thorn. Flowers are blue in the form of inflorecense. Unripe fruit is green with white striations on it and ripe fruit become yellow.

a) Macroscopic

Root well developed, long, ribbed, woody, cylindrical, pale yellowish-brown, 1-2.5 cm in dia., a number of secondary roots and their branches present, surface rough due to presence of longitudinal striations and root scars, fracture, short and splintery; no distinct odour and taste.

b) Microscopic

Root - Shows thin cork composed of 5 - 15 layers of thin-walled, tangentially elongated, rectangular cells filled with yellowish-brown content; cork cambium single layered; secondary cortex composed of 5 - 9 layers of thin-walled, oval and tangentially elongated cells; stone cells present in singles or in groups of 2-5 or more in this region; secondary phloem composed of sieve elements, parenchyma and stone cells, traversed by phloem rays; phloem parenchyma much abundant, thin-walled; stone cells present in outer phloem region in singles or in groups of 2-5, varying greatly in shape and size; phloem rays 1-3 cells wide, isodiametric to slightly radially elongated in inner phloem region and radially elongated in outer phloem region, occasionally stone cells also found in medullary rays; wood occupies bulk of root and composed of vessels, tracheids, fibres and xylem parenchyma, traversed by xylem rays, all elements being lignified, vessels occur singly or in groups of 2-5 with simple pits; xylem fibres moderately thick-walled with simple pits and pointed ends found in adundance; xylem parenchyma have simple pits or reticulate thickening; xylem rays uni to biseriate, thick-walled, cells radially elongated and pitted, microsphenoidal crystals of calcium oxalate as sandy masses and simple starch grains present in some cells of secondary cortex, phloem and medullary rays; simple and rounded to oval starch grains, measuring 5.5 -11.6 μ in diameter.

Powder - Cream coloured; shows groups of thin-walled, parenchymatous cells, aseptate fibres, vessels with simple pits, oval to elongated stone cells and simple, rounded to oval starch grains, measuring 5.5 - 11.6 μ in diameter.

Chemical constituents - Steroidal Alkaloids namely solanine, solanidine and Steroids.

Habitat-India mainly in Punjab, Himachal, Bengal and South India.

Properties: Guna- laghu, ruksha, tikshna Rasa- katu, tikta Vipaka- katu Veerya - ushna.

Doshakarma- Kaphavattashamaka.

Action and uses – It acts as a carminative, digestive, astringent and anthelmintic, hence indicated in indigestion, colic, anorexia and worm infection. Due to its kaphashamaka property indicated in rhinitis, cough, asthma, soar throat and hiccups. Its seeds have oxytocin, so it is indicated in ammenorrhoea, dysmenorrhoea and puerperal disorders. Seeds also promote virility.

Part used- Root, seed, fruit.

Dose- Decoction- 40 to 80 ml, powder- 0.5 to 1 g.

Main preparation- Dashmoolaristh, Brihatyadi kwath.

KAMAL

Botanical name- *Nelumbo nucifera* Linn.

Family- Nymphaceae.

Synonyms- Kamal, Padma, Nalin, Ambhoj, Arvind, Salilaj.

Botanical description- It is water living plant. Its root is dipped in mud. Leaves are shiny and round. Flowers are white or red, fragrant having diameter 10 to 14 cms. The plant bears flowers at the height of 30 to 35 cms from the stalk of flower. The ovary is a sponge like compartment, containing round and soft seeds are known as 'kamal kakari.

a) Macroscopic

Fruit pulp occurs mostly in broken pieces and sometimes entire, measuring about 4-5 cm in dia; semicircular, rough, hard, having longitudinal ridges and furrows; reddish brown; odour, aromatic; taste, sour.

b) Microscopic

Fruit Pulp - shows irregular, thin-walled, parenchymatous cells; numerous idioblast cells filled with reddish-brown content; stone cells, slightly triangular and oval, with concentric striations and narrow lumen, found in groups; a few fibro-vascular bundles distributed in the pulp; xylem vessels having spiral thickenings.

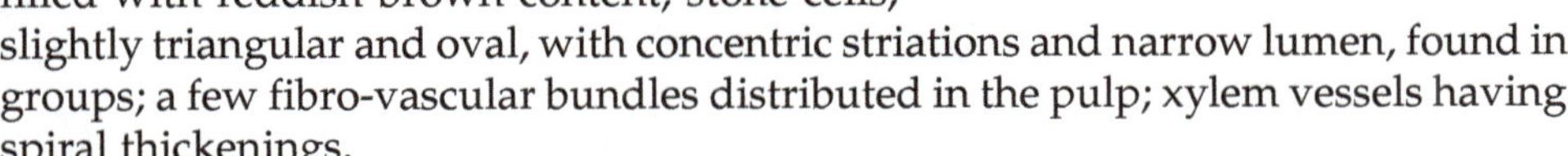

Powder - Reddish-brown; shows fragments of fibro-vascular bundles, stone cells, triangular to oval with concentric striations and narrow lumen, vessels and idioblast filled with cell content.

Chemical constituents- Citric acid and Mucilage.

Habitat- All over India, in lakes and ponds.

Properties: Guna- laghu, snigdha, pichchil Rasa- madhur, kashaya, tikta Vipaka- madhur Veerya - sheeta.

Doshakarma- Kaphapittashamaka.

Action and uses – It is a good brain tonic, it enhances the intellectual power and promotes sleep. It is cardiotonic and protects the heart from excessive heat. Due to pittashamaka property, it is indicated in various bleeding disorders.The seed of kamal is very nutritive. It is also indicated in urinary disorders caused due to pitta, fever and toxicity.

Part used- Root, seed, flowers.

Dose- Powder- 3-5 g.

Main preparation- Arvindasava.

VATA

Botanical name- *Ficus bengalensis* Linn.

Family- Moraceae.

Synonyms- Vata, Kshiri, Vanaspati, Avarohi, Skandharuh.

Botanical description- It is a big tree having huge and outspreading branches. Trunk is thick and grey in colour. Leaves are big, oval, thick, turgid and darkgreen having 3 to 5 veins. Fruits are red and round; they are not real fruit but contain innumerable minute flowers in them just like udumber.

a) Macroscopic

Mature stem bark grey with thin, closely adhered ashy white, light bluish-green or grey patches, bark fiat or slightly curve, thickness varies with age of tree : externally rough due to presence of horizontal furrows and lenticels, mostly circular and prominent, fracture short in outer two thirds of bark while inner portion shows a fibrous fracture taste, astringent.

b) Microscopic

Transverse section of mature bark shows compressed cork tissue and dead elements of secondary cortex consisting of mostly stone cells and thin-walled, compressed elements of cortex cork cells rectangular, thick-walled and containing brownish content, secondary cortex wide, forming more than half of thickness of bark, composed of large groups of stone cells and parenchymatous cells, stone cells vary in shape, parenchymatous cells thin-walled and somewhat cubical to oval few in number and occur between groups of stone cells, some of cells contain prismatic crystals of calcium oxalate, starch grains and tannin, secondary phloem composed of a few sieve elements parenchyma, fibres, stone cells and latex tube alternating with medullary rays, sieve elements compressed in outer region of bark while intact m inner region, few thick- walled phloem parenchyma occurring in between patches of phloem fibres and stone cells, stone cells similar to those present in secondary cortex, some phloem cells contain prismatic calcium oxalate crystals also, present in fibres forming crystal fibres, medullary rays 2-5 seriate, composed of thick-walled, circular to oval cells few cells also converted into stone cells and some have pitted walls, also containing plenty of starch grains, mostly rounded, rarely oval or semi-lunar in shape, simple as well as compound type, compound starch grains consist of 2-3 components, cambium composed of a few layers of small, rectangular, thin-walled cells.

Habitat- All over India.

Chemical constituents- Tannins, glycosides and flavonoids.

Properties: Guna- guru, ruksha Rasa- kashaya Vipaka- katu Veerya - sheeta.

Doshakarma- Kaphapittashamaka.

Action and uses – The latex of vata is applied on wounds, cracked soles, arthritis, otorrhoea, toothache, conjunctivitis and piles. The latex of vata is given orally to treat diarrhea, amoebic dysentery and bacillary dysentery. In leucorrhea and menorrhagia the decoction of bark is given. The latex of tree is given with sugar as a remedy for premature ejaculation.

Part used- Bark. Latex, leaf and fruit.

Dose- Powder- 3-6 g, decoction- 50-100 ml, latex- 5-10 drop.

Main preparation- Nyagrodhadi churna, Nyagrodhadi ghrita.

UDUMBAR

Botanical name- *Ficus glomerata* Roxb.

Family- Moraceae.

Synonyms- Udumber, Jantuphala, Yagyanga, Hemdugdha, Sadaphala.

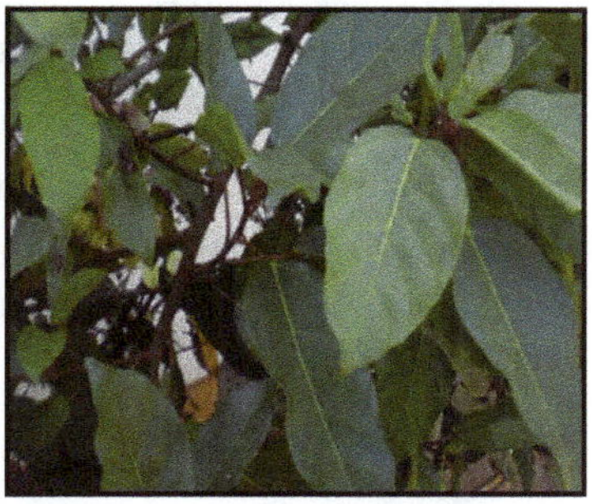

Botanical description- The tree of udumber is up to 60 feet height. The bark is reddish and smoky; leaves are oval shaped 5-7 inch long with tapering tips, having three veins. Fruits are green when raw and red when ripe.Fruits has small worms within them.

a) Macroscopic

Bark greyish-green, surface soft and uneven, 0.5-1.8 cm thick, on rubbing white papery flakes come out from outer surface, inner surface light brown, fracture fibrous, taste, mucilaginous without any characteristic odour.

b) Microscopic

Transverse section of bark shows cork, 3-6 layers of thin-walled cells filled with brownish content, cork cambium single layered, secondary cortex 6-12 layered, composed of thin-walled rectangular cells arranged regularly, a number of secondary cortex cells contain starch grains and some contain rhomboidal crystals of calcium oxalate, most of the cells filled with chloroplast giving green appearance, cortex a fairly wide zone composed of circular to oblong, thin-walled cells, containing orange-brown content, most of the cells filled with simple and compound starch grains, a number of cells also contain cubical and rhomboidal crystals of calcium oxalate, some cortical cells get lignified with pitted walls found scattered singly or in large groups throughout cortical region, secondary phloem a very wide zone composed of parenchyma with patches of sieve tubes, companion cells by medullary rays, phloem parenchyma circular to oval and thin-walled, phloem fibres much elongated, lignified, very heavily thickened and possess a very narrow lumen: medullary rays uni to pentaseriate widen towards peripheral region , a number of ray cells also get lignified and show pitted wall as described above, laticiferous cells also found in phloem region similar to parenchyma but filled with small granular masses, starch grains and rhomboidal crystals of calcium oxalate also found in most of phloem parenchyma and ray cells, cambium, when present, 2-3 layered, of tangentially elongated thin-walled cells.

Habitat- All over India.

Chemical constituents- In bark 14% tannin, wax and ash which contains silica and phosphoric acid.

Properties: Guna- guru, ruksha Rasa- kashaya, madhur Vipaka- katu Veerya - sheeta

Doshakarma- Kaphapittashamaka.

Action and uses – Its latex is used as emolument for swelling and pain. Its raw stem is used in the form of ointment for wound healing. Decoction of its bark is used as mouthwash and gargaling in pharangitis and stomatitis. Decoction of its bark is given in diarrhea and dysentery associated with bleeding per rectum and sprue (Grahani roga). In infantite diarrhea and teething problems its latex is given with sugar. The decoction of its bark is useful in leucorrhea and menorrhagia. In sperm debility latex is given with sugar preparation.

Part used- Bark, fruit, latex and root.

Dose- Powder- 3-6 g, decoction- 50-100 ml, latex- 5-10 drop.

Main preparation- Udumbersar.

PLAKSHA

Botanical name- *Ficus lacor* Buch-Ham.

Family- Moraceae.

Synonyms- Plaksha, Parkati, Pakar, Jati.

Botanical description- It is a big tree with greenish grey bark. Leaves are resembled to udumber leaves but are smaller. Its leaves have 4-10 pairs of veins. Fruits are greenish, round and small.

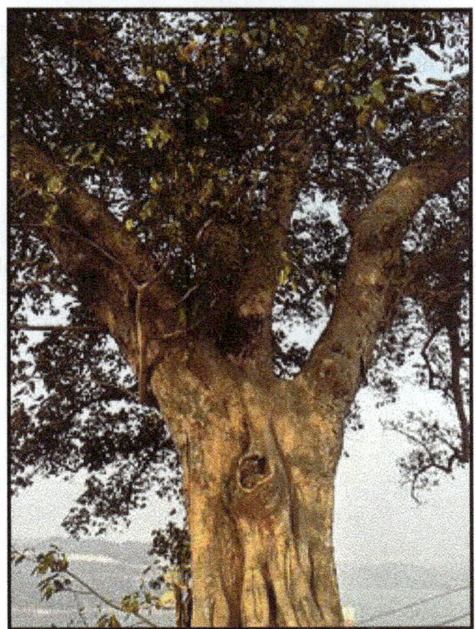

a) Macroscopic

Mature stem bark, 0.5 - 1 cm thick, greyish to pale brown, curved, rough due to presence of rhytidoma and scattered dark brown spots of exudate; rhytidoma 0.2 cm thick usually peels off, exposing light brown surface, exfoliation of cork and presence of shallow longitudinal and transverse fissures; fracture, laminated in outer part and fibrous in inner part; internal surface rough, pale brown; taste, slightly astringent.

b) Microscopic

Stem Bark -Mature bark shows rhytidoma consisting of alternating layers of cork, secondary cortex and phloem tissue; cork cells, thin-walled, 5-10 or more layered, rectangular, dark-brown; secondary cortical cells round and irregular in outline, dark brown, moderately thick-walled; tanniniferous cells, often in groups, having brown colour, sometimes containing mucilage and other materials found scattered in this zone; beneath this zone regular cork consisting of 4-12 rows of radially arranged, rectangular 136 cells followed by a zone of 2 - 4 layers of sclereids; secondary phloem consisting of sieve tubes, companion cells, phloem parenchyma, phloem fibres, crystal fibres, traversed by phloem rays; in outer and middle phloem regions phloem tissues get crushed and form tangential bands of ceratenchyma; phloem fibres arranged in tangential bands alternating with sieve tubes and phloem parenchyma; most of fibre groups contain prismatic crystals of calcium oxalate forming crystal sheath; in macerated preparation phloem fibres appear thick-walled lignified elongated with tapering or bifurcated ends; crystal fibres divided into a number of chambers containing a prismatic crystal of calcium oxalate in each chamber; phloem rays multiseriate 4 - 12 cells wide, 7 - 50 cells in height, straight; prismatic crystals of calcium oxalate found scattered in the secondary phloem tissues and phloem rays; starch grains simple or compound having 2 - 3 components, measuring 2.75 - 13.75 ì in dia., found scattered in phloem parenchyma and phloem ray cells abundantly; tanniniferous cells and secretory cavities also occur in secondary phloem.

Powder - Reddish-brown; shows numerous prismatic crystals of calcium oxalate, starch grains simple and compound with 2 - 3 components measuring 3-14 ì in dia., dark brown coloured cells, sclereids mostly in groups, thin-walled cork cells, numerous crystal fibres in group or singles.

Chemical constituents- Kinotannic acid and Gallic acid.

Habitat- All over India.

Chemical constituents- In bark 14% tannin, wax and ash which contains silica and phosphoric acid.

Properties: Guna- laghu, ruksha Rasa- kashaya Vipaka- katu Veerya - sheeta

Doshakarma- Kaphapittashamaka

Action and uses – The action of plaksh is hemostatic, anti-inflammatory and wound healing. It is indicated in bleeding, inflammation and wounds in the form of paste and powder. It is indicated in diarrhea and dysentery due to its kashaya rasa and astringent property. Its decoction is used for vaginal douche as well as orally in menorrhagia and leucorrhea.

Part used- Bark, latex and root.

Dose- Powder- 3-6 g, decoction- 50-100 ml.

MISHREYA

Botanical name- *Foeniculum vulgare* Gae.

Family- Umbelliferae.

Synonyms- Mishreya, Madhurika, Mishi, Madhura, Sugandha.

Botanical description- It is a fragransive herb of 5-6 inch in height. Leaves are devided in many parts. Flowers are racemose and yellow in colour. Fruits are small greenish or yellowish.

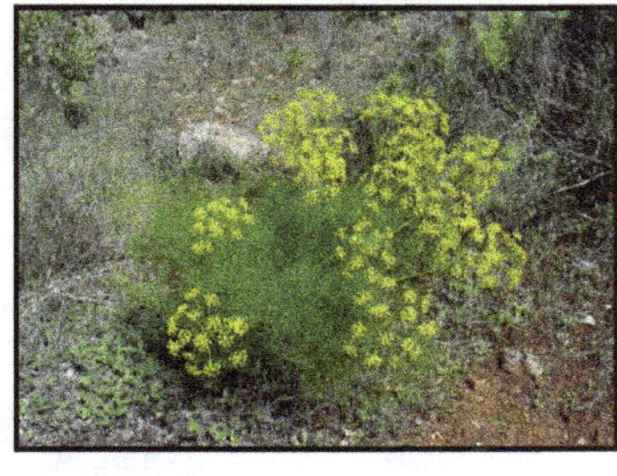

a) Macroscopic

Fruits, usually entire with pedicel attached, mericarps, upto about 10 mm long and 4 mm broad, five sided with a wider commissural surface, tapering lightly towards base and apex, crowned with a conical stylopod, glabrous, greenish or yellowish-brown with five paler prominent primary ridges , endosperm, orthospermous.

b) Microscopic

Transverse section of fruit shows pericarp with outer epidermis of quadrangular to polygonal cells with smooth cuticle and a few stomata, trichomes, absent vittae, 4 dorsal and 2 commissural extending with length of each mericarp, intercostal with an epithelium of brown cells and volatile oil in cavity, mesocarp, with much reticulate lignified parenchyma, costae, 5 in each mericarp, each with 1vascular strand having inner xylem strand and 2 lateral phloem strands separated by a bundle of fibres inner epidermis of very narrow, thin-walled cells arranged parallel to one another in groups of 116 5-7, many of these groups with longer axis of their cells at angle with those of adjacent groups (Parquetry arrangement), endosperm consists of thick-walled, cellulosic parenchyma containing much fixed oil, micro-rosette crystals of calcium oxalate, and numerous aleurone grains upto 5 ì in diameter, carpophore with very thick-walled sclerenchyma in two strands, often unsplit with two strands very close to each Other.

Habitat- All over India.

Chemical constituents- Seeds contain 3-4% volatile oil. 50 to 60 % Anethole is found as principle compound.

Properties: Guna- laghu, snigdha Rasa- madhur, katu, tikta Vipaka- madhur Veerya - sheeta.

Doshakarma- Vatapittashamaka.

Action and uses – It is brain tonic and useful for eye. It is act as deepan, pachan and anuloman. It is indicated in distention, pain in abdomen, dysentery and piles. In dysentery it acts as amapachak and anulomak of vata and it removes amadosha out of body. Due to its diuretic properties it is useful in urine retention, dysuria and burning micturation. In females it induced menstruation and increased breast milk secretion.

Part used- Seed, seed oil, root.

Dose- Seed powder- 3-6 g, oil- 5 to 10 drops.

CHANGERI

Botanical name – *Oxalis carniculata* Linn.

Family- Geraniaceae.

Synonyms- Amlapatrika, Changeri, Triparni.

Botanical description – It is a small shrub growing in moisture places. Its petioles are long. It has three conjugated leaves. Flowers are small, yellow. Fruits are oval shaped.

a) Macroscopic

Root - Dark brownish, thin, about 1-2 mm thick, branched, rough, soft; no odour and taste.

Stem - Creeping, brownish-red, soft, very thin, easily breakable; no odour and taste.

Leaf - Palmately compound, trifoliate; petiole-green, thin, about 3-9 cm long, cylindrical, pubescent; leaflet-green, 1-2 cm long, obcordate, glabrous, sessile or sub sessile, base cuneate; taste, somewhat sour.

Flower -Yellow, axillary, sub-umbellate.

Fruit - Capsules cylindrical, tomentose.

Seed -Tiny, dark brown, numerous, broadly ovoid transversely striate.

b) Microscopic

Root - Shows 3-4 layers of cork, composed of thin-walled rectangular cells, brownish in appearance; cortex, a wide zone, consisting of rectangular and oval, thin-walled parenchymatous cells filled with simple starch grains, yellowish pigment and tannin; inner cortical cells rectangular and polygonal, smaller in size than miter ones; cortex followed by thin strips of phloem consisting of sieve tubes, companion cells and phloem parenchyma, cambium not distinct; xylem consists of vessels, tracheids, fibres and xylem parenchyma; vessels cylindrical, pitted some with tail-like projection at one end; tracheids pitted with pointed ends; a few starch grains simple, round to oval measuring 3-11ì in dia., present scattered throughout the region.

Stem - Shows single layered epidermis, composed of rectangular to oval cells, some of which are elongated to become unicellular covering trichomes; cortex consists of 4-5 layers of thin-walled, circular and polyhedral parenchymatous cells; endodermis single layered of thin-walled rectangular cells; pericycle composed of two or three layers of squarish and polygonal sclerenchymatous cells; vascular bundles 6-7 in number, arranged in a ring, composed of a few elements of phloem towards outer side and xylem towards inner side; xylem composed of pitted vessels, tracheids, fibres and xylem parenchyma; central region occupied by pith composed of thin walled, parenchymatous cells, a few simple, round to oval starch grains measuring 3-11 ì in dia, scattered throughout the region.

Leaf

Petiole - Shows rounded or plano-convex outline consisting of single layered epidermis of rectangular or circular, thin-walled cells; cortex 3-4 layers of thin-walled, circular, oval or polygonal parenchymatous cells, generally filled with green pigment; endodermis single layered followed by 2-3 layers of sclerenchymatous pericycle, less developed towards upper side of petiole; vascular bundles 5 in number, arranged in a ring, consisting of phloem towards outer side and xylem towards inner side; centre occupied by a small pith; a few simple, round to oval starch grains, measuring 3-11 ì in dia., scattered throughout.

Lamina - Shows single layered epidermis on upper and lower surfaces, composed of rectangular cells; covering trichomes unicellular; palisade single layered composed of thin-walled, columnar cells, filled with green pigment; below palisade 2-3 layers of thinwalled, spongy parenchyma consisting of circular to oval cells filled with green pigment; stomata paracytic.

Powder- Greenish-brown; shows fragments of trichomes, parenchymatous, sclerenchymatous cells, fibres, epidermis showing irregular cell walls in surface view; a few simple, rounded to oval starch grains, measuring 3-11 ì in diameter.

Chemical constituents- Vitamin C, Carotene, Tartaric Acid, Citric Acid and Malic Acid.

Habitat- Throughout India.

Chemical Composition – It contains potassium and oxalic acid.

Properties: Guna - laghu, ruksha Rasa- amla Vipaka- amla Veerya- ushna.

Doshakarma- Kaphavatashamaka.

Action and uses – Its local application is helpful in inflammation, headache, moles and eye disorders. It improves the taste and appetite. It is a good liver stimulant and astringent, hence indicated in anorexia, heamorrhoids, dysentery and rectal prolapsed. It is cardiotonic and haemostatic, effective in heart diseases and bleeding disorders.

Part used- Whole plant

Dose- Juice- 2 to 3 teaspoons.

Main preparation- Shatapushpadi churna, Shatapushpa arka.

SHATAVARI

Botanical name – *Asparagus racemosus* Willed.

Family- Liliaceae.

Synonyms- Shatavari, Bahusuta, Bheeru, Narayani, Shataveerya, Varee.

Botanical description – It is a perennial herb creeper. It grows upward direction and having thorns on its stalk. Branches are triangular, smooth with straight line on it. Leaves are very small and in bunch. Thorns are curved and 0.60 to 1.25 cm in length. Flowers are small, white and fragrant, appear in bunches. Fruits are small round shape, 1-2 seeds, which become red when ripened. There are many thick oblong rootlets near the main root.

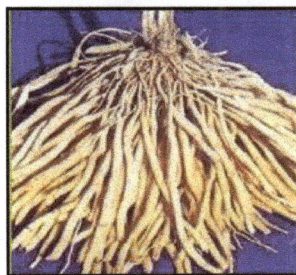

a) Macroscopic

Root tuberous, 10 to 30 cm in length and 0.1 to 0.5 cm thick, tapering at both ends with longitudinal wrinkles; colour cream; taste, sweetish.

b) Microscopic

Shows an outer layer of piliferous cells, ruptured at places, composed of small, thin-walled, rectangular asymetrical cells, a number of cells elongated to form unicellular root hairs; cortex comprises of 25 to 29 layers, distinct in two zones, outer and inner cortex; outer cortex consists of 6 or 7 layers, compactly arranged, irregular to polygonal, thick walled, lignified cells; inner cortex comprise of 21 to 23 layers, oval to polygonal, thin-walled, tangentially elongated cells with intercellular spaces; stone cells, either singly or in groups, form a discontinuous to continuous ring in the upper part of this region; raphides of calcium oxalate also present in this region; 2 or 3 layers of stone cells encirle the endodermis; endodermis composed of thin-walled parenchymatous cells; pericycle present below endodermis; stele exarch and radial in position; xylem consist of vessels, tracheids and parenchyma; xylem vessels have pitted thickening; phloem

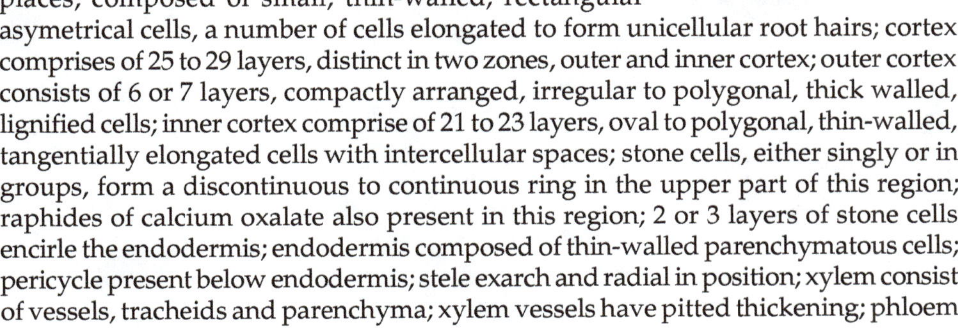

patches consists of usual element; pith composed of circular to oval parenchymatous cells, a few cells slightly lignified.

Powder - Yellowish-cream; fragments of lignified, thick-walled cells; vessels with simple pits, pieces of raphides, numerous, lignified, rectangular elongated' stone cells having clear striations with wide as well as narrow lumen and groups of parenchyma.

Habitat- Throughout India, but found especially in Northern India.

Chemical Composition – It fresh tuber is contain 52.5% water soluble ingredient, 33.3% fiber and 9% water. Water soluble ingredient contains sugar 7%.

Properties: Guna- guru, snigdh Rasa- madhur, tikta Vipaka- madhur Veerya- sheeta.

Doshakarma- Vatapittashamaka.

Action and uses – Shatavari is a brain tonic and pain reliever, so gives in epilepsy and vata disorder because its gives energy to brain and nerves. It is a good digestive, laxative and astringent hence indicated in loss of appetite, gulma, chronic colitis and hemorrhoids. It is used as galactagogue, foetal tonic and aphrodisiac in humans.

Part used- Tuber.

Dose- Powder- 3-6 gm with milk.

Main preparation- Shatavaryadi churna, Narayan tail.

SARPAGANDHA

Botanical name – *Rauwolfia serpentina* Benth. ex Kurz.

Family- Apocynaceae.

Synonyms- Sarpagandha, Dhavalavitap, Nakuli, Nagasugandha.

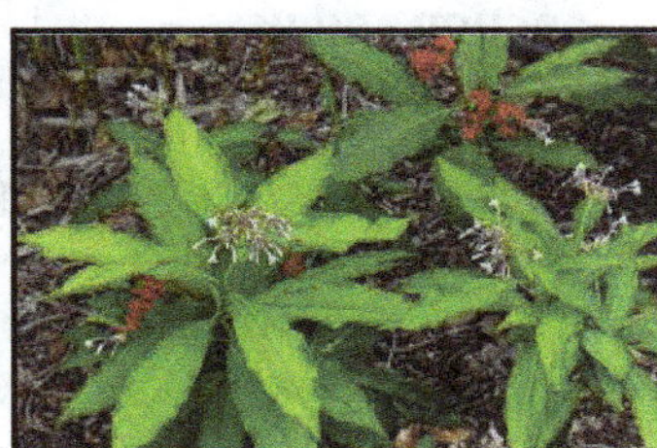

Botanical description -It is a shrub, 1 to 3 feet height. The bark of trunk is white with thin hairs on it. Leaves are light green on lower surface and dark green on upper surface, soft to touch. Flowers are white or pink in bunches. Fruits are small, fleshy pea shaped. Roots are finger sized thick.

a) Macroscopic

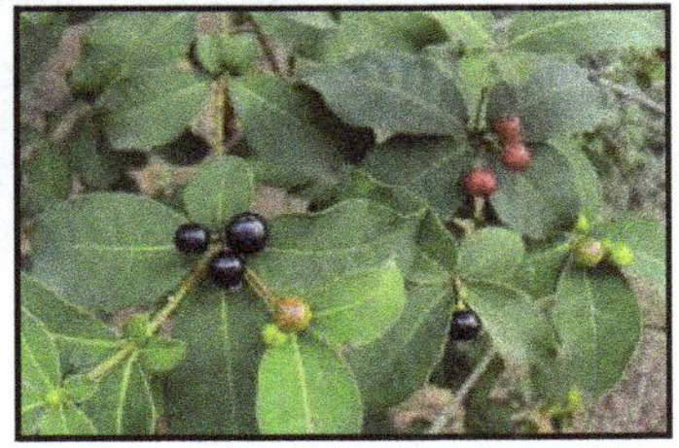

Pieces of roots mostly about 8 to 15 cm long and 0.5 to 2 cm in thickness, subcylindrical, curved, stout, thick and rarely branched; outer surface greyish-yellow to brown with irregular longitudinal fissures; rootlets 0.1mm in dia; fracture, short, slight odour and bitter taste.

b) Microscopic

Root- Root comprises of stratified cork of about 18 layers, of which the cells of 8 to 12 layers are smaller, suberized and unlignified; cells of remaining layers large, suberized and lignified; phelloderm parenchymatous, some cells packed with starch grains and prismatic and clusters crystals of calcium oxalate; secondary phloem tissue consists of sieve cells, companion cells and parenchymatous cell containing starch grains and crystals of calcium oxalate; phloem fibres absent; phloem parenchyma occasionally filled with granular substances; starch grains mostly simple but compound granules also occur with 2 to 4 components; individual granules spherical, about 5 to 15 ìm in 194 diameter, with well marked hilum simple or split in a radiate form; stone cells are absent (distinction from many other species such as *R. canescens*, *R. micrantha*, *R. densiflora*, *R. perakensis* and *R. vomitoria*); secondary xylem is traversed by well developed lignified medullary rays of about 1 to 5 cell wide but uniseriate rays are more prominent; vessels singly or in pairs; xylem parenchyma cells lignified; fibres present; cells of medullary rays thick walled also filled with starch grains and calcium oxalate prisms.

Powder - Coarse to fine, yellowish-brown, free flowing, odour slight, bitter in taste; characterized by spherical, simple to compound starch grains, calcium oxalate prisms and clusters; vessels with simple perforation, occasionally tailed; tracheids lignified; xylem fibres irregular in shape, occurs singly or in small groups, walls lignified, tips occasionally forked or truncated; wood parenchyma cells are filled with calcium oxalate crystals and starch grains; stone cells phloem fibres absent.

Habitat- All over India, mostly in Bihar, Bengal and Uttar Pradesh.

Chemical Composition – It contains salt, resin, potassium carbonate, phosphate, calcium and manganese. In the root of sarpagandha an yellow coloured alkaloid Serpentite and Serpentinine are found.

Properties: Guna- ruksha Rasa- tikta Vipaka- katu Veerya- ushna.

Doshakarma- Kaphavatashamaka.

Action and uses – Due to its analgesic, sedative and vatashamaka properties it is indicated in insomnia, irritative condition of nervous system. It is an appetizer, digestive, chologogue and anthelmintic, hence indicated in loss of appetite and worms. It reduces the heart rate and dilates the blood vessels with lowering of blood pressure. It is good known medicine of hypertension in Ayurveda. It is also used to increase the uterine contraction during labour.

Part used- Root.

Dose- Powder- 1 to 2 g for blood pressure - 3 to 6 g for insomnia.

Main preparation- Sarpagandha vati, Sarpagandhadi churna.

DARUHARIDRA

Botanical name - *Berberis aristata* DC.

Family - Berberidaceae.

Synonyms - Daruharidra, Darvi, Kateyaka, Pitadru.

Botanical description - It is an erect glabrous spinescent shrub, 3 - 6 m in height with obovate to elliptic subacute to obtuse, entire or toothed leaves. Flowers yellow in corymbose recemes. The fruits are bluish purple and small (known as jharishka). The stem is ash coloured from outside but is dark yellow coloured inside.

a) Macroscopic

Drug available in pieces of variable length and thickness, bark about 0.4 - 0.8 cm thick, pale yellowish-brown, soft, closely and rather deeply furrowed, rough, brittle, xylem portion yellow, more or less hard, radiate with xylem rays, pith mostly absent, when present small, yellowish-brown when dried, fracture short in bark region, splintery in xylem; taste, bitter.

b) Microscopic

Stem- Shows rhytidoma with cork consisting of 3-45 rectangular and squarish, yellow coloured, thin-walled cells, arranged radially; sieve elements irregular in shape, thin walled, a few cells containing yellowish-brown contents; phloem fibres arranged in tangential rows, consisting of 1-4 cells, each fibre short thick-walled, spindle-shaped, lignified having wide lumen; half inner portion of rhytidoma traversed by secondary phloem rays; phloem rays run obliquely consisting of radially elongated parenchymatous cells, almost all phloem ray cells having single prismatic crystals of calcium oxalate, a few cells of rhytidoma also contain prismatic crystals of calcium oxalate; stone cells also found scattered in phloem ray cells in groups, rarely single, mostly elongated, a few rounded, arranged radially, some of which contain a single prism of calcium oxalate crystals; secondary phloem, a broad zone, consisting of sieve elements and phloem fibres, traversed by multi seriate phloem rays; sieve elements arranged in tangential bands and tangentially compressed cells alternating with single to five rows of phloem fibres, phloem fibres short, lignified, thick-walled having pointed ends; secondary xylem broad consisting of xylem vessels, tracheids, xylem fibres and traversed by multi seriate xylem rays; xylem vessels numerous, small to medium sized, distributed throughout xylem region in groups or in singles, groups of vessels usually arranged radially; isolated vessels cylindrical with rounded or projected at one or both ends with spiral thickening; xylem fibres numerous, lignified, large, thick-walled with wide lumen, and pointed tips; xylem rays quite distinct, straight, multiseriate, consisting of radially arranged rectangular cells, each ray 30-53 cells high, 8-12 cells wide, a few ray cells containing brown contents.

Powder - Yellow; shows mostly fragments of cork cells, sieve elements, yellow coloured phloem fibres entire or in pieces, stone cells in singles or in groups, numerous prismatic crystals of calcium oxalate, xylem vessels having spiral thickening, thick-walled, lignified xylem fibres and ray cells.

Chemical composition- Berberin, oxyacanthine.

Habitat - Mostly found in Nepal, grown in Nilgirries and all over temperate Himalayas.

Properties: Guna - laghu, ruksha Rasa - tikta Veerya - ushna Vipaka - katu

Doshakarma - Kaphapittashamaka.

Action and uses - The plant is tonic, stomachic, astringent, antipyretic and liver stimulant and useful in treatment of jaundice, enlargement of spleen.

Part used - Root, stem, bark, fruit.

Dose: Decoction - 50-100 ml, fruits - 5-10 g.

Main preparation - Rasanjana, Darvyadi kwath.

INDEX OF BOTANICAL NAMES OF PLANTS

- *Ocimum sanctum* Linn.
- *Glycyrrhiza glabra* Linn.
- *Cuminum cyminum* Linn.
- *Plumbago zeylanica* Linn.
- *Mucuna pruriens* Bek.
- *Achyranthes aspera* Linn.
- *Calotropis procera* Ait.
- *Saraca ashoka* Roxb De Wild.
- *Emblica officinalis* Gaert.
- *Cassia fistula* Linn.
- *Ricinus cummunis* Linn.
- *Solanum virigianum* Linn.
- *Psoralea corylifolia* Linn.
- *Eclipta alba* Hassk.
- *Bergenia ligulata* Wall.
- *Pipper longum* Linn.
- *Areca catechu* Linn.
- *Boerhavia diffusa* Linn.
- *Pongamia pinnata* Pieree.
- *Aloe vera* Tourn. ex Linn.
- *Acacia catechu* Willed.
- *Commiphora mukul* Hook ex stocks.
- *Tinospora cordifolia* Willed Meirs.
- *Operculina turpethum* Silva Manso.
- *Aegle marmelos* Corr.
- *Terminalia belerica* Roxb.
- *Holarrhena antidyse-nterica* Linn.
- *Cynodon dactylon* Linn.
- *Terminalia arjuna* Wight. and Avn.
- *Zinziber officinale* Roscoe.
- *Woodfordia floribunda* Kurz.
- *Vitis vinifera* Linn.
- *Swertia chirata* Roxb.ex Flem.
- *Picrorrhiza kurroa* Royle ex Benth.
- *Elettaria cardemomum* Maton.
- *Convolvulus pluricaulis* Chois.
- *Embelia ribes* Burm.
- *Rubia cordifolia* Linn.
- *Piper nigrum* Linn.
- *Randia dumetorum* L.A.M.
- *Myristica fragrans* Houtt.
- *Acacia arabica* Wild.
- *Trachyspermum ammi* Linn.
- *Abrus precatorius* Linn.
- *Syzygium aromaticum* Merr Perry.
- *Terminalia chebula* Retz.
- *Pistacia intigerrima,* Stew. ex Brandis
- *Gmelina arboria* Linn.
- *Brassica compestris* Linn.
- *Mentha spicata* Linn.
- *Tamarandus indica* Linn.
- *Ficus religiosa* Linn.

- *Cassia angustifolia* Vahl.
- *Crataeva nurvala* Buch-Ham.
- *Acorus calamus* Linn.
- *Butea monosperma* Lam-Kutze.
- *Citrus lemon* Linn.
- *Azadirachta indica* Linn.
- *Vitex negundo* Linn.
- *Curcuma longa* Linn.
- *Mallotus philippinensis* Muell. Arg.
- *Centella asiatica* Linn.
- *Celastrus paniculata* Willd.
- *Pluchea lanceolata* C.B.Clarke
- *Valeriana wallichi* DC.
- *Cedrus deodara* Roxb-Loud.
- *Aconitum ferox* Wall.
- *Balliospermum montanum* Arg.
- *Onosma bracteatum* Wall.
- *Gloriosa superba* Linn.
- *Stereospermum suaveolens* DC.
- *Salmalia malabarica* Linn.
- *Albbizzia lebbek* Benth.
- *Oroxylum indicum* Vent.
- *Aquilaria agallocha* Roxb.
- *Quercus infectoria* Oliv.
- *Aconitum heterophyllum* Wall.
- *Ferula foetida* Linn.
- *Saussurea lappa* C.B. Clarke
- *Ureria picta* Desv.
- *Inula racemosa* Hook.
- *Moringa oleifera* Lam.
- *Abies webbiana* Lindl.
- *Strychnos nux-vomica* Linn.
- *Argyria speciosa* Sweet.
- *Mimosa pudica* Linn.
- *Mesua ferrea* Benth+Hook
- *Cissus quadrangularis* Linn.
- *Pueraria tuberosa* DC.
- *Premna mucronata* Roxb.
- *Semecarpus anacardium* Linn.
- *Solanum indicum* Linn.
- *Nelumbo nucifera* Linn.
- *Ficus bengalensis* Linn.
- *Ficus glomerata* Roxb.
- *Ficus lacor* Buch-Hum.
- *Foeniculum vulgare* Gae.
- *Oxalis carniculata* Linn.
- *Asparagus racemosus* Willed.
- *Rauwolfia serpentina* Benth Ez Kurzza.
- *Berberis aristata* DC.

REFERENCES

1. Acharya Agnivesha, Charaka Smahita, Edited by Brahmananda Tripathi, 3rd Edition, Chaukhamba Surbharti Prakashan, Varanasi, 1994.
2. Chunekar K.C.-editor, Acharya Bhava Mishra-author, Bhavaprakash Nighantu, 5th Edition, Chaukhamba Sanskrit Sansthan, Varanasi, 1979.
3. Vapalal vaidya, Nighantu Adarsh, first edition, Publication- Chaukhambha Sanskrit series, Varanasi, 1968.
4. Dr. P.N.V. Kurup, Dr. V.N.K. Ramadas, Shri. Prajapati joshi, Hand book of Medicinal plants, Central council for Research in Ayurveda and Siddha, New Delhi, 1979.
5. Gogte V. M., Ayurvedic pharmacology and therapeutic uses of medicinal plants, reprint 2009, publisher- Chaukhambha publications Ansari road, Daryaganj, New Delhi.
6. Ayurvedic pharmacopeia of India
7. www.google.co.in.
8. Acharya Priyavrat Sharma, Namaroopgyanam.

TULSI

YASHTIMADHU

JEERAKA

CHITRAK

KAPIKACHCHHU

APAMARGA

ARKA

ASHOKA

AMALAKI

ARAGVADHA

ERANDA

KANTAKARI

BAKUCHI

BHRINGARAJA

PASHANABHEDA

PIPPALI

PUGA

PUNARNAVA

KARANJA

KUMARI

KHADIRA

GUGGULU

GUDUCHI

TRIVRITTA

BILVA

VIBHITAK

KUTAJA

DURVA

ARJUNA

ARDRAK

DHATAKI

DRAKSHA

KIRATATIKTA

KUTAKI

ELA

SHANKHAPUSHPI

VIDANG

MANJISTHA

MARICHA

MADANAPHALA

JAIPHAL

BABBUL

YAVANI

GUNJA

LAVANG

HARITAKI

KARKATASRINGI

GAMBHARI (KASHMARI)

SARSHAPA

PODINA

AMLIKA (CHINCHA)

ASHWATTHA

SANAYA

VARUNA

VACHA

PALASHA

NIMBUK

NIMBA

NIRGUNDI

HARIDRA

KAMPILLAKA

MANDOOKAPARNI

JYOTISHMATI

RASNA

TAGAR

DEVADARU

VATSANABH

DANTI

GOJIHVA

LANGALI

PATALA

SHALMALI

SHIREESH

SHYONAK

AGARU

MAYAPHALA

ATIVISHA

HINGU

KUSHTHA

PRISHNAPARNI

PUSHKARMOOL SHIGRU

TALISHPATRA KUPILU

VRIDHADARU LAJJALU

NAGKESHAR ASTHISHRINKHALA VIDARI

AGNIMANTH BHALLATAKA BRIHATI

KAMAL VATA UDUMBAR

PLAKSHA MISHREYA CHANGERI

SHATAVARI SARPAGANDHA DARUHARIDRA

www.ingramcontent.com/pod-product-compliance
Lightning Source LLC
Chambersburg PA
CBHW050515190326
41458CB00005B/1549
* 9 7 8 9 3 8 7 0 5 7 1 8 0 *